FIRE ECOLOGY OF FLORIDA AND THE SOUTHEASTERN COASTAL PLAIN

UNIVERSITY PRESS OF FLORIDA

Florida A&M University, Tallahassee
Florida Atlantic University, Boca Raton
Florida Gulf Coast University, Ft. Myers
Florida International University, Miami
Florida State University, Tallahassee
New College of Florida, Sarasota
University of Central Florida, Orlando
University of Florida, Gainesville
University of North Florida, Jacksonville
University of South Florida, Tampa
University of West Florida, Pensacola

FIRE ECOLOGY OF FLORIDA AND THE SOUTHEASTERN COASTAL PLAIN

REED F. NOSS

University Press of Florida
Gainesville · Tallahassee · Tampa · Boca Raton
Pensacola · Orlando · Miami · Jacksonville · Ft. Myers · Sarasota

Published in the United States of America

First cloth printing, 2018
First paperback printing, 2024

29 28 27 26 25 24   6 5 4 3 2 1

Library of Congress Cataloging-in-Publication Data
Names: Noss, Reed F., author.
Title: Fire ecology of Florida and the southeastern coastal plain / Reed F. Noss.
Description: Gainesville : University Press of Florida, 2018. | Includes bibliographical references and index.
Identifiers: LCCN 2017032673 | ISBN 9780813056715 (cloth) | ISBN 9780813080772 (pbk.)
Subjects: LCSH: Fire ecology—Florida. | Fire ecology—Southern States. | Wildfires—Florida. | Wildfires—Southern States. | Forest fires—Florida. | Forest fires—Southern States.
Classification: LCC QH545.F5 N67 2018 | DDC 577.2/4—dc23
LC record available at https://lccn.loc.gov/2017032673

The University Press of Florida is the scholarly publishing agency for the State University System of Florida, comprising Florida A&M University, Florida Atlantic University, Florida Gulf Coast University, Florida International University, Florida State University, New College of Florida, University of Central Florida, University of Florida, University of North Florida, University of South Florida, and University of West Florida.

University Press of Florida
2046 NE Waldo Road
Suite 2100
Gainesville, FL 32609
http://upress.ufl.edu

I dedicate this book to the pioneering fire ecologists of the South, especially Roland M. Harper, Herman H. Chapman, Herbert L. Stoddard Sr., and Edward V. Komarek Sr. These men led the way globally in demonstrating fire as a fundamental ecological and evolutionary force. I also thank Professor William J. (Bill) Platt, who more than anyone taught me about fire in the southeastern Coastal Plain.

CONTENTS

FIGURES

TABLES

PREFACE

Like most ecologists trained in the latter half of the twentieth century, I learned in my university education that fire is an important ecological force. Precisely when I learned this, I don't recall, but it was not early enough. We should all be taught such fundamental facts about nature in primary school or earlier. In Ohio, where I grew up, the most fire-prone ecosystems—mostly prairies and oak savannas—had been almost entirely eliminated by agriculture more than a century earlier. The fire history of the likely fire-prone forests, such as oak-hickory, is poorly known. The movement to protect prairie remnants, create new "restored" prairies, and use fire to manage them did not progress very far until the 1960s and 1970s. I participated in my first controlled burn in the 1970s, on a small restored prairie near Yellow Springs, Ohio.

I was not exposed to large landscapes actively managed with fire until I visited Florida on class field trips as a graduate student at the University of Tennessee in 1977 and 1978. This fire management was oriented toward silvicultural and wildlife management objectives, but it had ecology in mind as well. On a plant ecology class trip to Tall Timbers Research Station, I learned that species, communities, ecosystems, and entire landscapes in Florida and the Coastal Plain are dependent on frequent fire for their survival. This new knowledge, for me, was revelatory. And fire itself was beautiful and exciting.

I am interested in fire because I cherish biodiversity: the full richness of life. Biodiversity in a place like Florida is nourished by fire. Every Florida naturalist knows that fire and water are the two major physical factors that control the distribution, abundance, and diversity of terrestrial and wetland species and natural communities here, usually even more important than geology and soils within regions of similar climate. Particularly in the case of fire, more is usually merrier. Species richness of plants in grasslands and savannas in the southeastern Coastal Plain, which is among the highest in the world at fine scales, increases with frequency of fire. Most of our endemic plants live in frequent-fire communities and possess fire-adaptive traits. Some fire-dependent birds such as the red-cockaded woodpecker and Florida grasshopper sparrow achieve greatest demographic success when their habitats are burned as often as they can carry fire, which is generally annually or biennially. These relationships suggest fire has been a central component of the evolutionary environment of species in this region for up to millions of years. The persistent myth that the history of fire here is mostly one of anthropogenic burning has no basis in fact. Fire is ancient in the southeastern Coastal Plain.

Many species, including humans, have developed relationships with fire during the 420 million years that fire has been on Earth. The relationship between humans and fire is complex. We depend on fire for many of our life activities, we are both thrilled and terrified by wildfire, we find fire aesthetically and emotionally appealing, and we have developed distinct branches of science and land management that concentrate on fire. Despite the long history of human use of fire, however, most people remain woefully ignorant of the role fire plays in Earth's ecosystems and the dependence of many species and entire ecosystems on fire. Even fire management professionals display a wide range of knowledge of fire ecology and evolutionary biology.

I was pleased to be invited by the University Press of Florida to submit a proposal for a book on fire ecology in Florida, which I expanded to include a large portion of the adjoining lower southeastern Coastal Plain that has climate, vegetation, and fire regimes similar to Florida's. This region east of the Mississippi River valley and extending northward to extreme southeastern Virginia is distinguished by the presettlement dominance of pine savannas and woodlands. A book on fire ecology that focuses on Florida and the southeastern Coastal Plain is needed because this region is one of the most fire-prone in the world. The science of fire ecology formally arose in Florida and southern Georgia (specifically at Tall Timbers Research Station and nearby quail-hunting plantations), and the relationships of plants, animals, and ecological processes to fire have been studied more intensively here than probably anywhere.

Despite the wealth of scholarly literature, however, information on fire ecology in Florida and the Coastal Plain has not been previously synthesized and compiled into a book. This is baffling, since other regions of North America where fire plays a prominent role in shaping evolution, vegetation, and species composition have books of their own, for instance,

Fire Ecology of Pacific Northwest Forests (Agee 1993), *Fire in California's Ecosystems* (Sugihara et al. 2006), and *Fire Ecology in Rocky Mountain Landscapes* (Baker 2009). Yet arguably no other region on Earth matches Florida and the southeastern Coastal Plain in the importance of frequent fire to the persistence of the native flora and fauna.

This book emphasizes the role of fire in the ecology and evolution of species, ecosystems, and landscapes of Florida and the southeastern Coastal Plain. My perspective is that of evolutionary fire ecology (Pausas 2015a). Instead of treating fires as discrete events, I concentrate on fire regimes, the heterogeneous patterns they create, and adaptations of species to these regimes. This book is not focused on the control or fighting of fire to protect human lives and property, which of course is sometimes needed, but not often in a well-burned landscape with natural levels of vegetative fuels. Controlled burns to restore and maintain the conditions under which native species evolved are critical in today's fragmented landscape, where lightning-ignited fires do not spread like they once did (and are usually quickly extinguished by firefighters) and where houses and busy highways often abut natural areas. I do not delve into the technical details of fire management, such as training, certification, and permitting procedures for fire managers. An experienced fire manager—which I am not—would be better suited to write a practical guide for controlled burning.

I sense that many fire managers would benefit from increased knowledge and understanding of the evolutionary ecology of fire. Fire managers tell me this is only a small part of their training, at best. I have sought to be reasonable, fair, and not overly judgmental about problems that I and many of my fellow ecologists see in real-world fire management. I recognize that practical constraints such as budgets, personnel, politics, weather, smoke, and public safety prevent controlled burning from perfectly mimicking a natural lightning-fire regime. Nevertheless, where I see weaknesses or errors in fire management, I point them out.

The integrity of fire-prone ecosystems and their native species and processes is my primary concern in this book. These ecosystems, species, fire regimes, and other processes have intrinsic value. They deserve our respect, admiration, and utmost care. Healthy ecosystems benefit people most of the time, but even when fire is inconvenient, we must still try to find a way to allow it to operate within its historic range of variability within our natural and seminatural ecosystems. Fire exclusion is one of the greatest causes of ecological degradation across North America and the world, and studies show that many of our most highly endangered ecosystems are fire-dependent (Noss et al. 1995). With frequently burned landscapes, but including adequate refugia for fire-sensitive species, life is enriched.

ACKNOWLEDGMENTS

Discussions with many fire ecologists, managers, naturalists, and biologists in person, by email, by telephone, and on Facebook greatly enhanced my knowledge of the topics addressed in this book. These intellectual contributors include, but are not limited to, Rick Anderson, Todd Angel, Paula Benshoff, Dave Breininger, Jim Brenner, Edwin Bridges, Bill Broussard, Dana Bryan, Brian Camposano, Susan Carr, Steve Christman, Jen Costanza, Jim Cox, Linda Duever, Brean Duncan, Robert Dye, Jen Fill, Cecil Frost, Sharon Hermann, Kevin Hiers, Pierson Hill, Richard Hilsenbeck, Ross Hinkle, Maynard Hiss, Ron Holle, Jean Huffman, Marty Main, Susan Marynowski, Bruce Means, Eric Menges, Paul Miller, Ed Montgomery, Steve (Sticky) Morrison, Steve Orzell, Bob Peet, Bill Platt, David Printiss, Zach Prusak, Jack Putz, Pedro Quintana-Ascencio, Kevin Robertson, Mike Ross, Latimore Smith, Bruce Sorrie, Jack Stout, Johnny Stowe, James Trager, Morgan Varner, and Alan Weakley. I am grateful to Leonardo Pedreros (Think & Engage, LCC) for preparing most of the illustrations (figures) for this book. I especially thank Eric Menges and Jean Huffman for their reviews of a draft of this book (and for revealing their identities), and the editors and staff at the University Press of Florida for bringing it to publication.

1

Fire in Florida and the Southeastern Coastal Plain

> Fire became a defining feature of the Earth's processes as soon as land plants evolved 420 million years ago and has played a major role in shaping the composition and physiognomy of many ecosystems ever since.
>
> He and Lamont (2017)

Florida and the southeastern Coastal Plain of North America (Fig 1.1) is one of the most fire-prone regions of the world (Scott et al. 2014). Fire here is "part of Nature's program" (Harper 1914). The pine savannas (Fig. 1.2) that define this region, by virtue of being the dominant presettlement terrestrial vegetation, depend on frequent fire to maintain their characteristic structure, composition, and richness. So, too, do many other natural communities here. This has been true for millions of years, which makes the southeastern Coastal Plain an ideal place to study the "emerging discipline of evolutionary fire ecology" (Pausas 2015a).

Florida and the lower Coastal Plain receive more lightning strikes and have the most frequent natural fire regime of any sizable region in North America. The climatic pattern of drought during the early lightning season (late April–June) predisposes the highly flammable vegetation that covers much of the region to burn regularly. The most fire-exposed landscapes have natural fire-return intervals averaging from one to three years (Frost 2006). This high frequency of fire has been confirmed by studies of fire scars in the growth rings of old pine trees and stumps (Huffman 2006; Huffman and Platt 2014). Such an evolutionary environment has shaped the physical traits and life histories of species native to this region. Restoring natural fire regimes or mimicking them through management is a paramount conservation goal here. Lacking sufficient fire, biodiversity in this global hotspot (Noss et al. 2015) collapses.

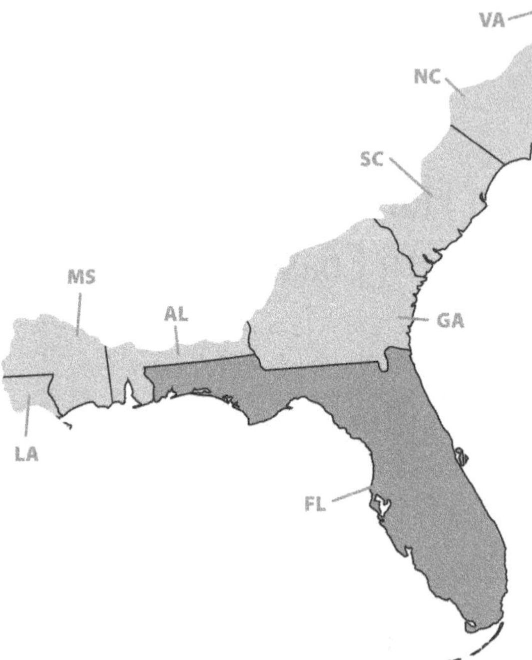

Figure 1.1. The study area for this book, the southeastern Coastal Plain, based on the distribution of pine savannas (Platt 1999), the dominant upland vegetation of the region prior to European settlement. Florida gets extra attention, so is shown in dark gray; the remainder of the region is in light gray.

Figure 1.2. Pine savannas, such as this mesic longleaf pine (*Pinus palustris*) flatwoods community in Apalachicola National Forest in the Florida Panhandle, are the dominant and characteristic natural terrestrial vegetation of the southeastern Coastal Plain. Photo by Reed Noss.

The Coastal Plain has a long history of lightning fire, possibly extending back to when it emerged from the late Cretaceous Sea around 100 million years ago. The Cretaceous, which had concentrations of atmospheric oxygen significantly higher than today, was a fiery time in Earth history, as evidenced by abundant fossil charcoal and fire-adaptive traits in pines from this period (He et al. 2012; Scott et al. 2014). By the middle Eocene, about 45 million years ago, fossils from the Coastal Plain, in what is now western Kentucky and Tennessee, include palms resembling modern *Sabal* and *Serenoa* (Graham 1999), which we know today are highly fire-adapted (e.g., cabbage palm [*Sabal palmetto*], scrub palmetto [*Sabal etonia*], and saw palmetto [*Serenoa repens*]).

The antiquity of fire is demonstrated by the southeastern Coastal Plain having one of the highest proportions in the world of endemic species and natural communities that are fire-dependent, in the literal sense that they would disappear in the absence of fire. For thousands of years this region also had a strong tradition of anthropogenic fire, which continues today among our dedicated cadre of professional fire managers. Fire here, unlike regions such as tropical moist forests that lack fire as a significant factor in their evolutionary histories, usually enhances the diversity of species and natural communities. Appropriately, Florida and adjacent southern Georgia are the birthplace of the science of fire ecology. The state of Florida leads the United States and probably the world in controlled burning ("prescribed fire") (Pyne 2016).

The main thrust of this book is on the interactions between fire and biota on ecological to evolutionary time scales, ranging from days to thousands and millions of years. Understanding the fire regimes to which our native species are adapted is fundamental to designing restoration and management strategies and practices that will allow persistence and continued evolution of these species and communities. Species have adapted through natural selection to specific fire regimes (Keeley et al. 2011). Therefore, allowing natural lightning fire regimes to operate wherever possible and simulating natural fire regimes with management will help to minimize extinctions.

In this chapter I consider the phenomenon of fire and why it is interesting to study from an ecological and evolutionary perspective. I examine relationships among climate, vegetation, and disturbance, and discuss how these relationships were misunderstood by mainstream ecologists in the context of the Coastal Plain for many decades. Most ecologists, for example, assumed that fire was a perturbation external to the ecosystem, which simply rebooted the predictable process of succession toward a stable climax forest. This climax state was thought to lack fire as part of its natural dynamics.

I embrace an alternative view of fire as an ecological driver that shapes the structure, composition, and function of ecosystems, and which enriches biodiversity at one or more spatial extents. In this view fire is intrinsic or endogenous to the ecosystem and is promoted by species in the community through coevolved vegetation-fire feedbacks. These feedbacks depend on key traits of species, such as high flammability of live or dead leaves and other tissues (Mutch 1970; Beckage et al. 2009; Fill et al. 2015a). The grasslands, savannas, woodlands, scrub, marshes, and other fire-dependent communities of the southeastern Coastal Plain are not successional stages, nor are they typical climax communities. They are dynamic nonsuccessional ecosystems. Finally, I review the fire environment of Florida and the southeastern Coastal Plain in terms of climate, weather, and landforms, which have influenced fire regimes and shaped our biota for millions of years.

Why Study Fire?

Earth is the only planet in the universe known to have fire, but if any other planet also has carbon-based terrestrial life, atmospheric oxygen concentrations above 13 percent, and at least occasional drying of fuels, it likely also has fire. Provide these ingredients, plus an ignition source, and fire is inevitable.

Approximately half of the terrestrial surface of Earth is regularly affected by fire (Willig and Walker 1999). Across the North American Coastal Plain, before conversion to agriculture and other human land uses, strongly fire-dependent ecosystems such as savannas and woodlands covered at least 55 percent of the region (Noss et al. 2015). Longleaf pine savannas covered roughly two-thirds of the southeastern portion of the Coastal Plain (Wahlenberg 1946). Fire certainly affected much more area—for example, by spreading into hardwood forests from their boundaries with more open habitats. Florida, with its highly incendiary vegetation and abundant lightning, probably had a higher proportion of its area affected by fire than any state in the country. A quick inspection of the *General Map of Natural Vegetation of Florida* by Davis (1967) shows that more than 90 percent of the land surface of the state—including both "uplands" and "wetlands"—was covered by fire-prone or fire-dependent vegetation prior to EuroAmerican settlement. Pine savannas, especially flatwoods, dominated the terrestrial vegetation of virtually all but the Everglades region and formed the matrix within which other communities were embedded.

With fire on Earth for more than 400 million years (see chapter 2), the evolution of many species and the development of natural communities were shaped by recurrent fire. Humans began to modify natural fire regimes as soon as they learned how to ignite their own fires and to extinguish fires they didn't like. This probably first happened about one million years ago. Especially since the Industrial Revolution and the invention of new tools to manipulate ecosystems, natural fire regimes have been disrupted through fire exclusion, human ignitions, intensive livestock grazing (which reduces grass cover and other "fine fuels"), road building (which creates firebreaks but also new points of ignition), logging, and creation of densely stocked tree plantations that are vulnerable to fire. Fire and related issues of mechanical treatment (e.g., thinning) to reduce the severity or spread of fire have become hot political topics, especially on national forests and other public lands in the American West, but increasingly in the South as well.

In Florida, despite better public acceptance of fire than in perhaps any U.S. state, the surge of newcomers from less fire-prone regions and the rapidly increasing urbanization and highway network now complicate controlled burning and smoke management. These changes raise serious questions about the extent to which fire will be part of our environment in the future. One thing we know for certain, however: without fire, our landscapes would be much poorer biologically and less attractive aesthetically. Without frequent burning, some natural communities would degrade to a condition where more severe and potentially dangerous fires are bound to occur. It is incumbent on naturalists and conservationists to serve as fire advocates and ambassadors and to use evidence-based arguments to promote rational fire management. Providing fodder for such arguments is one purpose of this book.

The Nature of Fire

Fire is a rapid oxidative chemical reaction known as combustion, in which oxygen and organic (carbon-based) substances are consumed, while generating heat, light, carbon dioxide, and other chemical by-products (Scott et al. 2014). Fire is basically photosynthesis in reverse. The source of ignition and the nature of the fuel are key considerations in fire ecology and fire management. In ecosystems, the primary fuels are cellulose and other carbohydrates in living and dead plant matter, and the ignition source may be lightning, volcanic eruptions, sparks from falling rocks, or humans. In the southeastern Coastal Plain, only the first and last of these potential ignition sources are relevant. Chemicals in living and dead plants either enhance or decrease flammability, and this varies greatly among species. Fuel must also be reasonably dry and arrayed in a structure that permits fire to access combustible material in the presence of sufficient oxygen and to move across the landscape. Because oxygen is consumed rapidly in the immediate vicinity of fire, wind is usually necessary to replenish oxygen and promote the spread of fire.

In the southeastern Coastal Plain lightning is the sole natural ignition source for fire, and on an annual basis it is not in short supply. Lightning is a sudden electrostatic discharge from the atmosphere, resulting from an imbalance between positive and negative charges. This imbalance often occurs when a warm air mass is mixed with a colder air mass and the atmosphere becomes polarized. The kind of lightning of interest in fire ecology is cloud-to-ground lightning, as opposed to within-cloud or cloud-to-cloud lightning. Lightning polarity (positive or negative) varies geographically and seasonally across North America, but in the southeastern United States negative polarity is dominant. A study of 16 years of lightning data in east-central Florida showed that 91.5 percent of cloud-to-ground lightning strikes were negatively charged, and these caused 93 percent of the lightning fires (Duncan et al. 2010).

Cloud-to-ground lightning accompanies thunderstorms, but sometimes emanates unexpectedly from small, innocuous-looking clouds in the absence of a storm or rain. Dry lightning (occurring in the absence of substantial rain) is more common in the southeastern Coastal Plain than in most regions. Lightning is one of the most energetically powerful phenomena on Earth, with each strike containing up to one billion volts of electricity. When a lightning bolt hits the ground, the temperature at that point increases instantaneously to approximately 30,000°C, more than five times hotter than the sun's surface. This rapid and extreme heating causes the surrounding air to expand and vibrate, producing thunder.

Fire is not inevitable when lightning strikes vegetation, despite the extraordinary temperature produced by each strike. For fire to occur, a sufficient accumulation of combustible material (fuel) must be present and dry enough to burn. Extremely wet vegetation will not catch fire, as the initial heat from a lightning strike will be lost through evaporation of water. Under somewhat less wet conditions, vegetation may burn only in a small patch, and the fire does not spread. A struck tree may be scorched without catching on fire. Damage to the tree occurs primarily as fluids in the vascular system of the tree boil along the path of the strike, generating steam and exploding cells in the wood. Strips or chunks of wood and bark may be blown from the tree immediately, or bark may peel off later. Lightning scars running or spiraling down trees are commonly observed in the southeastern Coastal Plain.

I've watched lightning-struck trees take months to die after a lightning strike. The effects of a strike may be delayed because a tree wounded by lightning often will be more susceptible to disease and decay over a period of months to years. Many lightning-struck trees that lack external evidence of a strike are nevertheless injured internally and below ground, sometimes fatally. The electrical current from a lightning strike typically passes down the bole of the tree and through the root system before dissipating in the ground. Savanna trees, such as longleaf pine, are surrounded by grassy ground cover and hence are proficient at converting lightning strikes to surface fires, a process aided by the highly incendiary fallen needles of the pines.

I once rushed out to see a neighbor's longleaf pine just after it was hit by lightning in an early June thunderstorm. In a ring with a radius of about 2–3 m around the bole of the tree, the grasses were on fire, even with light rain falling. Presumably the energy from the lightning strike traveled through the root system and ignited the grasses from below. The fire department arrived shortly thereafter to extinguish the fire, though it would have gone out naturally with the increasing rain. When a savanna tree dies after being struck, its offspring may have a better chance of germination and growth in the postfire environment than in a fire-free environment with no bare mineral soil for germination and with excessive competition for sunlight, water, and nutrients. Trees killed by lightning may remain standing for years (Platt et al. 1988a).

Fires do not burn only live or dead vegetation. If an accumulation of organic matter is present in the soil, as with a humic layer or peat, the soil itself may burn. Such "ground fires" may smolder for months or longer after the fire aboveground has gone out. When fire burns through thick layers of peat, which may happen when drought or drainage for agriculture or development lowers the water table in a peaty region such as the Florida Everglades, peat fires may burn for long periods of time and may not be extinguishable by water (Scott et al. 2014).

Because they are challenging to put out, peat or muck fires are a grave concern to fire managers, who usually avoid burning into wetlands where organic soils are prevalent. Periodic burning of organic soils, however, is a natural process that lengthens the hydroperiod (the annual time span that a wetland is inundated) and produces herbaceous wetlands with open-water patches that many native species of plants and animals require.

Intense fires, such as those typical of the late-spring drought in the lower Coastal Plain, are often needed to burn through wetlands and produce biologically desirable effects (Kirkman 1995). Prescribed burns conducted in winter when water levels are higher (especially north of central Florida) and air temperatures are lower often fail to pass through wetlands. Without sufficiently frequent and intense fires, many wetlands in this region become shallower and shrubbier, losing their open water. A categorization of fire types (e.g., surface fire, crown fire, and ground fire, or alternately, stand-maintaining, stand-replacing, and smoldering fire) is presented in chapter 4, along with an assignment of natural communities to fire regimes.

FIRE AS PART OF A DISTURBANCE REGIME

Ecologists usually consider fire a disturbance, with disturbance defined as "any relatively discrete event in time that disrupts ecosystem, community, or population structure and changes resources, substrate availability, or the physical environment" (White and Pickett 1985). Thus, fire ecology is appropriately a subdiscipline of disturbance ecology. White and Pickett (1985) note that disturbance "includes environmental fluctuations and destructive events, whether or not these are perceived as 'normal' for a particular system." Some ecologists describe frequent fire as more a stress than a disturbance. As ecologist J. Philip Grime describes, however, stress limits the quantity of biomass per unit of space and time by constraining its production, whereas disturbance limits biomass by partly or completely destroying it (Grime and Pierce 2012). Hence, even very frequent fire qualifies as a disturbance.

Fire and other disturbances are powerful forces for biological evolution, through their effects on the growth, reproduction, biomass, and mortality of organisms and on the interactions among species in a community. Many species experience improved fitness in burned or otherwise

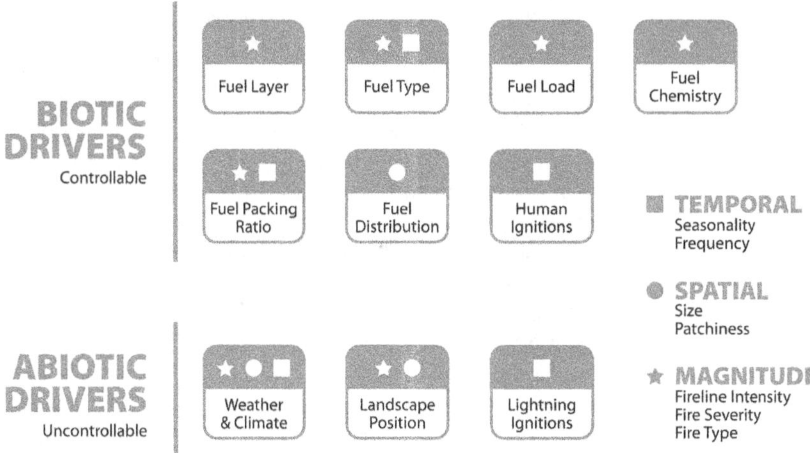

Figure 1.3. Temporal, spatial, and magnitude components of a fire regime, and the abiotic and biotic drivers that influence them. Based on information in Sugihara et al. (2006).

disturbed environments compared to undisturbed or less-disturbed systems (see chapter 3). Thus, over time, disturbances generate and maintain much of the biodiversity we find on Earth. As an agent of natural selection that shapes species' traits, fire can create an environment in which species and ultimately ecosystems become dependent on fire for their existence. These flammable ecosystems in turn promote fire, so the relationship is a classic positive feedback system (Fill et al. 2015a).

Along with hurricanes, floods, droughts, heat waves, volcano eruptions, earthquakes, landslides, avalanches, tsunamis, insect outbreaks, and other disruptive phenomena, fire is often a component of a region's *disturbance regime*—the sum of all disturbances at a given site and their interactions (Walker 2012). As explained by Turner (2010), "in contrast to a disturbance event, a disturbance regime refers to the spatial and temporal dynamics of disturbances over a longer time period. It includes characteristics such as spatial distribution of disturbances; disturbance frequency, return interval, and rotation period; and disturbance size, intensity, and severity." Every region and every ecosystem, over any given time span, has a natural disturbance regime. Fire-prone ecosystems are often said to have a characteristic fire regime, which has several components (Fig. 1.3; Table 1.1).

Table 1.1. Some components of a fire regime known or hypothesized to be important in evolutionary ecology

Component	Description
Frequency	The number of fires at a given location or typical of a natural community per unit time. May be described by "fire-return interval," the number of years between two successive fires in a specified area, often reported as a range (e.g., 1–3 years). Another, less satisfactory measure of frequency, but sometimes all that available data allow, is "time-since-fire." Changes in fire frequency alter competitive interactions of plants, with declining frequency generally favoring woody plants over grasses and associated forbs.
Seasonality	The time of year during which fires at a given location typically occur, or have occurred during a defined span of time. May be reported as months (e.g., May–June), seasons, or portions of seasons (e.g., "late spring"). The fire season is usually determined by the co-occurrence of ignitions and low fuel moisture, as produced by drought.
Intensity	The energy or heat released from a fire, typically expressed as the amount of energy per unit flame length (in kW/m). The temperature 1–3 m above the flame tip may be significantly higher than the temperature at the flame tip.
Severity	The effect of fire on vegetation or biomass, usually described as loss, damage, or change in vegetation cover or biomass. Can be measured remotely by satellite or aircraft or through ground observations. Burn severity, on the other hand, is related to the effects of fire on a range of ecosystem properties, and is often measured by the loss of organic matter in the soil.
Extent (Fire Spread)	The size of a fire event, often mapped and measured according to the total area within the outer perimeter of a fire. This measure is misleading to the extent that a fire is heterogeneous and includes unburned area or a range of fire severity patches.
Heterogeneity or Patchiness	The extent and particular manner in which the pattern produced by a fire event or series of events deviates from a uniform or homogeneous pattern. Burns that are more heterogeneous or patchy have a greater range of fire severities, often ranging from unburned through several levels of vegetation damage or mortality. Patchiness occurs across many grains, from fine-grained (small patches) to coarse-grained (large patches). Unburned or lightly burned patches in a heterogeneous mosaic are critically important as refugia for fire-sensitive species.
Patch Size	The area of patches within a fire mosaic burned at various severities (including unburned). Rarely measured.
Patch Shape	The shape (e.g., round, oval, linear) of a patch burned at a given severity, including the complexity or fractal dimension of its boundary. A shape approaching round provides more interior habitat, while elongated patches feature more edge, and jagged edges provide more complex ecotones. Rarely measured.

Fire events can be described according to these qualities, with the corresponding fire regime being the mean, range, or characteristic values of these qualities. Fire contributes to biodiversity to the extent that there is variability in these qualities; however, maximum "pyrodiversity" does not equate to maximum biodiversity (see chapter 5).

Consistent with the view of fire as a creative evolutionary force, with feedbacks to and from vegetation, some ecologists now question whether fires within the historic range of variability should be called disturbances. There is something unsatisfying about defining fire (or most other natural phenomena commonly considered disturbances) as a disturbance, given the common meaning of that term. Consider the dictionary.com definitions of "disturb:" "1. to interrupt the quiet, rest, peace, or order of; unsettle. 2. to interfere with; interrupt; hinder. 3. to interfere with the arrangement, order, or harmony of; disarrange. 4. to perplex; trouble." These largely pejorative descriptors of disturbance do not fit at all with what we know about the replenishing, regenerative, diversifying effects of fire and other natural disturbances. Rather, human activities such as exclusion or suppression of fire, or changing the frequency or season of fire away from the coevolved fire regime, may more accurately be considered disturbances because they disrupt community structure and can even cause extinctions (Noss and Cooperrider 1994). Fire in Florida is no more a disturbance than winter is in Minnesota.

Despite these terminological problems, I follow convention in considering fire part of a site's or region's disturbance regime, while acknowledging that, for fire-dependent ecosystems, lack of fire or human-induced changes in fire regimes more accurately fit the common definition of disturbance than fires within the historic range of variability. "Historic" in this sense refers to a time frame that corresponds to the evolutionary history of native species within the region.

Fire is different from other disturbances in at least three important ways. First, fire (like herbivores) needs organic material to propagate, which is not true for other disturbances such as floods, windstorms, and earthquakes. Second, fire regimes are strongly controlled by various aspects of climate and weather. In turn, they influence climate and weather as well as global carbon dynamics. Third, fire is more under the control of humans than most other disturbances, albeit not nearly as much as many people would like, especially in landscapes characterized by large high-severity fires, as in many mountain conifer, boreal, or Mediterranean-climate regions (Keeley et al. 2009). In these regions attempts at suppressing extreme fires ignited by lightning or humans are not only tremendously expensive and often result in casualties of firefighters, they are often futile. Some other disturbances can be partially controlled by humans. We can build dams, for example, to reduce the likelihood of flooding downstream (unless the dam fails). Still, fire is usually more under human control than hurricanes, earthquakes, tsunamis, droughts, and many other disturbances. Indeed, we have been trying to control and manipulate fire for our purposes—often with success—for a million years.

CHANGES IN FIRE REGIMES

By opening physical space for colonization and breaking the inertia of established biological communities, disturbances of various types may hasten the transition to new assemblages of species. With change in climate, with natural colonization or human-assisted introduction of plants that differ in their flammability from preexisting plant species, or with direct alteration of landscape conditions by humans, fire and other natural disturbance regimes change. Humans are notoriously capable of changing disturbance regimes, especially fire regimes, directly or indirectly through their activities. These changes often have unfortunate consequences for biodiversity, ecosystem function, and human well-being.

Under human influence fire regimes can change rapidly and radically, producing conditions outside the evolutionary experience of species in the community. Some species survive these changes because they can evolve relatively quickly. This may be because they have high genetic diversity or short generation times. Alternately, they may have sufficient phenotypic plasticity—that is, an ability to change phenotype (an organism's observable characteristics, such as morphology or behavior) in response to changes in the environment. Other species, slower to adapt to change or less flexible in their responses, decline and may ultimately go extinct when a fire regime is altered. Extremely rapid or pronounced change may exceed the capacity of species to disperse or adapt, resulting in nonresilient ecosystems that have compromised functions, reduced delivery of ecosystem services, and lost biodiversity.

In common with other disturbances, fire is an ecological filter that plants and animals either pass through or fail to pass through, depending in large part on their functional traits. Thus, fire determines the species composition and trait structure of communities, the abundances of species within communities, and the existence and distribution of communities across a landscape. Species in fire-prone ecosystems have evolved ways to avoid, tolerate, exploit, or promote fire, which enhances their Darwinian fitness, as defined by survival and reproductive success. Mutch (1970) proposed that flammability evolved to enhance competitive ability in the face of fire; this idea now has support at the level of individual selection, as discussed in chapter 3.

Over recent years, fire and other natural disturbances have captured the attention of the public and policy makers worldwide. Climatologists have long predicted that a global shift to a warmer climate will be marked by increased variability in weather and a greater frequency of extreme events such as severe hurricanes, floods, heat waves, and droughts (Smith 2011; IPCC 2012). These predictions are generally proving true. Hotter and drier conditions, for example, promote larger insect outbreaks and more intense stand-replacing fires in some types of forests (Westerling et al. 2006; Raffa et al. 2008; Hart et al. 2014). In other regions, such as the chaparral of southern California, fires apparently are not becoming more intense, but they are now more frequent, in large part due to fire-facilitating nonnative grasses (Keeley et al. 2009).

If we consider a sufficiently broad span of time, such as centuries to millennia, the droughts and extreme fires of recent years are not unprecedented. They remain within the historic range of variability. For example, fire frequency, severity, and associated debris flows vary with global and regional climate. Severe fires in ponderosa pine landscapes occur during warm, dry periods, such as the Medieval Climate Anomaly of 950–1250 C.E. (Pierce et al. 2004). Still, counting on disturbance regimes to remain within a historic range of variability in these highly uncertain times would be foolish. There will be many surprises.

Decades of excluding and suppressing fire have made many ecosystems more susceptible to intense fires outside the range of historic variability, at least over the last few centuries. Humans have excluded fires both directly and actively, and indirectly or passively through the creation of vast agricultural fields, cities, and highways that block the natural flow of fire across landscapes. It is a sad irony that these fire-excluded landscapes then often become more susceptible to catastrophic fire, which endangers human lives and property in addition to nonhuman species and ecosystem functions. Alternately—and this is a situation commonly observed in Florida and the southeastern Coastal Plain—ecosystem types that depend on frequent fire to maintain their natural composition, structure, and function may shift to an alternative stable state, which composed of less flammable plant material, is difficult to burn. Ecologists call such a change in ecosystem state a *regime shift*.

In many Coastal Plain ecosystems, we can identify two thresholds of fire behavior in response to fire exclusion: an early threshold, after only a few years in some cases, where a buildup of fine (e.g., grass thatch, pine needle duff) and intermediate (e.g., small woody stems) fuels increases fire intensity and severity, followed by a second threshold years to decades later, when the community flips to a new, less flammable or virtually nonflammable stable state (Myers 1985). Only the most severe fire weather (plus seed sources) or intensive management intervention might succeed in returning the community to its characteristic frequent-fire state.

The difficult and energy-requiring process of reversing a regime shift is summarized by the term "hysteresis." In ecology, hysteresis implies not only that two or more alternate stable states (A, B) exist for an ecosystem but also that proceeding backward from B to A (analogous to climbing up a steep hill) can be far more challenging than moving from A to B (sliding down a steep hill). Because of hysteresis, some regime shifts are essentially irreversible; for instance, visualize trying to ascend an overhanging cliff without climbing equipment.

Beyond the catastrophic disturbances that draw headlines are the more commonplace fires, floods, windstorms, and other natural events that are necessary and positive forces of nature. Diversity of species, habitats, and landscapes is promoted by regular disturbance, and, as noted above, plants and animals have evolved ways of coping with these events and often thriving after them. Nutrient cycling, soil fertility, and the various services that ecosystems provide free of charge to people are all dependent on natural disturbances such as fire.

Yet public understanding of the role of fire and other disturbances in nature is abysmal. Disturbances are almost universally considered "bad," and their ecological role is misunderstood and unappreciated. Even for students and researchers in the field of disturbance ecology, the literature is vast, complicated, and often contradictory. Perhaps the most accurate way to view fire is as an ecological driver: a major force that shapes species adaptations, ecosystem structure, and biogeochemical and ecological processes worldwide. Without fire, the world would be a less interesting place biologically.

Climate, Vegetation, and Disturbance

> Ecologists, biogeographers, and paleobotanists have long thought that climate and soils controlled the distribution of ecosystems, with the role of fire getting only limited appreciation.
>
> Pausas and Keeley (2009)

The overarching influence of climate on the distribution of vegetation was understood by many proto-ecologists. In 1792, German botanist Karl

Ludwig Willdenow published a synthesis of phytogeography, which included discussion of the pivotal role of climate in determining the distribution of vegetation on Earth. These ideas were expanded upon by Willdenow's student and colleague Alexander von Humboldt, one of the greatest naturalists of all time, who in 1805 described the elevation-related vegetation zones of the Andes and compared them to zones determined by latitude (Lomolino et al. 2006). In the late nineteenth and early twentieth centuries, Andreas Franz Wilhelm Schimper developed the concept of biomes, large regions of similar vegetation determined by the regional climate (Schimper 1903), and C. Hart Merriam defined life zones on mountains in the southwestern United States (Merriam 1894). Later, Holdridge (1947) developed a classification of global life zones along the three axes of mean annual temperature (controlled by latitude or elevation), precipitation, and potential water loss through evapotranspiration.

Natural disturbances were not explicitly considered in any of these climate-vegetation classifications or descriptions. Indeed, Holdridge's life zone concept failed to include savannas and grasslands, two major vegetation types that are largely maintained by recurring fire. Whittaker (1975) produced a simpler classification of the biome-scale relationships between vegetation and two climatic axes: mean annual temperature and mean annual precipitation. Temperate grasslands and tropical savannas show up in this classification, but they are portrayed as occurring in regions intermediate in rainfall between desert and forest. This is demonstrably not true for the wet southeastern Coastal Plain of North America, precisely because fire overrides the influence of precipitation in determining the existence and distribution of most major grasslands/savannas of this region (Noss 2013).

Contemporary understanding of savannas reveals that they are "bistable," that is, they could be either forest or savanna depending on one key variable: fire. Bond et al. (2005) showed that a "world without fire" would have twice as much forest and half as much grassland as presently occurs. Such would be a poorer world, because ancient grasslands globally are every bit as interesting and important as forests in terms of biodiversity and ecosystem services (Veldman et al. 2015a, 2015b; Bond 2016).

SUCCESSION AND CLIMAX THEORIES

Theories about ecological succession, an ostensibly predictable sequence of vegetation change following disturbance, are prominent in the history of ecology (McIntosh 1985). The hypothetical end point in a successional sequence is the climax stage. Climax theory, or what might be better called succession-to-climax theory, was developed by Frederic Clements and other ecologists in the late nineteenth and early twentieth centuries. This theory, in its "monoclimax" form proposed by Clements (1916), carried the assumption of climatic control of vegetation and the belief that only one "climatic climax" exists for a region. This climatic climax is composed of a single dominant life form (e.g., forest or grassland) and will eventually dominate a site and perpetuate itself over time.

Clements acknowledged the importance of disturbance—for example, stating that "even the most stable association is never in complete equilibrium, nor is it free from disturbed areas in which secondary succession is evident" (Clements 1916). Clements even conducted some of the first field research on fire ecology in North America, a study of wildfire in lodgepole pine (*Pinus contorta*) forests (Clements 1910). Nevertheless, Clements viewed disturbance as essentially an injury to the community "superorganism," which would be healed through ecological succession. He and many of his contemporaries also believed that, with rare exceptions, succession leads inexorably to a stable monoclimax, which is characteristic of each climatic region or biome. Fire and other natural disturbances, under this view, were events that reset the successional clock, which would then resume ticking toward a predictable climax. A fire-maintained community in this theory is never a true climax, but rather a "subclimax" or "disclimax."

An alternative concept to the monoclimax, and more realistic, is the polyclimax. The polyclimax idea is often attributed to Tansley (1939), but was built on the work of several earlier ecologists. Polyclimax theory does not assume that successional trajectories after disturbance will converge on one vegetation state. Rather, it proposes that several distinct climax states—including fire climaxes, edaphic (soil) climaxes, grazing-maintained climaxes, and human-maintained climaxes—could coexist as a mosaic within a given climatic region. Under both the monoclimax and polyclimax theories, as well as other theories of succession, disturbances are considered exogenous, that is, external to the biological community. Succession/climax theory leaves little room for biological properties of organisms, such as flammability, that might promote fire, with positive fitness consequences.

ALTERNATIVES TO TRADITIONAL CLIMAX

Some early twentieth-century ecologists were ahead of their time in recognizing the dynamic mosaics of patches in various stages of recovery from disturbance that characterize real landscapes. A splendid example is W. S. Cooper, who described the wind-disturbed forests of Isle Royale in Lake Superior as "a complex of windfall areas of differing ages . . . a mosaic or patchwork which is in a state of continual change" (Cooper 1913). In a monograph on the broad-sclerophyll vegetation of California, Cooper (1922) noted a similar effect of fire, stating that the great diversity of plant associations one finds in a single landscape "is in part due to slight habitat differences . . . but also in an important degree to the great number of species with restricted range and to the frequent occurrence of fires, which result in multitudinous combinations of species, depending upon which are able to survive or to repopulate the area burned." Recognition of how disturbance-generated patch dynamics maintains biodiversity blossomed later in the twentieth century (e.g., Pickett and Thompson 1978; Sousa 1979; White 1979; Sousa 1984; Pickett and White 1985).

Alternative views of climax communities that considered fire intrinsic to the community were voiced as early as the 1930s. For instance, H. H. Chapman (1932) boldly asserted:

> In the longleaf pine type of the south (and nowhere else in North America, to the writer's knowledge) fire at frequent but not necessarily annual intervals is as dependable a factor of site as is climate or soil. The conception of a climax type as one which has reached a stage of permanent equilibrium or perfect adaptation to these constant factors of site should include the longleaf pine type of the south, which presents by far the greatest area and most permanent characteristics of any climax to be found in the United States.

The condition that Chapman described for longleaf pine is what others (e.g., Daubenmire 1990) call a "fire climax," a state that is prevented from becoming a typical hardwood-dominated climax by recurring fire. E. Lucy Braun (1950), in her monumental *Deciduous Forests of Eastern North America*, considered the longleaf pine ecosystem the dominant vegetation in her "southeastern evergreen forest" region, but thought that it "is but an edaphic climax so modified and stabilized by recurring fires that it is considered a fire subclimax."

Ecologist Robert Whittaker (1953) maintained that many commonly recognized climax communities are actually mosaics of populations of various species distributed more or less individualistically (in the sense of Gleason [1926], who rejected the idea of communities as superorganisms) across complex environmental gradients. Whittaker (1953) recognized that fire incidence varies across environmental gradients, such as elevation, and "periodic burning is an environmental factor to which some climaxes are necessarily adapted." Furthermore, Whittaker (1953) described a spectrum of communities in natural landscapes that respond quite differently to fire, with some being essentially nonsuccessional communities for which fire does not behave as a typical disturbance: "in fire-adapted climaxes, fire either does not destroy the dominant populations or does not cause replacement of the dominant growth-form as in other climaxes."

Thus, Whittaker echoed a handful of earlier ecologists, such as Chapman (1932), in contending that fire-adapted communities in the southeastern United States can be quite stable and are not necessarily successional stages in a sequence leading to closed forest. Still, the prevailing paradigm of closed forest as the climatic climax of high-rainfall regions such as the southeastern Coastal Plain was hard to kill. This idea remained dominant at least through the 1960s and persisted in much ecological literature through the 1980s and beyond. The idea is still expressed in some popular writing today.

Ecologists and geographers influenced by the succession-to-climax paradigm, such as Wells and Shunk (1931), Quarterman and Keever (1962), Monk (1965), and Küchler (1964), described and mapped most of the southeastern Coastal Plain as "southern mixed hardwood forest" or "southern mixed forest," the presumed climax vegetation of the region. Pine savannas and other grasslands were considered transient successional stages dependent on disturbance by fire. These views were entirely consistent with those of the leading ecologists of that era. Eugene Odum's classic paper "The Strategy of Ecosystem Development" (Odum 1969) outlined a neo-Clementsian model of succession as an orderly, directional, and predictable trajectory toward a stable endpoint: the climax.

It is true that if one excludes fire from the pinelands and grasslands of the southeastern Coastal Plain, they usually transition relatively quickly (within decades) to hardwood forest. Nevertheless, except in topographically protected sites, fire would have maintained most of the landscape in pine savanna and embedded fire-dependent communities, such as depression wetlands. Roland Harper described this phenomenon in an early paper (Harper 1911), which unfortunately had little influence on mainstream ecology.

Another flawed but persistent assumption is that fire in this region is fundamentally anthropogenic. For example, although Van Lear et al. (2005) acknowledge that some fires in the southeastern Coastal Plain are ignited by lightning, they attribute most fires prior to EuroAmerican settlement to human agency: "Indeed, it can be argued that, at least in some places, the southeastern Coastal Plain prior to its discovery by Europeans was a cultural artifact largely molded and manipulated by Native Americans through their use of fire" (Van Lear et al. 2005). Popular books such as *1491* (Mann 2005) reinforce the myth that Florida and much of the Coastal Plain was "dominated by anthropogenic fire." No supporting data are provided because none exist.

The anthropocentric view of fire in the southeastern Coastal Plain is impossible to reconcile with the overwhelming evidence that shows great antiquity for pine savannas and other fire-dependent ecosystems that dominated the region. This evidence comes from diverse sources, including: (1) paleoecology (e.g., fossils of fire-associated taxa extending back tens of thousands to millions of years prior to human arrival in the Southeast less than 15,000 years ago); (2) phylogeny (e.g., monotypic genera and other ancient taxa associated with modern fire-prone ecosystems); (3) a high level of species endemism, especially paleoendemics, concentrated in fire-prone ecosystems; (4) fire-adaptive traits of plants and animals (especially old taxa, suggesting natural selection by fire long before humans appeared on the scene); and (5) the existence of major ecosystem types dependent on frequent fire (Platt 1999; Noss 2013).

The consequences of applying succession-to-climax theory to restoration and management of fire-dependent communities in the southeastern Coastal Plain are deleterious to biodiversity (Fill et al. 2015a). The succession model is flawed on several counts. First, it ignores evidence that characteristics of the vegetation and of individual plant species can create conditions that favor their persistence over other species or vegetation states. For instance, profound differences exist in the flammability of plant species associated with pine savanna vs. hardwood forest. Flammable species, especially longleaf pine (*Pinus palustris*) and some other pines, pyrophytic oaks such as turkey oak (*Quercus laevis*) (Hiers et al. 2014), as well as many C_4 grasses, exist in a coevolved positive feedback relationship with fire. They promote and facilitate fire. If fire is excluded, these light-demanding pyrogenic species gradually drop out of the community as fire-sensitive mesic hardwoods increasingly dominate the site. The hardwood forest eventually becomes an "asbestos community," an alternative stable state that burns infrequently and incompletely.

Traditional succession and climax theory have waned in their influence on ecology in recent decades. In their place is a more dynamic, nonequilibrium view of ecosystems, a view that has gained prominence with the accumulating evidence of individualistic changes in vegetation and species composition in response to climate change (Davis 1986; Gill et al. 2015). As summarized by ecologist Norman Christensen, "As we have learned more about succession in different places, initiated by different disturbances, and occurring in the context of other kinds of change, simple, deterministic, directional, and widely applicable models of succession have been replaced by much more complex, stochastic, and situation-specific constructs" (Christensen 2014). Still, the idea that closed-canopy hardwood forest is the natural climax vegetation of the Coastal Plain remains firmly embedded in the minds of many ecologists as well as in popular imagination. Many people tend to think of pine savannas and other southern grasslands as akin to old fields abandoned after agriculture, which if left alone will inexorably succeed to hardwood forest. It will take a concerted educational effort to correct this misperception.

The dominant fire-dependent vegetation of Florida and the southeastern Coastal Plain turns out not to be early successional communities, as long thought, but rather nonsuccessional communities that both require and promote frequent fire for their persistence. Their incredible diversity shows that fire, far from being a true destructive force, has been a central and positive component of their history.

The Role of Fire in Ecosystems

In this section I briefly summarize various roles that have been proposed for fire in ecosystems. Most of these topics will be treated in greater detail later in this book. These roles collectively suggest that fire is a major ecological and evolutionary driver of life on Earth.

FIRE ACTS AS AN HERBIVORE

A popular topic in the ecological literature is trophic cascades, in which higher trophic levels exert control over the populations, biomass, or functions of lower trophic levels. Carnivores, for example, often control the population size or spatial distribution of herbivores. If carnivore populations are depleted, herbivores expand and begin to have more significant effects on vegetation and other components of the ecosystem (Hairston et al. 1960; Terborgh and Estes 2010). Bond and Keeley (2005) noted that the effects of fire have been ignored in the literature of trophic cascades, even though fire's role is in many ways analogous to herbivory. Unlike other disturbances, such as windstorms or floods, but like herbivores, fire feeds on complex organic molecules and converts them to organic and mineral products. Bond and Keeley (2005) further suggested that fire and herbivores can be alternative consumers of vegetation, but that fire has much broader "dietary preferences," feeding on live or dead material that herbivores may avoid. Plants avoided by herbivores often fuel fires. When fire is excluded from fire-prone ecosystems, cascading changes in species composition (e.g., replacement of savannas by forests) can be expected.

Just as Hairston et al. (1960) predicted little competition among plants when herbivores are abundant, an analogous phenomenon occurs in species-rich grasslands when fire is frequent: many ecologically similar plant species coexist. By reducing plant biomass, fire reduces competition for key resources such as sunlight, soil water, and soil nutrients. A global study of the relationships between plant species richness and productivity in grasslands (Grace et al. 2016) shows that accumulation of plant biomass reduces species richness. Competitively superior plants dominate communities in the absence of fire or other disturbance. These superior competitors exclude other species, especially those that are shade intolerant.

FIRE STIMULATES REPRODUCTION AND REGENERATION

Many plant species are stimulated to germinate or flower in response to fire, sometimes only within a certain season of the year. Besides sexual reproduction, fire stimulates resprouting and vegetative growth of many plants. Animals, too, take cues from fire and from the vegetative conditions produced by fire. Some grassland birds are known to shift territories into very recently burned areas, where food, suitable nesting sites, or other resources are more available (see chapter 3).

FIRE MAINTAINS POPULATIONS OF FIRE-ADAPTED SPECIES AND THE COMMUNITIES THEY COMPOSE

Species vary tremendously in their responses to fire, from highly fire-sensitive to fire-dependent. Across a landscape, differences in topographic position, humidity, soil moisture, and other factors create a patchwork of communities with varying relationships to fire. The most fire-exposed sites burn frequently, such that fire-sensitive species are unable to invade, whereas ravine slopes, river bottomlands, sinkholes, and other landscape positions burn rarely and thereby offer zones of safety to fire-sensitive species. Thus, landscape position interacting with the components of the fire regime and the widely varying fire-sensitivities and flammabilities of plants allows for coexistence of many species.

FIRE CREATES A SHIFTING MOSAIC OF HABITAT CONDITIONS

Even in the most fire-exposed landscape positions, an individual fire virtually never burns every scrap of vegetation. Shifting winds, variation in microtopography, soils, fuel loads and other legacies of previous fires, fuel moisture, and other factors assure that every fire will be patchy to some degree. This fine-scale heterogeneity is extremely important, as it results in microrefugia being retained within the fire perimeter. These refugia are essential to the survival of fire-sensitive and relatively nonvagile animals such as many invertebrates and small vertebrates. The spatial heterogeneity of fire effects also fosters the coexistence of plants that differ in their optimal time-since-fire, in turn enriching plant species composition and vegetation structure.

The habitat mosaic produced by fire is not static. Each new fire creates a somewhat different patchwork, such that fuel-rich patches not burned during the previous fire will likely burn during a subsequent fire. Thus, fire creates a shifting mosaic of vegetation patches at multiple spatial grains, from fine to coarse, across the landscape. Over weeks to months, fire may be "stored" in the landscape, smoldering in woody debris or the organic soils of wetlands, only to flare up later.

FIRE RECYCLES NUTRIENTS AND AFFECTS WATER AND SEDIMENT DELIVERY THROUGH WATERSHEDS

Fire changes nutrient concentrations in soils and the atmosphere by volatilizing some elements (i.e., changing them from solids to gases) while also making many nutrients temporarily more available to plants. Indigenous cultures in tropical rainforest regions have long used fire—for example, through swidden agriculture—to increase nutrient availability to crops. The temperature of the fire mostly determines the proportion of nutrients in live and dead plant materials that are volatilized and temporarily lost from the system. With the abundant fine fuels that characterize Florida's grassy natural communities (see chapter 4), fireline intensity can be high, causing considerable temporary losses of nitrogen, carbon, and sulfur, as well as proportionately larger (but less in mass) volatilization of phosphorous, calcium, and potassium. However, the solid residue or ash that remains after a fire contains charcoal and soluble nutrients that raise soil pH and are readily absorbed by plant roots (Scott et al. 2014).

Most of the nutrients volatilized by a fire are recaptured by the ecosystem through redeposition of ash and through rainfall (Scott et al. 2014). The cumulative effects of volatilizing nitrogen and carbon in savannas are remarkably small, although concentrations of these elements are typically much greater under tree canopies than in open grassland (Coetsee et al. 2010). Intense heating of soils causes a temporary pulse of nutrient availability to plants, but can kill beneficial soil organisms such as nitrogen-fixing bacteria and mycorrhizal fungi, which are symbiotic with plant roots. Recovery of these organisms from fire is usually rapid, however (Scott et al. 2014).

FIRE IS IMPORTANT TO PEOPLE

Beyond its ecological and biological importance, fire is critical to human civilization, not just for staying warm and cooking food, but for clearing land for agriculture, promoting populations of favored plants and game animals, and controlling insect pests (i.e., you can experience the difference between an open pine sandhill and a hammock during the peak of mosquito season). Beyond these direct benefits to humans, fire and other natural disturbances are also intrinsically interesting and exciting, and they produce landscapes that, somewhat paradoxically, people find beautiful and sublime. This might be partly explained by humans having evolved in savannas. Carbon isotope ratios in dated soils indicate that our earliest bipedal ancestors inhabited open savannas dominated by C_4 grasses and with less than 40 percent woody plant cover (Cerling et al. 2011). These savannas were maintained by frequent surface fires. Hence, fire is worthy of our respect, appreciation, and understanding.

The Fire Environment of Florida and the Southeastern Coastal Plain

The dominant natural upland vegetation of the southeastern portion of the North American Coastal Plain, and extending into what are often considered wetlands, is pine savanna (Platt 1999). The distribution of pine savannas east of the Mississippi River therefore provides a suitable study region for this book. Ecologists recognize that, for regions with a savanna climate, fire is the primary factor that determines whether the vegetation at a given place and time is forest or savanna (Bond et al. 2005; Staver et al. 2011).

Savanna and forest are alternative stable states in seasonally dry climatic regions of the tropics and subtropics, and extending into warm temperate regions. That is, they both can occur under the same environmental conditions, and they maintain themselves through positive feedback. Savannas that occur today in regions wet enough to support forest may have formed during periods of drier or highly seasonal climate—for instance, in the late Miocene about six to eight million years ago (Keeley and Rundel 2005; Beerling 2007). The fire-savanna relationship is a positive feedback loop: fire promotes savanna, which due to its flammable grasses promotes more fire. When fire is excluded, most savannas, including our southeastern pine savannas, rather quickly convert to forest. Sadly, the savannas that dominated the landscape of the Coastal Plain were largely erased from our collective memory or were assumed to be artifacts of human activity such as burning by Indians or white settlers. These savannas were also falsely assumed to be successional stages that would develop into closed-canopy forests if humans just left them alone.

WEATHER AND CLIMATE

Florida and the southeastern Coastal Plain are climatically and meteorologically prone to fire, and apparently have been for millions of years. The pioneer fire ecologist Edward V. Komarek Sr. (1967) pointed out the principal factors favoring lightning-ignited fires in this region are extended dry weather (e.g., spring drought) combined with the amount, configuration, and moisture content of fuels (flammable live or dead vegetation). Both fuels and weather conditions in the southeastern Coastal Plain, especially the lower or southern portion, are ideal for promoting frequent fire. Fire managers in Florida, as in many other regions, pay close attention to weather conditions, as these strongly affect the probability of wildfire and define appropriate times and conditions for controlled burning. In this section I concentrate on Florida's fire environment, but with relevance to the remainder of the southeastern Coastal Plain.

Florida and its surroundings fall within the rainfall regime, both in terms of total annual precipitation and seasonality of precipitation, which defines the world's great savanna regions. Globally, a close association exists between fire activity and areas of intermediate primary productivity, such as savannas. Oddly, Florida was not recognized as a savanna region by many ecologists until quite recently. Staver et al. (2011) characterized savanna regions as having intermediate annual rainfall (100–250 cm, or 39–98 in) and mild seasonality (dry season shorter than seven months), and identified regions meeting these climatic criteria in South America, Africa, and Australia. They and most other savanna ecologists failed to recognize the Coastal Plain of the southeastern United States as another region with a savanna climate and vegetation, according to their criteria.

What Coastal Plain ecologists now take for granted about the fire-maintained savanna nature of our dominant vegetation has been recognized only recently by the global community of fire ecologists and savanna ecologists (e.g., Bond 2015, 2016; Veldman et al. 2015a). A bundle of myths and misconceptions, including succession-to-climax theory, prevented ecologists from outside the Coastal Plain, and even still some within it, from realizing the true nature of our vegetation and its determining factors (Noss et al. 2015).

Weather and climate both affect fire regimes. Weather is the day-to-day conditions of temperature, precipitation, humidity, wind, and storminess, whereas climate refers to the persistent pattern of weather conditions for a defined region. Most of Florida falls into a humid to subhumid subtropical climate regime, whereas south Florida (mostly south of Lake Okeechobee) is subhumid tropical in southwest Florida and humid tropical in southeast Florida (Biota of North America Program [BONAP], www.bonap.org/Climate%20Maps/climate48shadeA.png; PRISM Climate Group, http://www.prism.oregonstate.edu/). Most subtropical regions of the world are considerably more arid than Florida, but the marine influence from the Atlantic Ocean and the Gulf of Mexico keeps Florida and the outer Coastal Plain humid and wet for much of the year. Florida would be wetter still were it not for the Bermuda High, a cell of high atmospheric pressure in the Bermuda-Azores region of the North Atlantic. The Bermuda High varies in strength and position but keeps Florida from being consistently rainy by inhibiting the development of convective clouds from fall through early spring.

The weakening of the Bermuda High in late spring and summer leads to convective thunderstorms. If the western extension of the Bermuda High persists, drought can extend well into summer. In late summer and early fall ("hurricane season"), tropical low pressure storms from the Atlantic and Caribbean often bring heavy rain and wind to Florida (Chen and Gerber 1990). High and low daily temperatures are rather consistent across Florida in the summer, but winter brings frequent cold fronts that particularly impact north Florida, such that the north–south temperature gradient is much steeper in winter than in summer. The lower Keys have the mildest summers and winters in Florida. In contrast, the Panhandle, being less buffered by the moderating effects of the ocean, usually records both the highest and the lowest annual temperatures.

Annual rainfall in Florida meets the intermediate criterion for savannas, averaging 158 cm (62 in) in the Panhandle, 130 cm (51 in) in northern peninsular and central Florida, 146 cm (57 in) in southern peninsular Florida, and 109 cm (43 in) in the Keys. The lower Keys are particularly dry. The length of the dry season is variable in Florida, both geographically and from year to year, but like the major tropical savanna regions of the world, is usually no longer than seven months. North Florida, especially northwestern Florida (the Panhandle), has a bimodal pattern of dry and wet seasons: wet in winter and early spring, dry in mid to late spring and sometimes early summer, wet most of the summer, and then again dry in mid to late fall (Fig. 1.4). The wet winters and early spring (with rainfall peaking in March) in the Panhandle result from its proximity to winter storm tracks that move in from the north and northwest.

From central Florida southward, the rainfall pattern shifts increasingly to two distinct seasons: a dry season with mild temperatures from mid-fall through spring, and a hot wet season from very late spring or early summer through mid-fall, but again varying considerably from year to year in timing of the onset of wet and dry seasons and in the total amount and seasonal distribution of rainfall. Generally, proceeding down the Florida peninsula the dry season gets drier and the wet season gets wetter (Fig. 1.4). Key West, however, has the driest climate in the state and least seasonal variability in rainfall. The combined dry seasons in north Florida average about three (in northwestern Florida) to seven months, and the single dry season in central and south Florida lasts from six to eight months (Chen and Gerber 1990; Slocum et al. 2010).

The transition from the dry to wet season in late spring or early summer heavily influences the probability and spatial extent of fire, as this is the time of year when thunderstorm activity increases rapidly, while the vegetation is still relatively dry and the groundwater level low from the winter–spring drought. Most human-set fires also occur during this

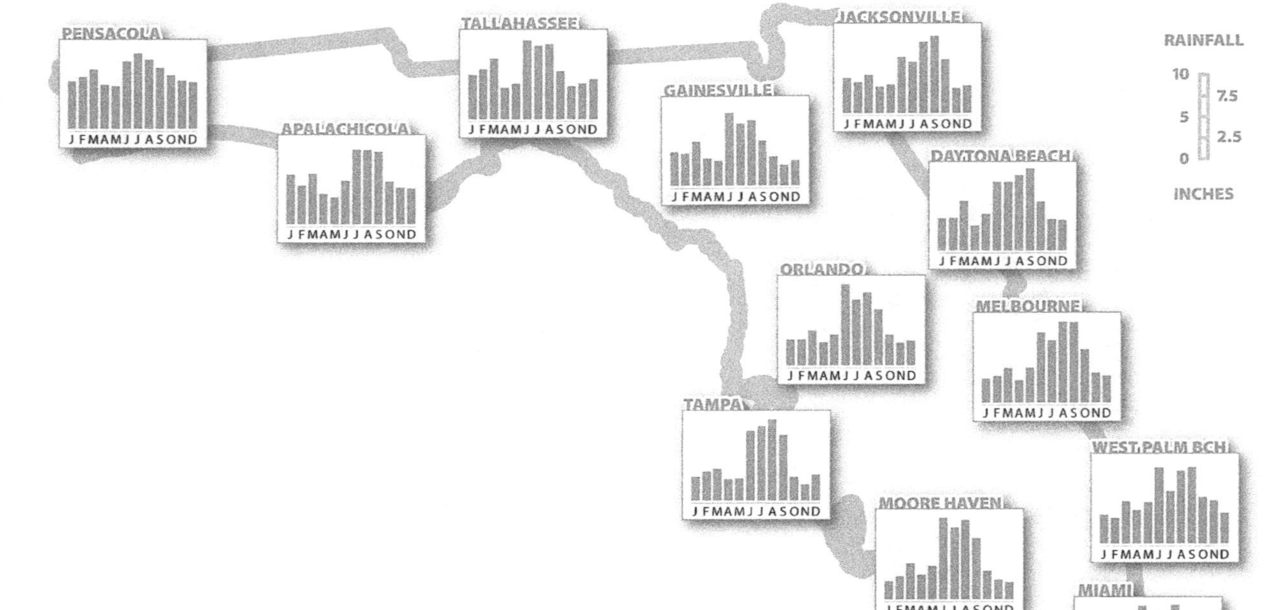

Figure 1.4. Seasonal variation in rainfall, 1981–2010 averages, for selected locations in Florida. North Florida, especially the Panhandle, has two rainy seasons (December–March and a stronger one in June–September) and two dry seasons (April–May and October–December). From around Orlando southward, a single wet season (June–September) and a single dry season (October–May) prevail. Data from http://www.usclimatedata.com/climate/florida/united-states/3179.

vulnerable period of low fuel moisture (Robbins and Myers 1992; Platt et al. 2015).

SEA BREEZE AND THUNDERSTORMS

Wind is another important element in fire weather, both through its direct effect on fire and through its effect on precipitation. Wind in Florida is highly variable over the course of a year and even within a single day, which keeps the job interesting for fire managers here.

An important phenomenon in Florida and the Coastal Plain, as in most near-coastal land areas globally, is sea breeze. Sea breeze is an onshore breeze caused by the land surface warming faster than the sea surface during much of the daylight hours. The land warms faster in the daytime and cools faster at night because it has a much lower heat capacity, the ability to store heat, than the sea. When the temperature of the land surface rises through the morning from the heat of the sun, the air above is warmed and rises. As the air rises it encounters lower atmospheric pressure and expands, which reduces air pressure over land near the coasts. The relatively higher air pressure above the sea then causes air to flow toward land, hence the sea breeze. The greater the difference in temperature between land and sea, the stronger the sea breeze.

Sea breeze becomes prominent along the coast by midmorning and then penetrates inland. For example, sea breeze picks up around noon at Oscar Scherer State Park, south of Sarasota and about 4 km from the Gulf of Mexico, and around 1:30–2:00 p.m. at Myakka River State Park, east of Sarasota and about 30 km from the coast. Fire managers engaged in a controlled burn like to secure the perimeters of their burn units with back fires (i.e., against the wind) and flank fires prior to the onset of the sea breeze, but then let the sea breeze carry a head fire (i.e., with the wind) through the heart of the unit (Robert Dye, personal communication). Meanwhile, as the rising air over land expands, it cools, causing moisture to condense and form clouds.

Some 70 percent of annual precipitation in Florida is attributed to sea breeze (Mogil and Seaman 2008). The sea breeze diminishes toward dusk and often shifts to an offshore wind during the night as the land cools faster than the sea, again due to the greater heat capacity of water. Sea breeze affects coastal land areas adjacent to both the Atlantic Ocean and the Gulf of Mexico. The Florida Keys lack such sea-breeze circulation and, accordingly, receive considerably less precipitation than nearby mainland (Mogil and Seaman 2008).

Sea breeze is relevant to fire in several ways. One major effect of sea breeze is to create a sea-breeze front, a phenomenon in which the cooler air from the sea creates a boundary with the warmer air over land. In concert with the condensation of moisture in rising and expanding air, the sea-breeze front stimulates the development of cumulus or cumulonimbus clouds and often convective thunderstorms, which can extend to heights exceeding 16 km (10 mi). These storm clouds are typically oriented in lines parallel to the coasts. The sea-breeze front progresses inland as warm air continues to rise and cooler air moves in to replace it, with the wind speed largely determined by whether the sea breeze is aligned with or opposes the prevailing wind, and by the degree of thermal contrast between land and sea.

During the summer wet season on the Florida peninsula, thunderstorms caused by powerful sea-breeze fronts converging from the Gulf of Mexico to the west and the Atlantic to the east are frequent in the afternoon. A collision of strong sea-breeze fronts from both directions can cause particularly intense thunderstorms down the center of the peninsula, with the area of convergence usually located toward the leeward coast of the prevailing wind (i.e., downwind). Appropriately, the central spine of Florida has been called "Lightning Alley" due to this collision of sea-breeze fronts. Thunderstorms, of course, include lightning and accompanying fire ignitions. Moreover, an existing fire will be intensified, if only temporarily, by the arrival of a sea-breeze front and its winds. As pointed out above, fire managers can use the arrival of the sea-breeze front to their advantage in controlled burns.

As the sea breeze is replaced by a land breeze at night, showers and thunderstorms often develop over the sea, only to dissipate as the sea breeze develops again the following morning. In the southeastern Coastal Plain north of the Florida peninsula, thunderstorms are also triggered by sea-breeze fronts, but there is less collision of sea-breeze fronts from the Gulf and the Atlantic. This difference is probably one reason why thunderstorm and lightning frequency decline markedly to the north of the Florida peninsula (areas of high lightning activity from the western Panhandle to eastern Louisiana notwithstanding).

The sea-breeze/land-breeze dynamic across the lower Coastal Plain is similar to that in the Florida peninsula. A study in southern Mississippi and eastern Louisiana showed prevailing winds as easterly–southeasterly in June and mostly southerly in July and August. July and August also can have a westerly land breeze from Louisiana. Precipitation typically begins by 11:00 a.m. central daylight time along the coast and is widespread throughout inland areas by afternoon, consistent with the sea-breeze effect, whereas precipitation offshore is primarily nocturnal (Hill et al. 2010). Urban land uses, now quite prevalent in much of Florida and along the coast of the Coastal Plain, create "heat islands" that influence convection and sea-breeze effects, significantly altering convective rainfall totals (Marshall et al. 2004) and presumably thunderstorm and lightning ignition patterns.

An interesting pattern occurs in the Sandhills region of the Carolinas and Georgia, on the inner edge of the Coastal Plain adjacent to the Piedmont. Thunderstorms occur more frequently in the Sandhills than in the immediately adjacent regions. One explanation for this pattern is that the Sandhills, with their drier soil and sparser vegetation than the Piedmont or most of the Coastal Plain, heat more rapidly, which causes convective thunderstorms to develop along the boundary between strongly contrasting soil types (Raman et al. 2005). This process is probably aided by sea breeze, which can reach as far inland as the Sandhills, and by the presence of denser vegetation on either side of the Sandhills, which supplies moisture to feed the storms (Raman et al. 2005).

Fog is a common phenomenon in Florida, most abundant in northern Florida and peaking during the winter. Heavy fog days, with visibility of one-quarter mile or less, decline from around 50 in Tallahassee to 1 in Key West. Fog conditions are highly relevant to fire management and particularly to the scheduling of prescribed burns. When smoke from fires combines with fog, driving conditions become perilous. Smoke from a prescribed burn in the Green Swamp Wildlife Management Area combined with heavy fog on 9 January 2008, resulting in a 70-car pileup on Interstate 4 that killed five people. Fog-smoke combinations are one of the arguments against prescribed burning in the winter, as opposed to in the natural lightning fire season (late April–July), a topic I explore later in this book.

PHYSICAL CHEMISTRY

Air temperature, relative humidity, and other physical and chemical factors related to weather and climate influence fire incidence and behavior. The Physical Chemistry Fire Frequency Model (PC2FM) developed by Guyette et al. (2012) (Fig. 1.5) predicts historic fire frequency (mean fire interval [MFI]) for the period 1650–1850 C.E. The model includes three main input variables: temperature, precipitation, and oxygen concentration as

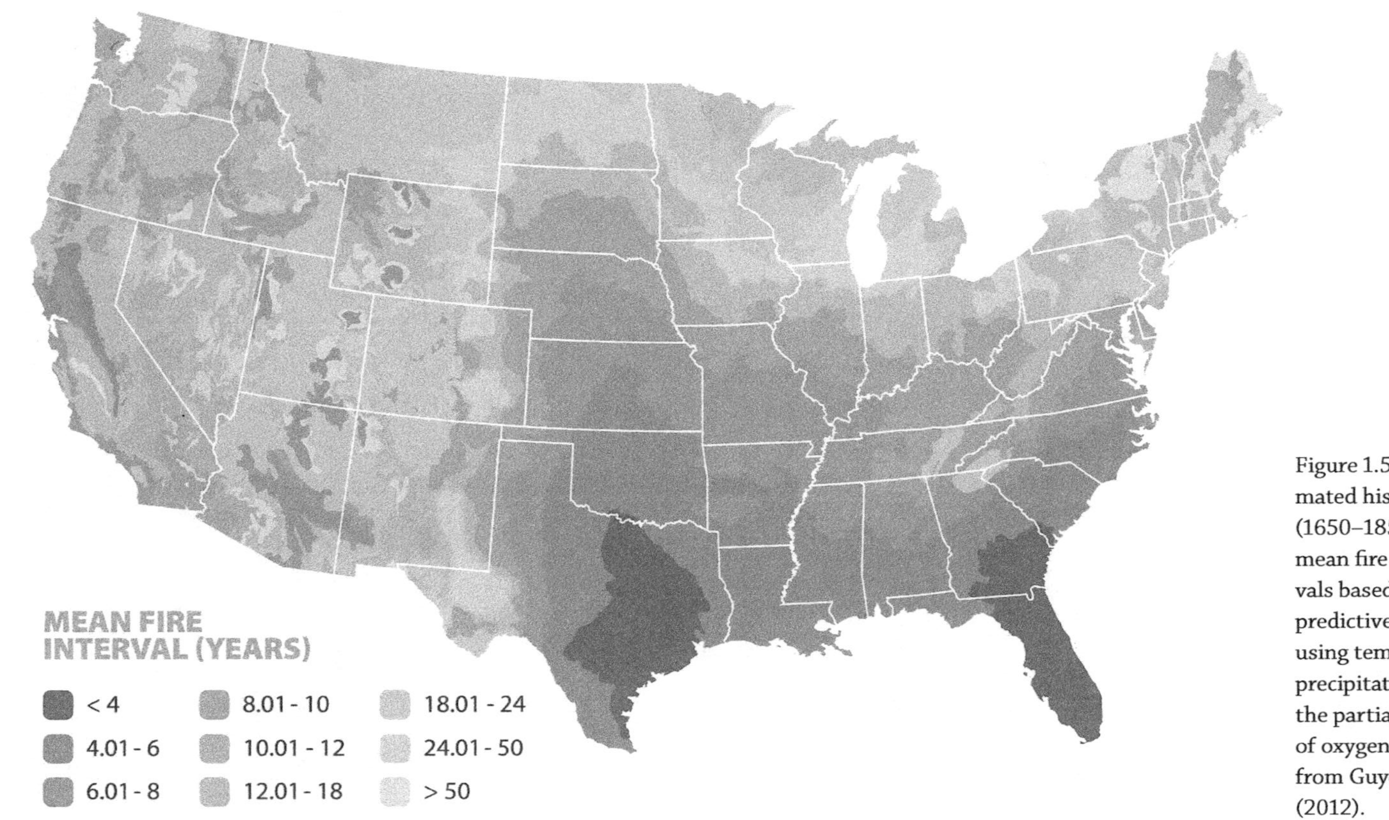

Figure 1.5. Estimated historic (1650–1850 C.E.) mean fire intervals based on a predictive model using temperature, precipitation, and the partial pressure of oxygen. Adapted from Guyette et al. (2012).

estimated by elevation. Each of these factors is known mechanistically to influence the frequency of fire, given an ignition source. Temperature influences reaction rates through kinetic energy, reactant concentrations (production and decay of fuels), and combustion. Higher temperatures promote increased fire; thus, the seasonality of fire is also related to temperature (more fire in warm seasons, all else being equal).

Precipitation serves as a proxy for fuel production, moisture, and humidity. As precipitation and humidity increase, higher temperatures are required to ignite and sustain fires. On the other hand, after a time lag higher precipitation increases the production and density of fuels, especially grasses and other fine fuels that promote frequent surface fires. Oxygen is required for combustion, and the low elevations of the southeastern Coastal Plain result in a high concentration (partial pressure) of oxygen. Wind, which is usually plentiful in Florida, often in the form of sea breeze, replenishes oxygen depleted by combustion, which sustains fire and enhances its spread across the landscape. Even without considering our high lightning frequency, it would be difficult to find a more climatically fire-prone region than Florida and the southeastern Coastal Plain.

In Guyette et al.'s (2012) model, perhaps surprisingly, fire ignitions are ignored. They explain that ignitions were not included in the model because data on lightning or anthropogenic ignitions are generally unavailable for historic periods; their model focused on climate, not ignitions; and anthropogenic ignitions, in particular, have varied greatly over time. They found that excluding topographic, vegetation, and ignition variables increased error at fine spatial scales, but improved predictability of their model at the broader spatial and temporal scales that were the focus of their study. In this model temperature has the strongest influence on fire, with fire incidence increasing as temperatures rise.

The weather conditions of Florida and the lower Coastal Plain vary from year to year, which influences the probability of fire. Long periods of drought were not uncommon historically in this region. Climatic variability and weather extremes, such as heat waves, droughts, and extreme precipitation events, are increasing with global climate change (Field et al. 2012; Wu 2015). A particularly long dry spell occurred in Florida between 1998 and 2002. Associated with this extended drought was increased wildfire activity. In the summer after the Super El Niño event in the winter of 1997–98, every county in Florida reported wildfires. The late spring and early summer of 1998 was unusually dry, with record-breaking high temperatures. The fires occurred in June and July, after two months of these extreme conditions. During the 1998–2002 period 1.5 million acres burned, and property worth $400 million was destroyed. The ecological effects of this increased fire activity were not thoroughly assessed and may well have been positive. Society pays more attention to the negative effects of fire than to the ecological benefits of a frequently burned landscape. Frequent burning has been shown to decrease the incidence and area of unwanted wildfire (Koehler 1993; Addington et al 2015).

Some of the interannual variability in weather and fire activity in Florida and the southeastern Coastal Plain is related to broad-scale weather patterns such as El Niño–Southern Oscillation (ENSO) and the Multi-Decadal Atlantic Oscillation (MDAO). In several regions fire occurrence is higher just after the switch to the La Niña phase of ENSO (Kitzberger et al. 2001), when abundant fine fuels produced during the moist El Niño phase are subsequently desiccated. Beckage et al. (2003) showed the La Niña phase of ENSO was characterized by lower dry season rainfall, reduced surface water levels, increased lightning activity, more fires, and more area burned in the Everglades. The El Niño phase, in contrast, was associated with increased dry-season rainfall, higher surface water levels, reduced lightning activity, and fewer and smaller fires. They reasoned that shifts between ENSO phases influence vegetation through periodic large fires, which lead to a prevalence of pyrogenic communities in the Everglades landscape.

LIGHTNING

Although physical chemistry predicts high fire frequency in the southeastern Coastal Plain even without explicit consideration of ignitions, fires cannot burn unless they are ignited. In a study of fire histories in national forests across the southeastern United States, an index of drought (the Palmer Drought Severity Index [PDSI]) explained more of the spatial and temporal variation in lightning fires than lightning flash density. In Florida and the southeastern Coastal Plain, however, lightning flash density was a strong predictor of the frequency of lightning-caused fires, with Florida having the highest density of lightning strikes and the most lightning fires. The national forest with the highest lightning fire frequency was Osceola National Forest, just south of the Okefenokee National Wildlife Refuge in Georgia, with 300 fires recorded in one decade (Mitchener and Parker 2005). These authors pointed out that many of the smaller national forests in the Southeast do not regularly experience lightning fires, which underscores the importance of large wild areas for sustaining natural disturbance regimes.

Over evolutionary time lightning probably was the sole ignition source that led to the development of fire-adaptive traits for species in the Coastal Plain. The only possible exceptions are young species that evolved during the less than 15,000 years that humans have inhabited the region and whose evolution may have been influenced by anthropogenic fire. Florida and the lower Gulf Coastal Plain have the most thunderstorms and the highest density of cloud-to-ground lightning flashes (Fig. 1.6) in the United States. Across the Coastal Plain, cloud-to-ground strikes average between 4 and 15 strikes per km^2 per year. The only place in North America with a higher lightning flash density (more than twice as high) than Florida is a relatively small area of northwestern Mexico on the western slope of the Sierra Madre Occidental (Murphy and Holle 2005).

From 2005 through 2014 Florida received an average of more than 1.2 million lightning strikes per year, about 200,000 more than any other state in the United States, and 47 people were killed by lightning (Allen 2015). Lightning ignited 265 building fires and house fires across the state in 2014 alone and led to more than 10,000 insurance claims.

With all the lightning strikes in Florida and the lower Coastal Plain, and with physical and chemical conditions ideal for fire, it is surprising that the greatest frequency of reported lightning fires 300 or more acres in size in the United States today is in the West, especially the Southwest and Intermountain West, according to the Federal Emergency Management Agency (FEMA]. This region has fewer cloud-to-ground lightning strikes (Fig. 1.6), considerably more topographic relief, and generally less flammable vegetation than in the Southeast.

Why does the West appear to have more lightning fires than the lightning-riddled southeastern Coastal Plain? Four non–mutually exclusive explanations for this paradox are: (1) Many lightning fires in the Coastal Plain are smaller than the 300-acre threshold FEMA uses for recording wildfires, especially because the southeastern landscape is more fragmented by roads, urban areas, and other human structures than most western landscapes, which limits the size and spread of fires. (2) As noted by Robbins and Myers (1992), many small lightning fires are unreported, especially during the busy lightning-fire season and if they are not perceived to create a fire-control problem. (3) Over the last several decades, spring and summer temperatures and the intensity of drought have been increasing across much of the western United States, in turn producing

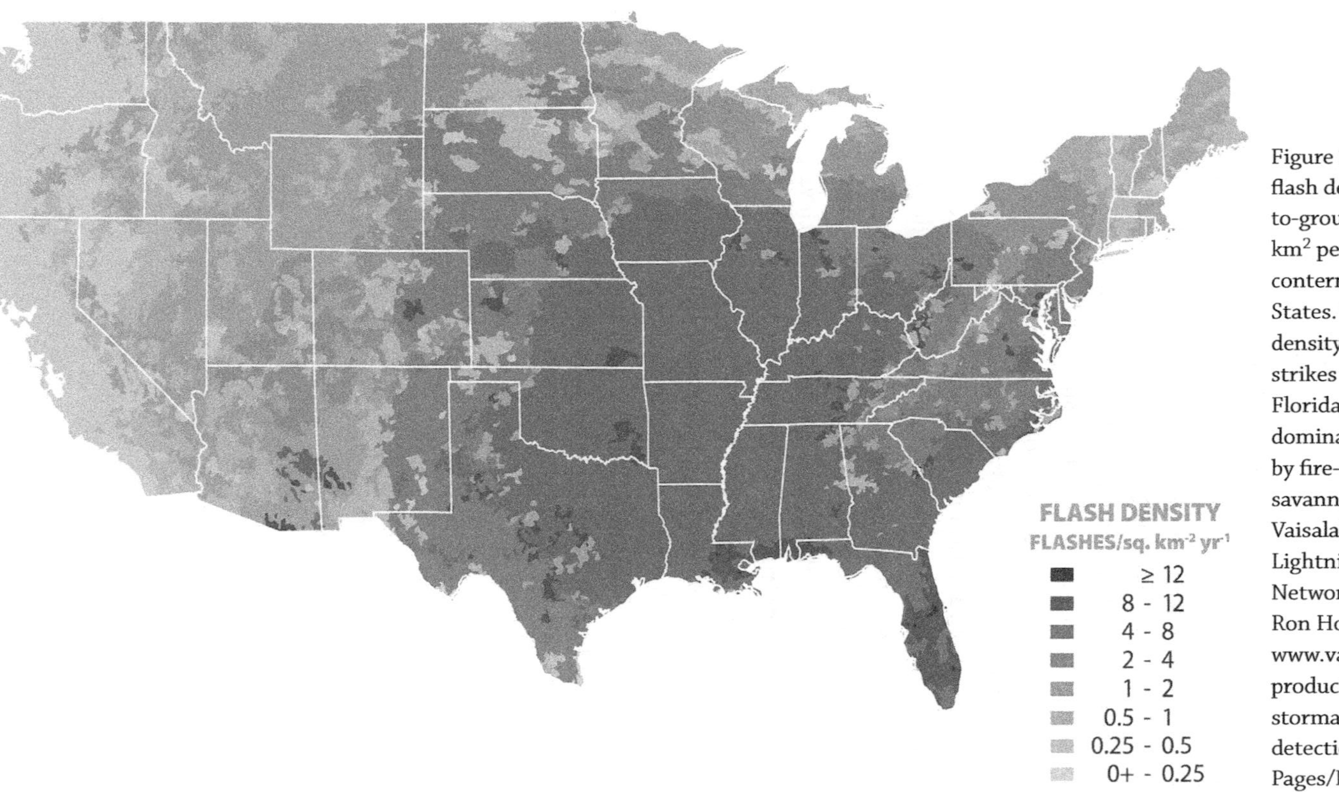

Figure 1.6. Lightning flash density (cloud-to-ground flashes per km^2 per year) in the conterminous United States. The highest density of lightning strikes is in central Florida, a region dominated historically by fire-dependent pine savannas. Data from Vaisala's National Lightning Detection Network, courtesy of Ron Holle. See http://www.vaisala.com/en/products/thunderstormandlightningdetectionsystems/Pages/NLDN.aspx.

more frequent large fires (Westerling et al. 2006; Liu and Wimberly 2015). This trend is less evident for the Southeast. (4) The states in the southeastern Coastal Plain burn more acres with prescribed fire than in any other region of the country or, probably, the world, which reduces the incidence of wildfire. Addington et al. (2015) show that prescribed fire and drought explain 80 percent of the variation in wildfire incidence and 54 percent of the variation in the areal extent of wildfires on Fort Benning, Georgia.

Lightning frequency is predicted to increase with global warming. A model in which the lightning flash rate is proportional to the convective available potential energy (CAPE) multiplied by rate of precipitation projects an increase in cloud-to-ground lightning flashes of 12 percent per degree Celsius of warming, an approximately 50 percent increase over the twenty-first century (Romps et al. 2014). Combined with warmer temperatures, higher rates of evapotranspiration, and drier fuels (Guyette et al. 2012), more lightning strikes will produce more fires. On the other hand, increasing levels of atmospheric CO_2 may promote growth of trees and other plants utilizing the C_3 photosynthetic pathway, thereby releasing them from control by fire (Midgley and Bond 2015). In chapter 3 we return to the fire environment subject with more detailed consideration of the frequency, seasonality, and intensity/severity of fire, components of the fire regime that have a strong influence on the evolution of adaptive traits of species. Chapter 5 provides additional discussion of managing fire into a future with new, possibly no-analog conditions.

2

History of Fire and Fire Ecology in Florida and the Coastal Plain

> Although anthropogenic fires have undoubtedly extended areas of flammable vegetation, there is now abundant evidence that natural fires occurred long before humans . . . and that flammable ecosystems predate anthropogenic burning by millions of years.
>
> Bond et al. (2005)

Any naturalist who finds the adaptations and life histories of organisms fascinating must, sooner or later, delve into the past or risk missing a big part of the story. By exploring deep history, a naturalist comes to understand something about the origin and function of traits, selective pressures in the environment that favored their evolution, how old they are, and which species (closely related or not) share similar traits. The extraordinary biodiversity of the North American Coastal Plain has been overlooked, in part, because of a false assumption of its geologic youth (Noss et al. 2015). In truth, as this chapter reviews, the terrestrial Coastal Plain extends back to the Cretaceous. Even its lowest-elevation subregion, the Florida peninsula, has been at least partially and continuously above the sea for many millions of years.

The history of fire in a region is a central consideration for determining how fire might have shaped the morphological features, life histories, and behaviors of species native to that region. Also valuable to learn is how long apparently fire-adapted species have coexisted. How long has fire been part of the evolutionary environment of species found in fire-prone ecosystems today? Did certain regions serve as refugia for fire-adapted species as the global climate shifted?

These kinds of questions are fundamental to the sciences of biogeography and evolutionary biology. They are also relevant to fire ecology, as answering them tells us about the role of fire as an ecological and evolutionary force and about the antiquity of fire in a region. In this chapter I show that lightning fire and fire-prone vegetation resembling what exists today have a long history in the Coastal Plain, vastly preceding the arrival of humans. Fire regimes were altered by humans thousands of years ago, but not as extensively as often assumed. Only recently have humans exerted strong control over fire, for better or for worse.

Global Fire in Deep Time: Paleozoic and Mesozoic Eras

Fire was on Earth long before what is now Florida separated from West Africa as Gondwana broke apart and the Atlantic Ocean formed at the end of the Paleozoic Era, about 250 million years ago. Recall that for a fire to start, three things must be present: oxygen, relatively dry fuel, and an ignition source. Lightning and other potential sources of ignition, such as volcanos, falling rocks, and meteorites, were presumably present virtually throughout Earth's history. Before photosynthesis evolved, however, there was no oxygen in the atmosphere to feed fire. Earth had enough oxygen in its atmosphere to support fire by the beginning of the Paleozoic, some 540 million years ago (Pausas and Keeley 2009). This oxygen was produced from the photosynthesis of marine algae, however, and fire doesn't carry well in water.

Suitable fuel to carry fire was lacking until terrestrial vascular plants evolved in the Silurian Period, beginning about 440 million years ago. Charred remains of stomata-bearing plants have been found from this time (Glasspool et al. 2004). Until recently charcoal-like plant fragments, known as fusain, were not fully acknowledged as resulting from wildfire. Only over the past three decades has the wildfire origin of fusain become well accepted and the study of fire in deep time developed (Scott 2000; Scott et al. 2014).

Studies of fire history over multimillion-year intervals demonstrate clearly the primary role of oxygen in both the ignition and spread of fire. High fire activity corresponds to periods of high atmospheric oxygen concentrations, and low fire activity corresponds to low levels of oxygen. At atmospheric oxygen levels around the current 21 percent, fire activity is relatively high, but not nearly as high as during the Carboniferous Period, when oxygen levels peaked at around 31 percent, or in the Cretaceous, when levels were approximately 23–29 percent (Pausas and Keeley 2009; Scott et al. 2014). During these high-oxygen periods, fires could have ignited and spread even with fuels much wetter than what will carry fire today.

After the initiation of wildfire in the Silurian, fire activity increased during the Devonian and through the Carboniferous (Mississippian and Pennsylvanian) Periods. The fire regimes during this span of time were similar to fire regimes today. For example, laminated charcoal deposits suggest surface fires at 3–35-year intervals in progymnosperm forests of the Devonian, 395 million years ago (Cressler 2001). Crown fire regimes with fire-return intervals of 105–1,585 years are inferred from charred growing tips of trees in tropical wetland forests of the Carboniferous (Falcon-Lang 2000). The rise in global fire activity from the Devonian through the Carboniferous is documented by microscopic charcoal in marine sediments deposited in ancient seas—for example, in what is now the Ohio River basin (Scott et al. 2014). This increase in fire was accompanied by changes in the structure of vegetation, with plants becoming taller and producing greater fuel loads over time. The evolution of seed plants in the late Devonian, as opposed to earlier reliance on spores in wet environments, permitted the spread of plants into drier habitats, which burned more readily.

Fire is highly improbable when atmospheric oxygen levels fall below 15 percent. Fire activity is low at levels of 15–17 percent, which occurred during periods such as the late Permian (coinciding with the largest mass extinction in the history of life, 250 million years ago) through the Triassic. During the Jurassic, which began around 200 million years ago, oxygen levels increased, and so did fire, as shown by charcoal deposits. Tree-ring patterns and charred fragments from gymnosperm forests of the Jurassic provide evidence of high-frequency surface fires (Francis 1984), perhaps not much different from what our Coastal Plain pine savannas experience today.

The Cretaceous (145–66 million years ago) saw further increases in oxygen and carbon dioxide levels, air temperatures, and fire activity, coinciding with the origin and spread of angiosperms (flowering plants, including grasses). Fire probably played a major role in the evolution and geographic expansion of angiosperms during this period (Bond and Scott 2010).

The spread of terrestrial vascular plants into more fire-prone habitats, which began during the Devonian and greatly accelerated much later, during the Cretaceous, paved the way for the evolution of fire-adaptive traits—that is, morphological and life-history features that allow plants and animals to survive and reproduce in the presence of fire. Evidence of likely fire adaptations in plant lineages, as determined by fossils and phylogenetic reconstructions, can tell us much about fire history. Fossils rarely provide direct evidence of fire-adaptive traits, but dating the origin and evolution of such traits from time-based molecular phylogenies offers insights into the role of fire as an agent of selection.

Fire adaptations in plant clades that arose prior to the Cretaceous are ambiguous at best. Beginning in the Cretaceous, however, the pines (family Pinaceae and especially genus *Pinus*) offer a compelling story of increasing adaptation to fire over time (He et al. 2012). Most species of *Pinus* living today occur in high-fire environments, and these trees show a marvelous collection of fire-adaptive traits (see chapter 3). The Pinaceae probably originated in the mid-Triassic, around 237 million years ago. Northeastern North America and western Europe, which were connected as one landmass at that time, is the most likely region of origin of *Pinus* (Millar 1998). The oldest fossil pine, *Pinus mundayi*, was recovered from lower Cretaceous rocks 133–140 million years old in Nova Scotia, Canada. *Pinus mundayi* was preserved in charcoal and shows resin ducts, which in modern pines produce flammable terpenes (Falcon-Lang et al. 2016).

The first unambiguous fire-adaptive trait seen in pines, thick bark, appears during the early to mid-Cretaceous, around 126 million years ago. This is the earliest evidence for any taxa for fire acting as an evolutionary force (Pausas 2015a). As shown by He et al. (2012) in their dated phylogeny of Pinaceae, the origin of thick bark in pines coincides with a period of high atmospheric oxygen and frequent surface-fire activity. This suggests these pine-dominated communities were open-canopied savannas or woodlands.

The analysis by He et al. (2012) suggests that other fire-adaptive traits in pines, such as serotiny associated with crown fires in more closed-canopied communities, evolved later in the Cretaceous (Fig. 2.1). Thus, *Pinus* diverged during the Cretaceous into two distinct adaptive peaks, one associated with stand-maintaining surface fires and the other with stand-replacing crown fires. Studies of trait evolution in taxa older than pines would likely push the date for fire as a selective pressure farther back in time (Pausas 2015a).

Not only oxygen, but atmospheric carbon dioxide (CO_2) was high during the warm greenhouse conditions of the Cretaceous, which would have led to high primary productivity and biomass. This translates into high fuel loads. Atmospheric CO_2 fertilization leads to higher rates of photosynthesis and accumulation of above- and belowground biomass. Although high

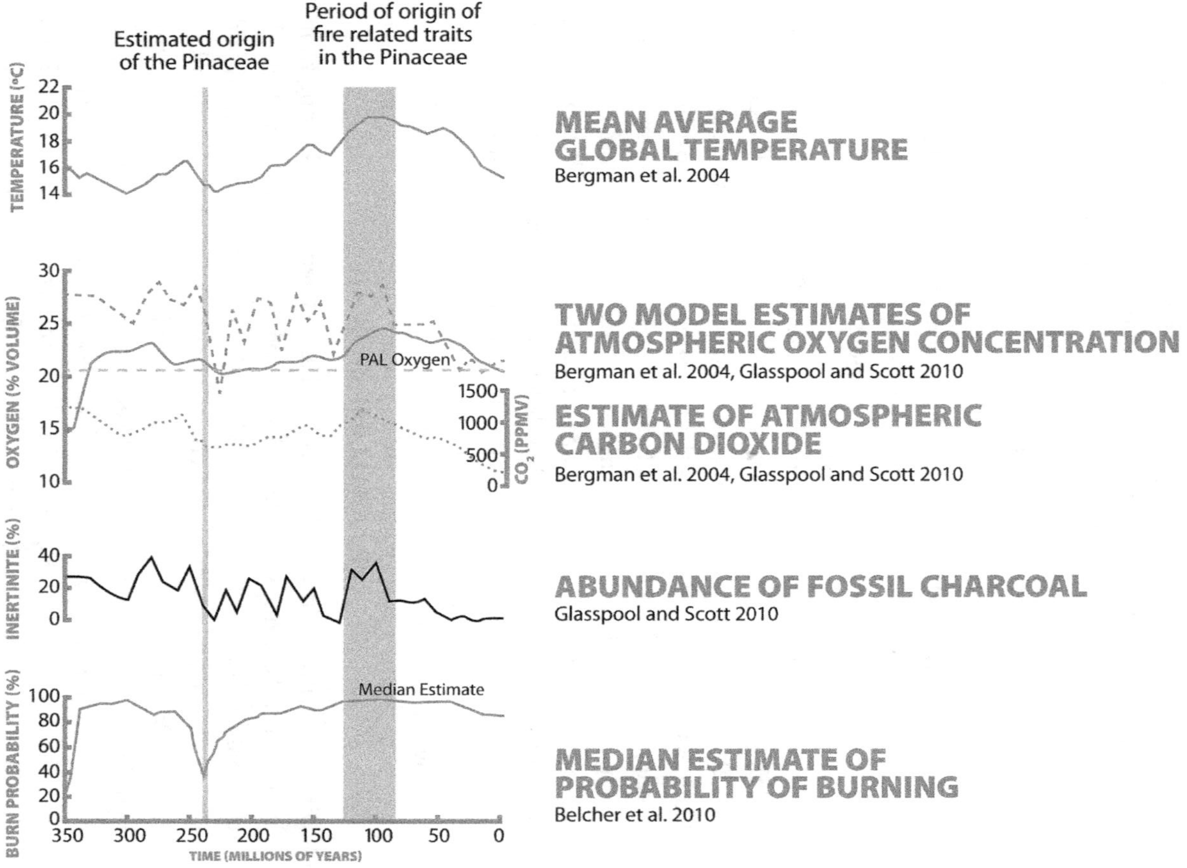

Figure 2.1. Period of origin of fire-related traits in the Pinaceae in relation to mean annual global temperature, estimated atmospheric oxygen concentrations, estimated atmospheric carbon dioxide, abundance of fossil charcoal, and estimated probability of burning over the last 350 million years. Dashed horizontal line indicates the present atmospheric level (PAL) of oxygen. Sources of data are shown under each parameter. Modified from He et al. (2012).

fire activity temporarily increases atmospheric CO_2 levels and leads to higher air temperatures, charcoal is a very stable, decay-resistant form of carbon: a sink. Thus, the accumulation of buried charcoal from many fires can potentially reduce CO_2 levels and result in global cooling (Scott et al. 2014).

Early angiosperms in the understories of fire-prone pine savannas and woodlands probably increased the amount of fine fuels and the flammability of the stands, as well as potentially increasing charcoal accumulation. These angiosperms included grasses, most of which carry fire well, although burned woody species produce more charcoal (Pausas and Keeley 2014a). Although fossil evidence of grasses in North America before the Paleocene does not yet exist, grasses evolved during the Cretaceous.

Ecosystems and Fire in the North American Coastal Plain from the Late Cretaceous through the Neogene (66–2.6 million years ago)

We now concentrate on the close of the Mesozoic Era (the Late Cretaceous) and the unfolding of the Paleogene and Neogene Periods of the Cenozoic Era. This was the span of time during which the North American Coastal Plain emerged from the sea and the fire-prone ecosystems of the region developed, spread, and diversified in species. The species that invaded the emerging Coastal Plain, beginning in the Late Cretaceous and accelerating later as more of the region was exposed, presumably brought with them some fire-adaptive traits that developed in fire-prone communities in their regions of origin, such as the Appalachians, the Great Plains, and the West Indies. Over time these traits were shaped to match the unique fire environment of the Coastal Plain.

THE CHARCOAL PROBLEM

Direct evidence of fire, in the form of charcoal deposits, is lacking from some intervals during which fire-adapted vegetation was present. Indeed, charcoal evidence of fire is sparse through much of the Cenozoic. Bond (2015) showed "a fundamental conundrum in the geological record of fire. . . . [W]e live in what seems to be a very flammable world where fire is influential in structuring global vegetation. Yet the inertinite [fossil charcoal component of coal] record suggests we are living in a world of very low fire activity" (Fig. 2.2). The inertinite content of coal has been considered one of the best proxies of historic fire activity, since it presents

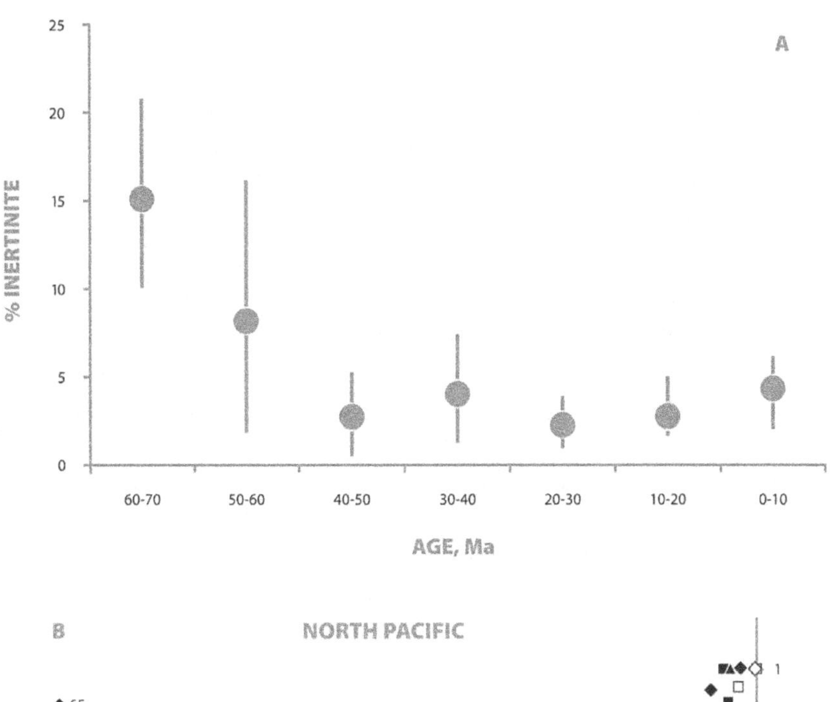

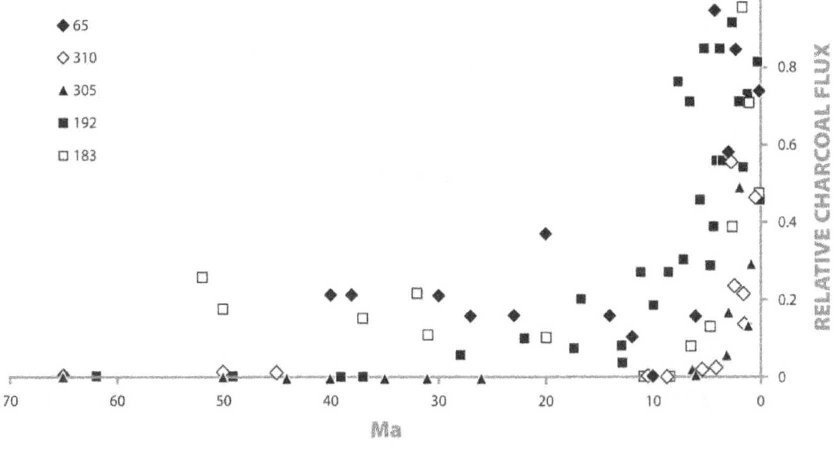

Figure 2.2. (*A*) The abundance of inertinite (the fossil charcoal content of coal) from the late Cretaceous through the Cenozoic, the last 65 million years of Earth history. By this measure commonly used to estimate historic fire activity, fire appears rare throughout the Cenozoic. However, inertinite is an imperfect measure of fire activity, especially in frequent-fire ecosystems such as savannas. (*B*) Ocean cores from the North Pacific, on the other hand, show a spike of charcoal from 10 million years ago until present, indicating a fiery environment over this period. Symbols refer to localities (not noted here), and values are relativized to the maximum charcoal recorded at each site. Adapted from Bond (2015).

a near-continuous record from the Silurian to the present (Glasspool and Scott 2010). However, inertinite is formed in peatlands, which are uncommon in such landscapes as seasonally dry savannas or shrublands (Keeley et al. 2012; Bond 2015), the ecosystems that dominate much of Florida and the Coastal Plain.

Although charcoal is present in some sediment cores from lake bottoms in what are now fire-dependent pine savanna and scrub landscapes in Florida, its abundance is low. This may be because grassy ecosystems, which dominate Florida and much of the southeastern Coastal Plain (Noss 2013), produce far less charcoal than woody vegetation (Pausas and Keeley 2014a; Bond 2015). Also, dense, fire-resistant vegetation that typically develops next to water bodies (from which sediment cores are derived) filters out much of the charcoal arising in the adjacent frequently burned uplands (Aleman et al. 2013; Bond 2015). Macrofossils that might indicate fire adaptations, such as serotinous cones, are rare in upland ecosystems, because opportunities for fossilization are virtually nonexistent.

Due to these biases, we are forced to make inferences about past fire environments based on indirect evidence, which is less than ideal. Nevertheless, both phylogenetic studies and fossils can tell us much about past fire activity, despite lack of corroboration by charcoal deposits. When fossils suggest vegetation in the past that closely resembles current vegetation that we know is fire-prone, it is reasonable to surmise that fire played an important ecological and evolutionary role in those ancient ecosystems.

THE LATE CRETACEOUS AND PALEOGENE (PALEOCENE, EOCENE, OLIGOCENE)

From the middle of the Cretaceous to the early Paleogene a vast inland sea extended from the Gulf of Mexico northward through the western lowlands, all the way to the Arctic Ocean (Graham 1993; DiPietro 2013). Although portions of what is now the Coastal Plain may have been above the sea intermittently during the Mesozoic (DiPietro 2013; Blakey http://cpgeosystems.com/index.html), fossil evidence of terrestrial vegetation in the region begins in the Cretaceous and expands during the Paleogene. Cretaceous macrofossils in the Coastal Plain include fir (*Abies*), pine (*Pinus*), birch (*Betula*), holly (*Ilex*), tulip-tree (*Liriodendron*), and oak (*Quercus*), all of which continue into the richer microfossil (palynomorph or fossil pollen) record of the Paleogene (Gray 1960). A mass extinction event approximately 66 million years ago—the last really big one until the present day—marks the close of the Cretaceous and the beginning of the

Paleogene. The nonavian dinosaurs went extinct at the end of the Cretaceous, and Earth's climate swung from greenhouse heat to cooler conditions, only to become warmer and wetter again in a few million years.

The Paleogene extended from 66 to 23 million years ago and includes three epochs: the Paleocene (66–59 million years ago), Eocene (59–34 million years ago), and Oligocene (34–23 million years ago). The Paleocene plant fossil record from the Coastal Plain is meager, but it suggests broad-leaved tropical forest along the Gulf Coast. The southeastern Coastal Plain, our study area, was still entirely inundated until the middle Eocene, and most of it apparently until the Oligocene or later. Fluctuations in sea level continued to inundate considerable portions of Florida and the southeastern Coastal Plain at various times, most recently during the Pleistocene interglacials (see below). Temperature and humidity fluctuated during the Paleogene, but conditions were warm and humid during much of the period. The transition from the Paleocene to the Eocene was marked by one of the most rapid increases in global temperature known in Earth history: ca. 5–8°C over a few millennia (Bowen et al. 2015). Sea level rose sharply during this period, only to fall again as the climate cooled, beginning around 49 million years ago and continuing through the middle and late Eocene and into the Oligocene.

The extensive Eocene fossil record from the Coastal Plain is divided into three major floras: the Wilcox (lower Eocene), Claiborne (middle Eocene), and Jackson (upper Eocene) (Berry 1924). Several plant fossils from the Eocene Coastal Plain hint at fire in the environment. For example, grass spikelets occur in the Wilcox Flora from western Tennessee. The Claiborne Flora (named after the Claiborne Bluffs along the Alabama River in southern Alabama) is represented at several sites. *Ephedra* (Mormon tea) is first recorded in the Southeast from the Claiborne Flora. *Ephedra*, which today is distributed largely in arid western North America but is still extant in the West Gulf Coastal Plain, was likely associated with well-drained, deep sandy soils (Graham 1999). These soils create physiologically xeric conditions even with high precipitation, much like pine sandhill and scrub in Florida today. *Ephedra* can resprout after topkill, a potential adaptation to fire.

Pines, which well before this time had developed a suite of fire-adaptive traits (He et al. 2012), are also represented in the Claiborne Flora in southern Alabama (Gray 1960). Although the Eocene is considered a mostly wet period, and charcoal deposits from marine cores are rare (Scott et al. 2014; Bond 2015), many low-latitude regions were seasonally dry, as the

Coastal Plain is today (Millar 1998). Seasonally dry savannas and woodlands are typically fire-prone. The southeastern United States appears to have served as a refugium for pines during the Eocene (Millar 1998), which suggests fire may have been part of this environment.

Sea level dropped substantially from the middle Eocene through the Oligocene in association with a trend toward a generally cooler and drier climate and lower CO_2 concentrations. Graham (1999) noted that the middle Eocene transition to cooler conditions was "an important time in the modernization of North American plant communities." At Claiborne Flora sites in western Kentucky and Tennessee (then and now part of the Coastal Plain) legumes are prominent. These and other taxa, including palms (*Serenoa* and *Sabal*), pines, *Ephedra*, and oaks (*Quercus*), suggest an active fire regime. Graham writes that "on sandy flats, but removed from the tidal influence, *Pinus* with an understory of *Sabal-Serenoa*-type palms and *Ephedra* were present." This description is reminiscent of the fire-prone vegetation of modern-day Florida. As pointed out by Bond (2015), however, "the plants that built new fire-dependent biomes were present many millions of years before they began to expand under increased fire activity," which happened largely in the late Miocene.

Common mythology maintains that terrestrial Florida is only a few tens of thousands of years old, at most, and vegetation similar to today's developed only recently. The origin of this myth is uncertain, but it is pervasive and has led to many false notions about the biogeographic and evolutionary history of Florida and the Coastal Plain (Noss 2013; Noss et al. 2015). For example, historian Stephen Pyne (2016) claims "the Pleistocene Ice Ages raised Florida from the ocean," and "Florida is among the most recent of American landscapes, even younger than much of the land that emerged from beneath the Wisconsin glaciation." That would make terrestrial Florida less than 20,000 years old! In fact, Florida has yielded terrestrial fossils more than three orders of magnitude older (see below). Solid evidence shows Florida emerged from the sea tens of millions of years ago. Moreover, the modern flora and fauna of Florida are not substantially different from what existed through much of the Pleistocene, and many of the taxa have been here for many millions of years longer.

Florida was long thought to have been completely submerged throughout the Eocene, continuing into the Oligocene. The oldest terrestrial vertebrate fossils are from a middle Oligocene (ca. 28 million years ago) site along Interstate 75 near Gainesville, the "I-75 Local Fauna" (Patton 1969). Thus, paleontologists thought the Florida peninsula first became terrestrial during the Oligocene. Recently, however, land plant microfossils were discovered in limestones of the Avon Park Formation from the middle Eocene, ca. 56–44 million year ago, the oldest exposed stratigraphic unit in Florida (Jarzen and Dilcher 2006).

These finds, from quarries in the Gulf Hammock area of Levy and Citrus Counties, represent the earliest evidence of a landmass in Florida. The fossils include pollen and spores from 21 terrestrial plants, as well as marine taxa. Other discoveries from these sites, including seagrasses and remains of whales, crocodiles, sea turtles, and dugongs, suggest a shallow marine lagoon, perhaps near the mouth of a river. The terrestrial plants include *Pinus*, *Quercus* (or a close relative), and *Ephedra* (Jarzen and Dilcher 2006). As with fossils from the middle Eocene Claiborne Flora elsewhere in the Coastal Plain, these fossils indicate potentially fire-prone vegetation. Apparently much of the southeastern United States was semideciduous tropical dry forest during the middle Eocene, but the beginnings of warm-temperate deciduous forest and "various edaphic and fire-controlled pine woodlands" can be traced to this time (Graham 1999).

By the late Eocene, the trend toward elimination of tropical species and expansion and diversification of temperate species in the Coastal Plain was well under way. During the Oligocene global cooling, sea level fell significantly, and by 30 million years ago the Mississippi Embayment retreated rapidly southward (Graham 1993). The Great Plains transitioned from forest and woodland to more open vegetation, approaching "a savanna in the modern sense" (Graham 1999). Vertebrate fossils indicate these savanna conditions extended southeastward to Florida, where, for instance, the I-75 Local Fauna includes a horse, an oreodont (an artiodactyl related to camels), a tortoise, and other animals associated with open environments (Patton 1969). The late Oligocene was marked by a warming trend that extended into the Miocene, but with rainfall staying relatively low.

THE NEOGENE (MIOCENE, PLIOCENE)

The Miocene Epoch (23–5.3 million years ago) represents the early Neogene Period. By the early Miocene remnants of the Paleogene tropical flora were probably limited to southern coastal areas, but pine woodlands and hardwood forests were extensive across the Coastal Plain. Vertebrates associated with open and potentially fire-maintained ecosystems included horses, a rhinoceros, a horned cameloid, an oreodont, a peccary, a pocket mouse, a beardog, a Gila monster, and a tortoise. All these taxa have been found at Thomas Farm in Gilchrist County, Florida (Webb 1990), one of the premier sites for fossil vertebrates in North America. The most common bird in the Thomas Farm collection is a small ground-dove (*Columbina prattae*) (Hulbert 2001). The extant common ground-dove (*Columbina passerina*) in the Coastal Plain is associated primarily with fire-dependent ecosystems, especially xeric oak scrub, scrubby flatwoods, open pine flatwoods, and coastal strand (Stevenson and Anderson 1994). By the middle Miocene, *Geochelone* (a giant tortoise), a giant ground sloth, and elephant-size proboscideans had joined the vertebrate fauna. Altogether, the early and middle Miocene vertebrates of Florida suggest a mix of subtropical forest and savanna. Pines increased in abundance across much of North America during the Miocene, and the direct ancestors of many modern pines apparently arose during this epoch (Millar 1998).

Global temperatures sharply decreased and became more seasonal in the middle Miocene between 15 and 10 million years ago, a trend that continued into the late Miocene (Graham 1999). During the middle Miocene one of the most profound vegetation transitions in Earth's history began taking shape. Several lines of evidence suggest a global increase in grasslands and savannas at the expense of forest, beginning with expansion of C_3 ("cool season") savannas and woodlands in the middle Miocene, followed by an explosion of C_4 ("warm season") grasses about eight to six million years ago in the late Miocene. These changes are accompanied by a clear trend of mammalian tooth structure from low-crowned (brachydont) teeth adapted to browsing on woody plants, to medium-crowned (mesodont) teeth adapted to a diet of woody plants and herbs (including grasses), to high-crowned (hypsodont) teeth adapted to grazing on grasses with their abrasive high-silica content (Janis et al. 2002).

The diversity of ungulates in North America peaked in the middle Miocene, when a mix of species with all three types of teeth inhabited C_3 woodlands and savannas with no modern analogs (Janis et al. 2002). Fossil vertebrates in Florida from the middle Miocene indicate a mix of subtropical forest and open habitat. The subtropical forest (which may have been a tropical forest in the process of being invaded by temperate species) is well represented by the Alum Bluff Flora along the Apalachicola River. In contrast, open habitats, which again were likely fire-maintained, are indicated by fossils of giant tortoises and grazing hypsodont mammals, such as a rhinoceros and seven species of horses (Webb 1990). Antilocaprids (pronghorns) are present but rare in middle and late Miocene deposits in Florida (Hulbert 2001).

During the late Miocene charcoal in marine sediment cores increased by up to a thousandfold, signaling a pulse in fire activity globally (Beerling 2007). This increased fire activity corresponds closely with the expansion of C_4 grasses and a surge of plant speciation as fire-sensitive forests were largely replaced by grasslands and savannas (Bond 2015). A shift to a monsoonal climate with dry winters and wet summers, like the Florida peninsula experiences today, promoted frequent fires and expansion of C_4 savannas (Keeley and Rundel 2005; Bond 2015). A wet summer allows for abundant growth of fine fuels, while a dry winter or spring makes the fuels more combustible.

We can be confident about the shift in dominance from C_3 to C_4 grasses in the late Miocene because plants with C_3 versus C_4 photosynthetic pathways contain different ratios of heavy and light stable isotopes of carbon in their tissues. These isotopic signatures are retained in the fossil tooth enamel of the herbivores that ate them, and these fossil teeth show a pronounced trend toward increasing C_4 isotope signatures in the late Miocene (Cerling et al. 1993, 1997). Based on carbon isotope analysis of mammal teeth, MacFadden and Cerling (1996) estimated C_4 grasses arrived in Florida between 9.5 and 4.5 million years ago. The expanding late Miocene grasslands were maintained by frequent fire. In addition to the huge increase in charcoal in deep ocean sediments, major increases in charred fossil grass cuticles occurred then (Morley and Richards 1993). By the late Miocene, faunal connectivity was established between the Gulf Coastal Plain and the Great Plains, which probably represents an eastward expansion of a savanna corridor that had previously been confined to the Rio Grande Trench (Webb 1977).

The late Miocene shift to C_4 grasslands was accompanied by a reduction in the numbers of brachydont and mesodont mammal species and an increase in hypsodont mammals in North America (Janis et al. 2002). Vertebrate fossils in Florida—for example, from the Love Bone Bed near Archer—suggest a mix of stream-bank vegetation, riverine forest, and savanna, with the last of these indicated by abundant cranes, tortoises, an extinct pronghorn antelope, rhinocerotids, camelids, proboscideans, and 10 species of hypsodont horses (MacFadden and Hulbert 1990; Webb 1990). Only in sub-Saharan Africa can one find such a large assemblage of savanna ungulates today. A reconstruction of the paleoclimate of Florida in the late Miocene suggests annual rainy and dry seasons, much like a modern African savanna (MacFadden and Hulbert 1990) and modern Florida. Such a climate is highly conducive to frequent surface fires.

During this period, tropical savannas spread near-synchronously in Africa, Asia, and the Americas (Keeley and Rundel 2005).

Dated phylogenetic evidence for multiple lineages of plants and animals associated with fire-dependent ecosystems is sparse in the southeastern Coastal Plain compared to Africa and South America, where excellent reconstructions exist (Simon et al. 2009; Maurin et al. 2014; Pennington and Hughes 2014). One intriguing study (Germain-Aubrey et al. 2014), however, traced the origins of four plants endemic to scrub in central Florida: scrub plum (*Prunus geniculata*), Lewton's milkwort (*Polygala lewtonii*), silk bay (*Persea humilis*), and scrub holly (*Ilex arenicola*). Although previous authors assumed that divergence of scrub endemics from their ancestors took place during the Pleistocene, when northern species were presumably pushed southward during glacial periods and then isolated on islands by high sea levels during interglacials, dated phylogenies of these four taxa show much earlier origins in the Miocene or Pliocene (Germain-Aubrey et al. 2014).

Although these fire-adapted plants evolved from ancestors with distributions in eastern North America, other members of central Florida scrub and sandhill communities, such as the Florida scrub-jay (*Aphelocoma coerulescens*) and the unusual plant *Ziziphus celata,* have their closest ancestors far to the west. They probably spread to Florida via the Gulf Coastal Corridor. A key point here is that assembly of our present-day pyrogenic communities drew from multiple regions and started longer ago than previously thought.

Feedbacks between vegetation and fire were essential to the global expansion of grasslands. As cooler, drier, and more seasonal conditions increased through the Miocene, grasses were highly competitive over trees and other woody vegetation. Due to their high flammability, C_4 grasses promoted increases in fire frequency, which in turn favored the continued spread of C_4 grasslands and savannas—a classic positive feedback loop (Beerling 2007; Pausas and Keeley 2009). Moreover, black smoke aerosols and tropospheric heating reduce regional cloud cover and the size of cloud droplets and hence their propensity to fall as raindrops. These changes reduce overall regional precipitation but make individual rainfall events more intense, including more extreme convective thunderstorms with increased lightning activity, in turn igniting more fires (Beerling and Osborne 2006).

The Pliocene Epoch (5.3–2.6 million years ago) began with the cool conditions of the late Miocene and the continued abundance of C_4 grasslands/

savannas over much of the globe. The vertebrate fauna shows continuity through this interval and includes a diverse assortment of savanna ungulates as well as inhabitants of subtropical forest. Webb (1990) writes: "More than any other time interval recorded in Florida, the early Pliocene supported several relicts that had already become extinct in well-documented contemporaneous faunas of the Great Plains," such as several horses and a large cameloid, *Kyptoceras*. Thus, Florida (and perhaps adjacent areas) played the critical climatic refugium role that distinguishes so much of its history and is reflected in its wealth of ancient endemic taxa.

By the mid-Pliocene atmospheric CO_2 increased to near modern levels, accompanied by high temperatures and sea level (Graham 1999). An incomplete fossil record obscures the fauna and vegetation of Florida during much of the middle Pliocene. The Atlantic and Gulf of Mexico coastline extended farther inland than at any time since the Eocene. Estimates of middle Pliocene sea level have varied widely, with corresponding uncertainty about the area of Florida and the Coastal Plain remaining above the sea. The amount of exposed land has obvious pertinence for survival of ancient plant and animal lineages. Mid-Pliocene (3.6 million years ago) sea level appears to have been 35 ± 18 m (17–53 m) above present (Dowsett and Cronin 1990). The lower portion of this range (ca. 25 m [82 ft]) now appears most likely, but "we cannot, at present, place robust limits on the maximum height of mid-Pliocene sea level" (Rowley et al. 2013).

To put these sea-level estimates into the perspective of current elevations in Florida, the highest spot in the state is Britton Hill near the Alabama border, at 105 m (344 ft) above mean sea level. In the Peninsula, the highest point is Sugarloaf Mountain on the northern Lake Wales Ridge, at 95 m (312 ft). Much of northern Florida and the central Florida ridges are higher than 50 m (164 ft). Therefore, considerable land area in Florida and the Coastal Plain remained above the sea during the Pliocene and continuously until present. Pyrogenic communities remained in Florida throughout this span of time, as evidenced by the persistence of narrow-endemic scrub plants that evolved earlier (Germain-Aubrey et al. 2014). Still, the gap in the terrestrial fossil record in Florida during the mid-Pliocene may suggest above-water refugia were relatively small and isolated—ideal conditions for allopatric speciation of endemics!

Pines increased and deciduous hardwoods decreased in the Coastal Plain during the middle Pliocene (Graham 1999), likely accompanied by an increase in the incidence of fire. Regarding middle Pliocene Florida, Graham (1999) comments that the trend in vegetation was "toward pine, grass, rosette palms, and scrub vegetation that could cope with physiologically dry deep sandy soil." Many equids (horses) and antilocaprids (pronghorns) went extinct through the course of the Pliocene (Janis et al. 2002). Florida vertebrate fossils from the late Pliocene, however, show abundant grazing and some browsing species suggestive of subtropical savanna and indicating relative stability and continuity of fire-prone vegetation. These vertebrates include horses, peccaries, a llama, a proboscidean (elephant relative), a small pronghorn (*Capromeryx arizonensis*), and the armadilloid *Glyptotherium*, among others (Webb 1990; Hulbert 2001).

By the late Pliocene the Gulf Coastal Corridor of semiarid habitat (presumably savanna/grassland and scrub) that had been established in the middle to late Miocene was continuous to Florida. This biogeographically significant connection between southeastern and western North America—and on to Central America via the Isthmus of Panama—persisted at least into the early Pleistocene (Webb 1990). The corridor apparently reappeared during subsequent glacial periods when the lower Gulf Coastal Plain was broad and extended southward to present-day central Florida (Morgan and Emslie 2010; Noss 2013).

The Gulf Coastal Corridor allowed many semiarid western taxa and some neotropical taxa to spread eastward. Many disjunct species in Florida and the southeastern Coastal Plain, whose closest relatives (in some cases the same species) occur today only in the West, bear witness to this phenomenon. Among the many fire-adapted species in this primarily western assemblage are pocket gophers, tortoises, diamondback rattlesnakes, crested caracara, burrowing owls, scrub-jays, harvester ants, grama grasses, yuccas, and wild-buckwheats (Noss 2013). Many of these species persist today in Florida and other portions of the lower Coastal Plain in physiologically xeric habitats on deep sands, particularly sandhill and scrub.

A fossil bird from the late Pliocene in Florida and Texas, *Titanis walleri*, was a flightless predator larger than an ostrich. It almost certainly was associated with open, probably fire-maintained habitats. This species is a member of the extinct South American family Phorusrhacidae. It must have spread northward along the Panamanian Land Bridge and then eastward along the Gulf Coastal Corridor. Also during the late Pliocene and continuing into the Pleistocene, several representatives of two ruminant families, Cervidae and Bovidae, dispersed into North America from Asia across the Bering Land Bridge. The most abundant cervid was a deer (*Odocoileus*) virtually indistinguishable from the extant and widespread white-tailed deer (*O. virginianus*) and adapted primarily to browsing. The first known North American bovid was an unknown species of *Bison*, a grazing animal associated with large expanses of grassland. Modern bison (*B. bison*) interact with fire by concentrating their grazing on recently burned sites (Fuhlendorf et al. 2009). By the middle to late Pleistocene, bison became one of the most widespread and characteristic large ungulates in North America, including the Coastal Plain (Hulbert 2001; Froese et al. 2017).

Ecosystems and Fire in Florida and the Southeastern Coastal Plain during the Quaternary (2.6 million years ago-present)

The Quaternary Period includes the Pleistocene and Holocene Epochs, plus what some geoscientists and ecologists now recognize as a new epoch, the Anthropocene. Although pollen and other microfossils provide an excellent record of vegetation history from the late Pleistocene through the Holocene, especially in Florida, this record is poor for the early and middle Pleistocene. Direct evidence for the vegetation during this span of time is therefore scant. As for earlier history, however, other kinds of evidence about past ecosystems come from plant and animal macrofossils. Florida, with a richer Cenozoic fossil vertebrate assemblage than anywhere in eastern North America (Webb 1990), provides ample opportunities to interpret past vegetation and fire history.

THE ORIGIN OF ESSENTIALLY MODERN PINE SAVANNAS

Webb (1990) opined "the two most impressive biogeographic features of the early Pleistocene in Florida are continuation of the 'great American Interchange' with South America and extensive development of longleaf pine habitats." In fact, the history of longleaf pine in the South has been controversial and remains unsettled, largely because the fossil pollen of pines here cannot yet be identified reliably to species (although DNA analysis of fossil pollen has identified some pines elsewhere to the species level). We only know the pollen grains belong to genus *Pinus*, although northern taxa such as jack pine (*P. banksiana*) tend to have smaller pollen grains, and southern pines larger pollen grains (Watts 1980a). With few exceptions, pine macrofossils are unfortunately lacking.

Many scientists, myself included, believe it prudent to assume savanna-associated pines in Florida and the southeastern Coastal Plain included longleaf pine or a direct ancestor since perhaps the Oligocene, when "savanna in the modern sense" (Graham 1999) became established in the region. Or conceivably longleaf pine or its ancestor extends as far back as the middle Eocene, when "*Pinus* with an understory of *Sabal-Serenoa*-type palms" (Graham 1999) reminiscent of modern pine flatwoods or scrub communities occupied what was then the outer Coastal Plain. Whatever pine species did occur in the Coastal Plain during Eocene to Pleistocene times (probably cycling up and down in abundance and shifting in distribution throughout this period), they were apparently ecological analogs of modern pines, if not the same taxa. These ancient pines were almost surely fire-adapted, with longleaf pine or its ancestor (or analog) adapted to surface-fire regimes characteristic of savannas.

Some scientists, however, have interpreted the rapid increase in pine pollen in lake sediments, which occurred most recently 7,000–5,000 years ago, as indicative of the time when longleaf pine suddenly entered the southeastern Coastal Plain. In an analysis of fossil pollen and plant macrofossils in sediments from Goshen Springs in southern Alabama, Delcourt (1980) showed replacement of early Holocene oak-hickory forest by "widespread southern pine forest on the sandy uplands of the Gulf Coastal Plain" 5,000 years ago, when the essentially modern climate regime was attained. This and similar findings have been interpreted as signifying the origin of the longleaf pine ecosystem of this region. As we shall see, however, comparable spikes in pine pollen occurred periodically over at least the last 61,000 years in peninsular Florida, cycling out of phase with spikes in oak pollen (Grimm et al. 2006). Why would longleaf pine abruptly replace older and ecologically equivalent taxa of pines 7,000–5,000 years ago?

Another problem with the hypothesis of recent origin of longleaf pine is evidence that fossil bones of animals closely associated with longleaf pine today, and often considered longleaf pine specialists, date back at least to the early Pleistocene, some 1.5–2 million years ago. Fossils of herpetofauna from this time—for instance, in the Inglis IA fauna—indicate a species assemblage nearly identical to that found in contemporary longleaf pine communities (Means 2006). The red-cockaded woodpecker (*Leuconotopicus* [*Picoides*] *borealis*), a specialist on longleaf pine over most of its range, is documented in fossils from Rock Springs Run, near Orlando, from Pamlico or post-Pamlico time 180,000–120,000 years ago (Webb and Wilkins 1984).

Schmidtling and Hipkins (1998) analyzed allozymes (different forms of proteins) from longleaf pines in several locations within the species' current range, and discovered allozyme diversity was higher in the western portion of the range than in the east. They interpreted these data as suggesting longleaf pine entered the southeastern Coastal Plain from a single refugium in southern Texas or northeastern Mexico in the very recent past—after the Pleistocene. This interpretation begs the question of what ecologically similar pine was present in the Coastal Plain before this time, and how it came to be replaced by longleaf.

A plausible alternative scenario to that offered by Schmidtling and Hipkins (1998) is that the Florida peninsula was another refugium for longleaf pine during periods when pines were less abundant across the remainder of the Coastal Plain. Overpeck et al. (1992) concluded southern pines were restricted to Florida during the late Pleistocene until about 9,000 years ago, "when southeast pine forests, similar to modern forests, spread northward." Living on a peninsula with restricted dispersal and gene flow, populations in the Florida peninsula would be expected to show reduced genetic diversity relative to "mainland" populations. Moreover, Schmidtling and Hipkins (1998) incorrectly state that "during the late Pleistocene, longleaf pine (*Pinus palustris*) was undoubtedly absent from the lower Coastal Plain of the southeastern United States and the area was dominated by a type of boreal forest." I address problems with this interpretation below. Available evidence suggests that longleaf pine or a direct antecedent has been a dominant component of the lower southeastern Coastal Plain, at least peninsular Florida, for millions of years. Thus, the distinctive fire ecology of longleaf pine savannas has great antiquity.

EARLY TO MIDDLE PLEISTOCENE

The Pleistocene began with a continuation of climatic and vegetation conditions of the late Pliocene and the continued dispersal of megafauna from Asia across the Bering Land Bridge into North America. Atmospheric CO_2 levels were declining, ushering in the first of at least 11 major glacial-interglacial cycles and many minor cycles (Richmond and Fullerton 1986). The minor, or briefer, cycles include stadials (cooler periods of glacial advance) and interstadials (warmer periods of glacial retreat) and are embedded within the major cycles. For example, Greenland ice cores indicate 24 interstadials during the 100,000 years of the most recent glaciation, the Wisconsinan (Wilson et al. 2000). Cooler periods with lower atmospheric CO_2 favor C_4 grasses because of their ability, thanks to the PeP-carboxylase enzyme, to more efficiently concentrate CO_2 in their cells for photosynthesis. Ironically, this adaptation also served C_4 grasses well during hot, dry periods of high moisture stress because, being more efficient at concentrating CO_2, they open their leaf pores (stomata) less widely for gas exchange. With smaller stomatal openings, less water is lost from the plant.

The glacial cycles led to pronounced changes in sea level and the amount of land exposed in Florida and the Coastal Plain. Contrary to simplistic portrayals, Florida and the rest of the lower southeastern Coastal Plain were never completely inundated during Pleistocene interglacials and interstadials. Sea levels during the various interglacials of the Pleistocene appear to have been all lower than during the mid-Pliocene high sea-level stand of 17–53 m above present (Dowsett and Cronin 1990). Large portions of the southeastern Coastal Plain remained terrestrial throughout the Pleistocene. The highest Pleistocene inundation apparently occurred 400,000 years ago, when the global sea level (which is very close to sea level surrounding Florida) was 6–13 m above present (Raymo and Mitrovica 2012). During the most recent interglacial, 120,000 years ago, sea level was 6.6–8.3 m higher than today (Muhs et al. 2011).

During high sea-level stands of the Plio-Pleistocene in peninsular Florida, fire frequency may have been reduced. The smaller the island, the lower the probability of lightning striking and igniting fuels, and fire would not spread beyond that island. Modern islands in the southeastern Coastal Plain have longer historical fire-return intervals than nearby mainland sites (Huffman et al. 2004; Henderson 2006; Huffman 2006). The Plio-Pleistocene islands of the Lake Wales Ridge that remained above the sea during the interglacials preserved relict taxa and promoted allopatric speciation and high endemism (James 1961; Sorrie and Weakley 2001).

Little direct evidence of vegetation composition and fire history exists for the southeastern United States during the early and middle Pleistocene (Graham 1999). Microfossil (pollen, spore, charcoal) records are poor for this period, and plant macrofossils are few. Once again, however, vertebrate fossils provide a picture of what these ecosystems were like. The presence of northern bobwhite (*Colinus virginianus*) bones in the early Pleistocene Inglis IA fauna from the Big Bend Coast area of Florida (Hulbert 2001) indicates a fire-prone habitat, as this species is one of the most fire-dependent of all modern birds in the Coastal Plain (Stoddard 1931; Cox and Widener 2008). Red-headed woodpecker (*Melanerpes erythrocephalus*) bones from this same site also indicate fire, as this species in the southeastern Coastal Plain primarily inhabits pine savannas and pine-oak woodlands with open understories or sparse shrub layers (Frei et al. 2015).

The herpetofauna of the Inglis IA site, indicative of longleaf pine savanna, also has semiarid western affinities, although three snake genera—*Drymarchon* (indigo snake), *Rhadinea* (pine woods snake), and *Tantilla* (crowned snake)—probably originated in Central America and followed the Panamanian Land Bridge to the Gulf Coastal Corridor (Meylan 1982; Webb 1990). All these taxa are found in high-fire environments (e.g., sandhill, flatwoods, and scrub) today and presumably were during the Pleistocene.

The continued importance of the Gulf Coastal Corridor and the persistence of savanna/grassland are indicated by fossils of a jackrabbit (*Lepus*), a western pocket gopher (*Thomomys*), and a small ground sloth (*Nothrotheriops*) primarily associated with the Great Basin, in the early Pleistocene Leisy Shell Pit site by Tampa Bay (Webb et al. 1989). The Gulf Coastal Corridor appears to have weakened during the middle Pleistocene, based on the disappearance of pronghorns, jackrabbits, and other western taxa from Florida fossils then. Nevertheless, pine savanna and scrub-associated vertebrates remained abundant; many of these taxa are considered Florida or southeastern Coastal Plain relicts: *Cemophora* (scarlet snake), *Aspidoscelis* (whiptails/racerunner), *Plestiodon* (skinks), *Pituophis* (pine snake), *Masticophis* (coachwhip), and *Aphelocoma coerulescens* (Florida scrub-jay), among others (Webb 1990).

LATE QUATERNARY (LATE PLEISTOCENE TO HOLOCENE)

Analyses of fossil pollen and other small particles recovered from cores of lake bottoms can tell us much about past vegetation, climate, and, by inference, fire regimes. In contrast to the early and middle Pleistocene, the fossil pollen record of the late Pleistocene and Holocene in Florida, and some other areas in the Coastal Plain, is superb—the best in eastern North America. Paleobotany and palynology (the study of pollen, spores, charcoal, and other particles) are not perfect sciences. Biases exist, for instance, in how pollen is dispersed, deposited, and preserved. Nevertheless, the ability to reconstruct paleovegetation from pollen and spore diagrams has improved dramatically over recent decades (Graham 1999).

Early cores from lake bottoms in the Coastal Plain extended back, at most, to the Last Glacial Maximum around 20,000–30,000 years ago (e.g., Watts 1971; Whitehead 1973; Delcourt 1980; Watts 1980a, 1980b), so inferences about vegetation change were mostly limited to full glacial and postglacial history. Now, however, we have continuous sedimentary sequences extending much farther back in time. Lake Annie at Archbold Biological Station on the Lake Wales Ridge, with a record extending back 37,000 years, was the first site in the southeastern Coastal Plain to have a continuous pollen record from the Pleistocene to present (Watts 1975).

Recent climate/vegetation reconstructions in the Coastal Plain have led to corrections of some earlier interpretations and to an improved picture of Late Quaternary climate and vegetation in the region. Deevey (1949), based largely on reports of macrofossils of northern taxa, including spruce (*Picea*) in Late Quaternary deposits from the Tunica Hills of southeastern Louisiana and adjacent Mississippi (31° N latitude) (Brown 1938), speculated that boreal forest extended southward to the lower Coastal Plain during full-glacial times. The spruce in the Coastal Plain during these times was identified as white spruce (*Picea glauca*), but more likely was the extinct *Picea critchfieldii* (Eric Grimm, personal communication), which was common in forests in the Mississippi Alluvial Valley during the late Pleistocene.

Abundant spruce and tamarack (*Larix*) pollen was subsequently found in late Pleistocene sediments in the Tunica Hills (Delcourt and Delcourt 1977, 1996), as well as in central Texas (Graham 1999). This was not pure boreal forest, however, as Deevey (1949) concluded, but rather a no-analog mix of boreal and temperate taxa. Macrofossils at Tunica Hills previously identified as northern white cedar (*Thuja occidentalis*) were corrected to Atlantic white cedar (*Chamaecyparis thyoides*), a warm-temperate to subtropical Coastal Plain species (Delcourt and Delcourt 1996). Graham (1999) suggests that "boreal elements migrated southward partly along the Mississippi River Valley system with cold-air drainage and increased moisture from fog. This brought stands of *Picea* within pollen dispersal range of Central Texas but not extensive boreal forests along the entire Gulf Coast."

Jack pine (*Pinus banksiana*) and spruce pollen are abundant in sediments at White Pond in Columbia, South Carolina (34° N) from the Last Glacial Maximum (Watts 1980b). Pollen from these taxa was previously recorded in lake sediments in the Coastal Plain of North Carolina and Virginia (see citations in Watts 1980b). Williams et al. (2000) interpret this biome as open conifer woodland, rather than boreal forest. Oak and hickory expanded at White Pond around 15,000 years ago, while jack pine and spruce were still present (Watts 1980b). Jack pine has many adaptations to stand-replacing fire, such as serotinous cones and branch retention, similar to sand pine (*P. clausa*).

Pollen of spruce (*Picea*), along with a northern aquatic plant (*Najas gracillima*), was found within sediments dated at 14,000–12,000 years ago at Camel Lake within the Apalachicola National Forest in the Florida Panhandle (30° N) (Watts et al. 1992). Camel Lake is 64 km (40 mi) from the Gulf Coast today, but with sea level 14,000–12,000 years ago about 70 m lower than today (Donoghue 2011), the east–west Gulf Coast here extended southward to the modern-day central Florida peninsula, in the vicinity of Tampa Bay (Delcourt and Delcourt 1981; Russell et al. 2009). The Camel Lake spruce pollen does not suggest boreal forest, because this pollen is associated with abundant pollen of temperate hardwoods, chiefly hickory (*Carya*), which peaked at that time, as well as oak and beech (*Fagus*). Given the spruce component, this was a forest community with no analog in the modern Coastal Plain, though these taxa occur close together today in the southern Appalachians and to the north in New England and southeastern Canada.

During the Last Glacial Maximum, a sharp climatic boundary, the Polar Frontal Zone, created an abrupt ecotone between boreal and temperate forest at around 33° N latitude (Delcourt and Delcourt 1984). In the present southeastern United States, this latitude extends from northern Louisiana just south of the Arkansas border (within the Coastal Plain), eastward through Birmingham, Alabama, and Atlanta, Georgia, both north of the Coastal Plain, to just north of Charleston, South Carolina. Jackson et al. (2000) place the Last Glacial Maximum boreal-temperate ecotone around 34° N latitude, with some temperate trees as far north as 35° N.

The vegetation of the broad southeastern Coastal Plain during the late Pleistocene ranged from cool temperate to temperate, warm temperate, and subtropical vegetation, the latter two mostly in peninsular Florida. In north-central Florida, cores from 14,360 to 12,760 years ago in Sheelar Lake, adjacent to present-day Gold Head Branch State Park, show a rich mesic temperate forest (Watts and Hansen 1994). Common tree taxa were oak, hickory, beech, elm (*Ulmus*), Atlantic white cedar or red cedar (*Juniperus*), and hornbeam (*Ostrya* or *Carpinus*). These tree genera still occur in naturally fire-protected hammock or swamp communities in north and (marginally) central peninsular Florida, often referred to as the "Hammock Belt." These data and information from other sites suggest that northern Florida, including the Panhandle (e.g., Camel Lake), served as a refugium for mesic temperate species during the late Pleistocene, after which these species expanded their ranges northward as the Holocene unfolded (Watts and Hansen 1994).

Particularly informative is the spectacular pollen record from Lake Tulane in south-central Florida; at 61,000 years, this is the longest continuous pollen record from anywhere in eastern North America. As found earlier at Lake Annie and other lakes in peninsular Florida, the most prominent pattern from Lake Tulane is the striking cycles of pine and oak abundance, out of phase with each other (Fig. 2.3). When oak peaked, pine usually was of low abundance, and vice versa (Grimm et al. 1993, 2006). At times, however, such as ca. 10,000 years ago, oak and pine appear equally abundant.

Based on the large size of the pollen grains, Jackson et al. (2000) surmised that most of the pines in peninsular Florida during the Last Glacial Maximum, and likely during other Pleistocene-Holocene pine cycles, were "southern" pines, which could have included longleaf (*P. palustris*), slash (*P. elliottii*), South Florida slash (*P. densa*), shortleaf (*P. echinata*), or loblolly pine (*P. taeda*). Palynologist Eric Grimm (personal communication) suspects the pines in northern Florida during the Last Glacial Maximum were probably "northern southern pines," such as loblolly and shortleaf, whereas longleaf was probably confined to farther south in the peninsula.

A puzzling aspect of the pine-oak cycles evident in the late Pleistocene and Holocene pollen record from peninsular Florida (Fig. 2.3) is the correlation of grass (Poaceae), ragweed (*Ambrosia*), other herbs, and hickory with oak abundance. The oak phases had been assumed to correspond to something like modern-day Florida scrub, a major type of which (oak scrub) is dominated by several scrubby oak species (especially *Quercus geminata, Q. myrtifolia, Q. chapmanii,* and *Q. inopina*) and is maintained in relatively short stature by recurring fires at approximately 5–20-year intervals. Grasses, however, are not abundant in Florida scrub because they are shaded out by the often dense oaks and other shrubs. They also may be inhibited by allelopathic chemicals produced by Florida rosemary (*Ceratiola ericoides*), plants of which are often surrounded by gaps of bare sand. *Ceratiola* is known to exclude other species of scrub plants through allelopathy (Hunter and Menges 2002; Hewitt and Menges 2008). Grasses are much more abundant and dominate the herbaceous layer in modern pine savannas.

One intriguing hypothesis for the close association of oaks and grasses in the late Quaternary pollen record of central Florida is that this

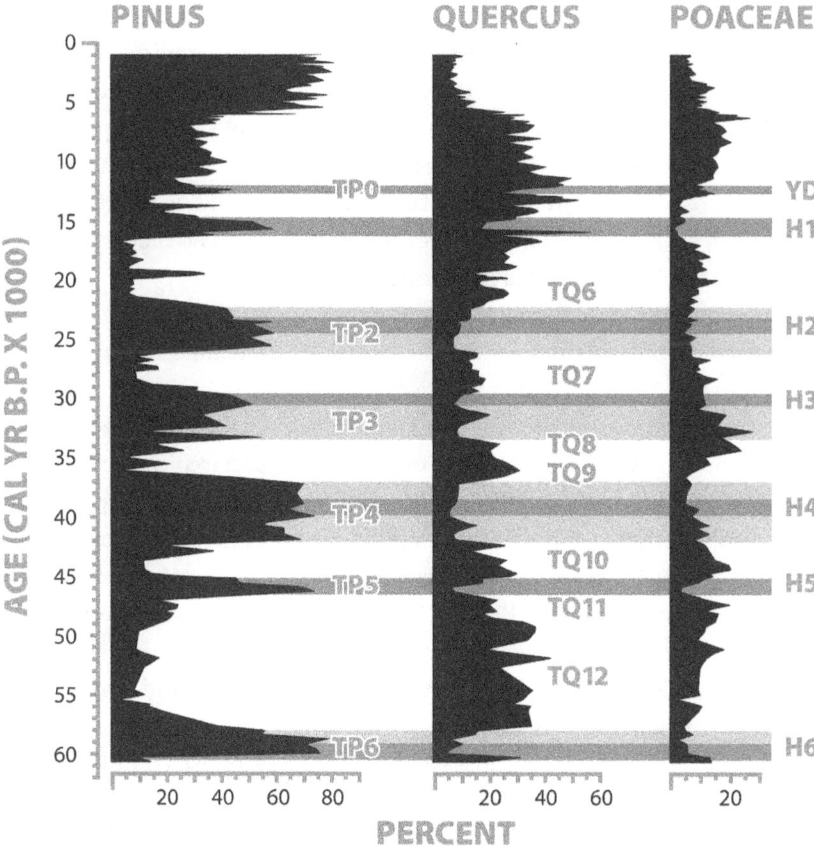

Figure 2.3. Fossil pollen record of three taxa from Lake Tulane on the Lake Wales Ridge in central Florida. Pine (*Pinus*) and oak (*Quercus*) abundance cycled out of phase in response to glacial/interglacial cycles over the 61,000 years represented in this core (see text). During some time periods, however, such as ca. 10,000 years ago, oak and pine were about equally abundant. Grass (Poaceae) abundance is temporally correlated with oak abundance, as is hickory (*Carya*, not shown here). "TP" refers to pine phases, "TQ" to oak phases, "YD" to Younger Dryas, and "H" to Heinrich events. Adapted from Noss (2013) after Grimm et al. (2006) and Grimm (personal communication).

combination of species represents Florida dry prairie (better called subtropical hyperseasonal grassland; Steve Orzell and Edwin Bridges, personal communication). This community, endemic to south-central Florida, burns naturally even more frequently, at one-to-two-year intervals, than many pine savannas. The high fire frequency may be explained by the very flat, relatively undissected terrain of this region of Florida, which allows lightning-ignited fires to spread across broad landscapes (Bridges 2006a). A dominant and characteristic species in subtropical hyperseasonal grassland, especially in its mesic and wet-mesic phases, is dwarf live oak (*Q. minima*). Could it be that much of the oak pollen in the pollen record is from dwarf live oak and represents a grassland rather than scrub? Another possible interpretation of the grassy oak phases is that they represent oak savanna (Putz 2015), a vegetation type with no modern analog among natural communities in Florida.

The typically highly flammable C_4 grasses were much (40 percent) more abundant than C_3 plants during the Last Glacial Maximum than during the Holocene, perhaps owing to drier conditions and lower atmospheric CO_2 concentrations during the glacial period. C_4 plants declined during the warmer and wetter Pleistocene pine phases associated with other stadial phases (Huang et al. 2006). During the Last Glacial Maximum, the biologically richest (and also fire-dependent) savanna biome on Earth, the Cerrado of Brazil, also was drier, with treeless grassland expanding at the expense of savanna (Behling 2002). Nevertheless, suitable climatic refugia for more mesic species persisted in the Cerrado during this time (Souza et al. 2017). Such a scenario is also likely for Florida, given the sustained presence of pine during oak-grass phases. This convergence may indicate similar climatic pressures and organism responses for Florida savannas and the Cerrado during the late Pleistocene.

Grass abundance generally correlates with oak abundance in the central Florida pollen record, but so too does hickory. This hickory is very likely *Carya floridana*, the scrub hickory endemic to peninsular Florida scrub (Williams et al. 2004; Grimm et al. 2006). Although we will not know the details of these species associations until fossil pollen DNA analysis is conducted, the oak-grass phases in the pollen record of central Florida may have represented a mosaic of oak scrub, oak savanna, and hyperseasonal grassland vegetation. Given the species associations, fire frequency was probably moderately to very high in these ancient communities.

Another intriguing aspect of the late Pleistocene pollen record in peninsular Florida is a counterintuitive correlation with paleoclimate. Early analysis of the pollen data in association with climate suggested that the pine phases represented periods of cool, wet climate, and the oak phases periods of warm, dry climate (Grimm et al. 1993). More intensive analysis of pollen assemblages, paleoclimate records, and past lake levels revealed the pine phases were warm and wet, whereas oak phases were dry and cooler (Grimm et al. 2006). Surprisingly, the warm-wet pine phases in Florida coincided not with globally warmer interstadial periods, but rather the opposite; they occurred during stadial periods, when global temperatures were far cooler and glaciers advanced. The pine peaks coincide closely with the massive iceberg discharges called Heinrich events, which are associated with the termination of stadial periods in the North Atlantic (Fig. 2.3).

A climate-pollen inference model shows that November temperatures and summer precipitation both increased in peninsular Florida during stadial periods, including the Younger Dryas, and their terminating Heinrich events (Donders et al. 2011). A compelling explanation for this unexpected pattern is that heat was retained in the Gulf of Mexico via a strengthened Loop Current, which recirculated heat through the Gulf. This strengthening was aided by a persistent Atlantic Warm Pool and an increase, during summer, of easterly trade winds. With the Gulf Loop Current so strengthened, less heat was transported to the north by the Florida Current and Gulf Stream, which arise from the Loop Current. When the Gulf Stream heat conveyor belt is weakened during stadial periods, heat is retained in the south (Grimm et al. 2006; Donders et al. 2011). This is one of several physical and climatological phenomena that lend climatic stability to Florida and the lower southeastern Coastal Plain.

Whereas some scientists (e.g., Hiers et al. 2000) see the vegetation cycles of the late Pleistocene in the southeastern Coastal Plain as evidence of climatic instability, the researchers who uncovered this pattern (e.g., Grimm et al. 2006), among others, see it as indicative of a relatively stable climate throughout this period during which most other regions of North America were climatically turbulent. The climatic stability of the southeastern Coastal Plain has allowed it to accumulate and retain a wealth of endemic species over a long period of time, with most of these endemics associated with fire-dependent ecosystems (Russell et al. 2009; Noss 2013; Noss et al. 2015). As a general global pattern, climatic stability fosters both the long-term persistence of paleoendemics and the retention of recently evolved neoendemics, such that global hotspots of endemism correspond closely to climatic refugia (Jansson 2003; Sandel et al. 2011; Harrison and Noss 2017).

A potential source of error in vegetation reconstructions based on fossil pollen and spores is the inclusion of species represented only by a small number of pollen grains, which may have dispersed from hundreds of miles away. Palynologists typically exclude pollen from taxa known to disperse long distances and present only in small quantities at a site, to avoid a misleading interpretation of the vegetation surrounding a site during a particular period. For example, spruce pollen has been found in lake cores farther south in Florida, but only in the Panhandle at Camel Lake did it reach "significant" quantities—8 percent of the total pollen in the record from 14,000 to 12,000 years ago (Watts et al. 1992). The cutoff points are somewhat arbitrary and can vary among researchers.

Williams et al. (2000) addressed the potential long-distance pollen error by raising the thresholds for inclusion of pollen from species that are "generally over-represented in the pollen record in order to reduce the chance of misassignments" of local vegetation to past biomes. The raised thresholds were *Pinus* 5 percent, *Quercus* 2.5 percent, and 1 percent for other taxa, in contrast to the standard practice of setting all thresholds at 0.5 percent. Williams et al. (2000) also avoided interpolation of biome types between sites, which can produce errors. This use of raised pollen thresholds assigned plant communities from the Last Glacial Maximum from sampled lakes in the southeastern Coastal Plain to "broadleaved evergreen warm mixed forest" (Florida Panhandle), "open conifer woodland" (Florida peninsula and South Carolina), "cool mixed forest" (southeastern North Carolina), and "cool conifer forest" (northeastern North Carolina).

Thus, the biome assignments of Williams et al. (2000) are consistent with the boreal/temperate ecotone at around 33–34° N suggested by Delcourt and Delcourt (1984) and Jackson et al. (2000). These studies are also consistent with the conclusion of Grimm et al. (2006) that the late Pleistocene climate and vegetation of central Florida was very similar to the climate and vegetation of today, with no major distributional shifts and no northern taxa identified in the paleoflora. Hence, the fire-prone pine savanna, scrub, and open grassland communities that dominated the uplands of Florida when the first Europeans arrived, along with their embedded wetlands, probably existed here in similar form for tens of thousands to potentially millions of years earlier.

Fauna of the Late Pleistocene

The fauna of the late Pleistocene in the Coastal Plain is more abundantly documented with fossils than for any preceding period. This record reflects a continuation of grassy, scrubby, fire-prone vegetation, as well as forest (Webb 1990). Bison first entered North America around 195,000–135,000 years ago, then rapidly spread across the continent (Froese et al. 2017). Although the taxonomy is still not entirely clear, at least two species of bison, *Bison latifrons* and *B. antiquus*, inhabited the southeastern Coastal

Plain during the Pleistocene. *B. latifrons* was twice the size of the modern bison and went extinct by 15,000 years ago. The smaller *B. antiquus* probably descended from *B. latifrons* and persisted until the late Pleistocene or early Holocene. A related species, the steppe bison (*B. priscus*), persisted into the early Holocene in Europe, Asia, Beringia, and North America, and is the most likely ancestor of the modern bison, *B. bison* (Guthrie 2006). Megaherbivores, including mastodons and mammoths, were apparently abundant in Florida and the Coastal Plain during the late Pleistocene. Carbon isotope analysis of their teeth indicates mastodons were primarily browsers of woody vegetation, whereas mammoths were grazers, consuming chiefly C_4 grasses (Koch et al. 1998).

Besides the fire-dependent red-cockaded woodpecker mentioned earlier, birds of fire-maintained habitats in late Pleistocene Florida include the greater prairie chicken (*Tympanuchus cupido*), now restricted to the Great Plains; the Florida scrub-jay; the burrowing owl (*Athene cunicularia*); and two species of caracara, *Caracara plancus* (southern caracara, closely related to our modern crested caracara, *Caracara cheriway*) and *Milvago chimachima* (yellow-headed caracara, now distributed from Costa Rica south to northern Argentina and Uruguay). With sea level 120 m or more below present during the Last Glacial Maximum, Florida was at least twice its current size, so this diverse Coastal Plain fauna would have been distributed more broadly, with colonization and recolonization of populations less constrained by the peninsula effect (i.e., increasing dispersal limitation toward the tip). The large birds *Gymnogyps* and *Teratornis*, as well as the giant tortoise *Geochelone*, probably inhabited open savanna environments and accompanied the mammalian megafauna into extinction at the end of the Pleistocene (Webb 1990).

Megafauna Extinctions and Fire

A well-known but still enigmatic phenomenon of the late Pleistocene is the decline and eventual extinction of the vast majority of megafauna: large mammals, large birds, and giant tortoises. Biologists and paleontologists have debated the cause of the extinctions for more than two centuries (Cooper et al. 2015), with the leading contenders being climate change and "overkill" by humans hunting the megafauna with increasingly efficient methods and technology. Although the debate continues, overkill and climate change are not mutually exclusive causes of the extinctions. Most of the recent literature points to hunting by humans as a major cause (Koch and Barnosky 2006). Climate change, however, is emerging as a significant factor explaining extinctions across diverse taxa and regions.

Using ancient DNA and radiocarbon data from 31 detailed time series of species extinctions and replacements (when populations were replaced by others of the same species or genus), Cooper et al. (2015) show that extinction and replacement events coincide with Dansgaard-Oeschger interstadial warming events. Although some previous studies suggested abrupt cooling (stadial periods) may have caused extinctions, Cooper et al. (2015) found extinctions were apparently absent from the cold periods of the Last Glacial Maximum and the Younger Dryas. (As noted earlier, the climate was warmer and wetter in peninsular Florida and perhaps other portions of the lower southeastern Coastal Plain during these globally cold periods.) Instead, populations and species were lost during the same abrupt episodes of global warming that caused shifts from pine phases to cooler and drier oak-grass phases in Florida. The extinctions occurred during the interstadials leading up to the Last Glacial Maximum as well as during those that followed.

The study by Cooper et al. (2015) leaves more questions than answers for Florida, given that late Pleistocene temperature changes here, both warming and cooling phases, were opposite the global pattern (Grimm et al. 2006; Donders et al. 2011). If rapid warming during interstadials caused the global megafaunal extinctions, then what explains the disappearance of megafauna in Florida, where the interstadials were somewhat cooler and drier? Cooper et al. (2015) were interested in broad general patterns, not local events, so they combined data across immense geographical areas, one of which was "North America (south of Late Pleistocene Laurentide Ice Sheets)." Their 31 time series included no Florida species or sites (Cooper et al. 2015 supplementary materials). Possibly the megafaunal extinctions in Florida also occurred during interstadials in response to increased aridity, and therefore less forage, rather than temperature change. Alternately, the extinctions in Florida may have occurred during the warm and wet stadial events. Or perhaps there was no association with either stadials or interstadials. This is a topic for further research.

Human hunters are not off the hook in the disappearance of megafauna. Climate change and human impacts are intertwined in the late Pleistocene extinctions. Recent analyses agree that human hunters likely fragmented populations of megafauna, perhaps by concentrating their hunting on key movement corridors, which in turn disrupted metapopulation structure and made subpopulations more vulnerable to continued hunting as well as climatic stresses (Guthrie 2006; Haynes 2013). The rapid "blitzkrieg" model of overhunting is not supported by recent analyses. For example, species that survived the Pleistocene in Alaska and the Yukon, such as bison (*Bison priscus*), elk (*Cervus canadensis*), and moose (*Alces alces*), increased in numbers through the period of human colonization and before horses (*Equus ferus*) and mammoths (*Mammuthus primigenius*) went regionally extinct. Given the lack of information specific to Florida and the Coastal Plain on the timing and causes of megafaunal extinction, multiple working hypotheses should be entertained and tested.

At this point the reader might be wondering how the late Pleistocene megafaunal extinctions pertain to the subject of this book. The extinctions are relevant to fire ecology because they have been explicitly linked to changes in fire regimes, albeit not yet for our study region. As in many regions (Bond 2015), a puzzling aspect of late Pleistocene lake sediment cores in Florida and the lower Coastal Plain is the low abundance of charcoal. Jackson et al. (2000) point out that "we have little indication of the nature and roles of disturbance, particularly fire, in structuring [late Pleistocene] vegetation; the only stratigraphic charcoal records (Lake Tulane, FL and Clear Pond, SC) show low charcoal influx." So here was all this seemingly fire-dependent pine, oak, and grassy vegetation during the late Pleistocene, but little charcoal to directly implicate fire.

Besides the biases mentioned earlier that affect charcoal records, such as filtering of upland-derived charcoal by dense vegetation along streams and lakes (Bond 2015), and the crucial fact that grassland/savanna fires produce much less charcoal than shrubland and forest fires (Pausas and Keeley 2014a), there is another possible explanation. The abundance of large grazing and browsing herbivores in the late Pleistocene may have been so great that their consumption of palatable hardwoods, as well as grasses, reduced both fine and coarse fuel loads, resulting in relatively low fire activity.

Gill et al. (2009) examined fossil pollen, spores of the dung fungus *Sporormiella* (linked to megaherbivore abundance), and charcoal from a lake in Indiana and several sites in New York studied by Robinson et al. (2005). Megafaunal populations declined steadily between 14,800 and 13,700 years ago, perhaps due to human hunting, before the final extinctions took place 13,500–13,000 years ago. The demise of megaherbivores was closely followed by a spike in charcoal in sediments. This spike indicates increased fire activity, which probably reflects increased fuel loads, particularly woody fuels. These changes were followed at around 11,900 years ago by a sharp increase in fire-tolerant pine, followed by a drop in pine and a pulse of oak at the beginning of the Holocene 11,700 years ago. Gill et al. (2009) acknowledge that their study "does not conclusively resolve the debate over climatic versus human causation (or both) of the North American megafaunal extinctions." Trends in charcoal influx from 35 lake-sediment records across North America show high fire activity during intervals of rapid warming at 13,900, 13,200, and 11,700 years ago, which do not coincide closely with changes in human population or with the timing of megafaunal extinctions (Marlon et al. 2009). Only one of these sites (Lake Tulane), however, was in the Coastal Plain.

A study in Australia reveals a chain of events comparable to that documented in eastern North America. Humans arrived much earlier in Australia, perhaps 49,000 years ago; by 40,000 years ago they were widespread across the continent. Australia's megafauna included giant marsupials, monotremes, reptiles, and birds, which were extinct by 40,000 years ago, very likely due to human hunting (Rule et al. 2012). A high-resolution 130,000-year record of pollen, *Sporormiella*, and charcoal shows that the rapid megafaunal extinction triggered a replacement of mixed rainforest by fire-prone grassy and sclerophyll vegetation. Rule et al. (2012) relate the change in vegetation to the dual action of reduced herbivory and perhaps landscape-scale burning by humans, although they reason that "the extended trajectory of the rise in charcoal and its close matching with falling *Sporormiella* suggest . . . that relaxation of herbivory directly caused increased fire, presumably by allowing the accumulation of fine fuel."

A rebound of fine fuels and increase in fire frequency following extinction of megaherbivores anywhere might be transitory. Megaherbivores actively maintain grasslands—an example of niche construction—with some species of proboscideans (elephant relatives) actively destroying trees by stripping their bark and pushing them over (Means 2006). With megaherbivores gone, grasslands often transition quite quickly to shrublands or forests with reduced fire frequency but often higher-intensity fire due to increased woody fuel loads. Because woody fuels produce more charcoal than grasses do, the accumulation of woody fuels following disappearance of megaherbivores is probably most responsible for the observed charcoal spike (Pausas and Keeley 2014a).

Unlike the activity of native megaherbivores, except in patches of temporarily concentrated activity, intensive grazing by livestock commonly reduces grass cover and the frequency of surface fires, producing a higher density of small trees, ladder fuels, and increased incidence of high-intensity, stand-replacing fire. This problem has been well documented in dry forests (e.g., ponderosa pine, *P. ponderosa*) of the western United States since the early twentieth century (Leopold 1924; Covington and Moore 1994; Allen et al. 2002). Although livestock do not as easily remove grass cover in high-precipitation regions such as the southeastern Coastal Plain, I have observed many sites in Florida where surface fires do not carry well because of the sparse fine fuels and abundant bare ground caused by heavy grazing and trampling by cattle.

Could a fire-herbivore-fuel relationship like that documented by Gill et al. (2009) and Rule et al. (2011) also have existed for the southeastern Coastal Plain in the late Pleistocene? Again, region-specific research is needed, given our unique climatic history and vegetation. Nevertheless, a plausible scenario is that the currently fire-dependent species and ecosystems of the southeastern Coastal Plain evolved and assembled in high-fire environments, perhaps extending back to the Eocene or Oligocene, but at least to the Miocene. During subsequent periods of high megaherbivore biomass (which are not precisely known, because abundance of fossils and high species richness do not necessarily imply high biomass), fire incidence may have declined due to intense grazing, browsing, and the consequent paucity of both fine and coarse fuels. The preference of some bird species in Florida, such as crested caracara, burrowing owl, and Florida sandhill crane (*Grus canadensis pratensis*), for heavily grazed or otherwise short lawns to taller-statured native grasslands (Morrison and Humphrey 2001) suggests these birds coevolved with large herbivores and may have tracked their herds as they moved across the ancient landscape (Noss 2013).

One of the few megaherbivores that did not die out during the late Pleistocene extinctions was bison (*Bison priscus* and *B. bison*). Although the Pleistocene and Holocene history of bison east of the Mississippi River remains enigmatic (Rostlund 1960; Belue 1996; Noss 2013), at times of abundance bison and fire likely interacted in a relationship known as "pyric herbivory." In this relationship bison follow fire across the landscape, concentrating their grazing on the palatable and nutritious new growth of grasses that emerges after fire. As shown in the Great Plains today, pyric herbivory creates a shifting mosaic of disturbance patches, enriching landscape heterogeneity and biodiversity of several taxa (Fuhlendorf et al. 2009). To my knowledge no rigorous studies of pyric herbivory have been conducted in the southeastern Coastal Plain. Such studies would be useful for determining linkages between management of fire and livestock.

UNFOLDING OF THE HOLOCENE

Humans entered North America in the late Pleistocene, but greatly expanded during the Holocene. The Holocene began 11,700 years ago, according to a time scale revision in 2009 by the International Union of Geological Sciences. Earth had been warming rather continuously since the Last Glacial Maximum 27,000–20,000 years ago, but that was interrupted by several brief cooling periods. The most recent and pronounced of these cool phases was the Younger Dryas, which lasted from about 12,900 to 11,700 years ago. The end of the Younger Dryas now marks the boundary between the Pleistocene and the Holocene. Early Holocene warming was rapid and strong enough to cause one of the highest rates of sea level rise in geologic history, up to 45 mm per year (Donoghue 2011), which equals 4.5 m (nearly 15 ft) per century, some 20 times the current rate and faster than all but the most extreme forecasts of sea-level rise for the next century or two.

The Holocene pollen record from peninsular Florida shows a continuation of the pine-oak cycles that characterize the late Pleistocene. Oak and grass dominated the early Holocene. The most recent pine expansion, which may have involved several species of southern pines, began as early as 8,000 years ago in northern Florida and as late as 5,000 years ago in central Florida (Watts and Hansen 1994). We remain in the most recent pine phase today. By the mid-Holocene the climate was similar to today's, with predominantly summer precipitation associated with thunderstorms and high lightning frequency (Watts and Hansen 1994). Based on the work of Grimm et al. (2006), we can assume that similar warm and wet (but probably seasonally dry) periods with high lightning-fire frequency probably characterized most or all the pine phases over the last 61,000 years and perhaps much longer (i.e., beyond the available continuous pollen record).

Although warm and dry periods, particularly droughts, have been associated with increased fire activity in many regions and at several temporal scales (Westerling et al. 2006; Pausas and Fernández-Muñoz 2012; Liu and Wimberly 2015; Platt et al. 2015), extended droughts can have the opposite effect because they reduce the production of fine fuels that carry fire. In grasslands of southern Africa, wetter and cooler periods over the last 170,000 years caused increased burning due to shifts in seasonality toward increased summer rainfall and productivity of grasses and the resulting development of high fine-fuel loads (Daniau et al. 2013). On an annual scale, wet summers with high primary productivity followed by winter or spring drought that dries fuels produce the most combustible vegetation and the most fire activity (Keeley and Rundel 2005; Bond 2015).

Influence of Humans on Fire Regimes

The influence of humans on fire regimes is complex and varies over time and space. Human modification of fire regimes not only can affect the structure and composition of ecosystems across the landscape, it can also change selection regimes and the evolutionary trajectories of species. People can alter fire regimes by modifying the key variables that affect fire activity (Bowman et al. 2011). For example, changes in land cover such as agricultural development or clear-cutting would tend to increase wind speed, in turn potentially increasing the spread of fire in adjacent natural communities. In contrast, habitat fragmentation and the creation of anthropogenic barriers (including those same clear-cuts or agricultural fields) would disrupt fuel continuity and reduce the spread of fires. Intensive livestock grazing also lowers fine fuel loads and continuity, reducing not only fire spread but the intensity and severity of fires. Deliberate fire suppression obviously also reduces fire spread.

Human ignitions, either intentional or accidental, tend to increase the number, frequency, and extent of fires, unless the natural fire regime leaves little fuel to burn (see below). Road networks are notorious sources of human ignitions in the modern world (Keeley et al. 2009). On the other hand, especially for surface fire regimes with relatively low flame heights, roads are often effective fuel breaks that stop the spread of fire. In central Florida, as little as 10 percent disruption of land cover by linear features such as roads can result in as much as 50 percent decline in fire extent (Duncan and Schmalzer 2004).

Globally, the first evidence of fire use by humans vastly predates the Holocene. People may have begun to cook food as early as 1.9 million years ago, although regular domestic use of fire probably did not begin until 100,000 or fewer years ago (Bowman et al. 2009). Evidence from a cave in South Africa suggests use of fire for cooking and staying warm one million years ago (Berna et al. 2012). Hunter-gatherers used fire to manage landscapes for various purposes, including hunting of game and nurturing of favored plants, for at least tens of thousands of years before present (Pyne 2001). One can imagine how important fire must be to people who lack modern tools for clearing vegetation. Pausas and Keeley (2014a) suggest early people used fire "for a wide range of purposes, including clearing ground for human habitats, facilitating travel, killing pests, hunting, regenerating plant food resources for both humans and livestock, warfare between tribes, and even for stimulating precipitation."

Use of fire was one of the hallmarks of human civilization, allowing people to eat a far greater range of foods, improve their nutrition, colonize colder environments, and manage entire landscapes to construct a unique human niche. The importance of Indian burning in the Americas is suggested by a massive reduction in burning following colonization by Europeans and decimation of native populations (albeit this does not seem to be the case for the southeastern Coastal Plain; see below). In the Neotropics this depopulation period is marked by decreased charcoal levels in soils and lake sediments and changes in stable isotopes of carbon in Antarctic ice cores (Pausas and Keeley 2014a).

One of the more controversial topics in American fire ecology is the relative role of humans vs. lightning in driving fire regimes prior to European settlement. One camp holds that humans transformed fire regimes on landscape, regional, and even continental scales, hence influencing selective pressures, adaptive traits, and community structure across North America. Another camp insists that lightning alone can account for most landscape-scale fires prior to European arrival. Still other researchers attempt to discern the relative influence of humans and climate (including lightning) in fire regimes for particular regions and periods of time, rather than generalizing wildly.

Changes in fire regimes clearly are linked to changes in climate. Nevertheless, Pausas and Keeley (2014a) review several ways in which fire regimes could change substantially in the absence of climate change. Humans are responsible for many of these changes through such activities as hunting, farming, introduction of flammable (or, conversely, nonflammable) nonnative plants, and habitat fragmentation (Table 2.1). Humans, however, may have had less influence than climate on overall fire incidence in the Americas since human colonization. For example, although a distinct decline in charcoal deposition after 1500 C.E. has been attributed to the decline of Indian populations, this decline is evident globally, not just in the Americas. Moreover, the decline of fire activity (as measured by charcoal, an imperfect measure of surface fire activity) closely coincides with the Little Ice Age (Power et al. 2012). A broad synthesis of Holocene fire records does not exclude a role for human activity in promoting fire,

Table 2.1. The various ways in which humans can influence fire regime parameters by modifying the key variables that affect fire activity

Fire variable	Natural influences	Human influences	Fire regime parameters
Wind speed	Season Weather Topography Land cover	Climate change Land cover change	Fire spread
Fuel continuity	Terrain type (slope, rockiness, aspect) Rivers and water bodies Season Vegetation (type, age, phenology)	Artificial barriers (roads, fuel breaks) Habitat fragmentation (fields) Exotic grasses Land management (patch burning, fuel treatments) Fire suppression	Fire spread
Fuel loads	Tree, shrub, and grass cover Natural disturbances (e.g., insect or frost damage, windthrow) Herbivory Soil fertility Season	Grazing Timber harvests Exotic species establishment Fire suppression Fuel treatments Land use and land cover (deforestation, agriculture, plantations)	Fire intensity and severity
Fuel moisture	Season Antecedent precipitation Relative humidity Air temperature Soil moisture	Climate change Land management (logging, grazing, patch burning) Vegetation type and structure (species composition, cover, stem density)	Fire intensity and severity
Ignition	Lightning Volcanoes Season	Human population size Land management Road networks Arson Time of day Season Weather conditions	Number and spatial and temporal patterns of fires

From Bowman et al. (2011).

but suggests a stronger influence of climate, at least until the late Holocene in eastern North America (Marlon et al. 2013).

INFLUENCE OF INDIGENOUS PEOPLE ON FIRE REGIMES IN FLORIDA AND THE SOUTHEASTERN COASTAL PLAIN

> For many years, the importance of fire use by American Indians in altering North American ecosystems was underappreciated or ignored. Now, there seems to be an opposite trend.
>
> Barrett et al. (2005)

> The southeastern Coastal Plain appears to be one region where frequent natural lightning fires, facilitated by the occurrence of dry lightning, may have been adequate to sustain pyrogenic vegetation, pine forests, savannas, and grassland and scrub, with or without Indian burning.
>
> Abrams and Nowacki (2015)

I devote considerable space to this issue because it is among the most contentious in North American fire ecology. Prevailing but untested assumptions of large Native American populations and their control of fire regimes in the southeastern Coastal Plain have misled a generation of fire scientists and managers about the long-term evolutionary and ecological role of fire in this region. The continuing debate appears, in part, to reflect ideological differences about the role of humans in nature. Ideology has an emotional component. Hence, arguments about people vs. nature as the main control of fire activity are unlikely to be resolved by anything but the most powerful evidence.

Accumulating information strongly suggests the southeastern Coastal Plain stands out as a region where climatic control of fire regimes has overshadowed human control until very recently, after European settlement and long after most species native to this region evolved. This does not mean earlier human use of fire was unimportant ecologically. Nevertheless, the evidence suggests that the spatial and temporal extent of this influence was limited in this region.

Humans are most likely to change fire regimes when they colonize regions that contain vegetation susceptible to burning, but which are ignition-limited—that is, with low lightning activity (Pinter et al. 2011). The logic is inescapable. Florida and the southeastern Coastal Plain have highly pyrogenic vegetation and are decidedly not ignition-limited (see chapter 1). Prior to European settlement and widespread landscape transformation, which disrupted fuel continuity and the spread of fire, this region was perhaps one of Earth's biomes "fully saturated with natural fires" (Pinter et al. 2011), or at least close to saturation. Although pre-Columbian humans used fire for various purposes in the Coastal Plain, the flammable vegetation and high lightning incidence suggest only rarely would enough fuel have been available and sufficiently connected over large areas for humans to ignite landscape-scale fires.

At the time of European arrival, most of the southeastern Coastal Plain away from the immediate coast and major waterways had extremely low population density, perhaps among the lowest in North America (Larson 1980). This low population density was observed by the first Spanish explorers to enter the region, and therefore was not the result of introduced diseases, although those diseases subsequently reduced native populations even further.

Exaggerated claims that human use of fire created and maintained grasslands and other fire-dependent vegetation globally—for example, "Most if not all of the great grasslands of the world, from the Serengeti Plains to the prairie bioregion of the contiguous United States, were maintained with fires set by native peoples" (Anderson 2007)—have little basis in fact. Rather, the influence of humans varied markedly by region, mostly depending on the frequency of lightning strikes, the combustibility of the vegetation, and the density and cultural attributes of the human population.

Abrams and Nowacki (2015) reviewed evidence for climate vs. humans as sources of fires during the Holocene in the eastern United States. Climate dominated human activity as a control of fire regimes and vegetation during the early and middle Holocene. During the late Holocene, humans became the main source of fires in the northern and central regions (including the tallgrass prairie) of eastern North America due to the general lack of dry lightning. In contrast, "in the southeastern coastal plain . . . lightning (including dry lightning) is frequent, the vegetation (conifers, ericaceous shrubs, grasses) is highly flammable, and frequent and reasonably large fires may ensue" (Abrams and Nowacki 2015). Similarly, humans had a relatively low impact on fire regimes in tropical African savannas until recently, when fire extent declined with increased livestock grazing, road density, and human population density (Archibald et al. 2009).

Paleoindians

Humans did not enter North America until the very late Pleistocene. The dates of Paleoindian artifacts and other signs of human activity around the Americas remain controversial, but the earliest well-accepted dates fall between 14,000 and 15,000 years ago. These early Paleoindians, after migrating to North America in probably several waves, spread extensively and rapidly across the continent and on to South America. They reached Florida quickly.

The site with the earliest evidence of Paleoindian activity in southeastern North America, 14,550 years ago, is the Page-Ladson sink in the Aucilla River of the Florida Panhandle (Webb 2006; Halligan et al. 2016). Among the artifacts discovered at this remarkable site are a stone knife and a mastodon tusk with circular knife cuts. Other megafauna remains found in sediments at the site contemporaneous with human artifacts include camelid and bison, indicating savanna/woodland interrupted by forested wetlands. The latter may have been primary habitat for browsing mastodons in this region (Newsom and Mihlbachler 2006). Annual cycles of tusk growth from a mastodon here "suggest a sub-tropical climate with one moderately severe dry season per year, and a secondary, less severe dry interval, each bracketed by times of relief from moisture stress" (Fisher and Fox 2006), a climate much like today's and conducive to fire.

Remains of mammoths also were found at Page-Ladson and other Aucilla River sites. With their high-crowned teeth, mammoths were primarily grazers of C_4 grasses here as elsewhere (Newsom and Mihlbachler 2006). Although growth rings in mammoth and mastodon tusks from this and other sites have led some researchers to suggest seasonal latitudinal migrations of both species, perhaps as far north as the southern Appalachians and back (Fisher and Fox 2006), other evidence such as strontium isotope ratios from tooth enamel suggests mammoths "moved only locally in Florida, while mastodons appear to have migrated across distances of at least 120 km into granitic terrain," and therefore possibly to the Appalachians (Hoppe and Koch 2006). Paleoindians likely accompanied mastodons, and perhaps other megafauna, on their seasonal migrations (Milanich 1998).

Sea level was around 95 m lower in the postglacial climate of 14,550 years ago, although its rapid rise beginning around 14,000 years ago (Donoghue 2011) suggests the Paleoindians lived during a time of rapid climate change. Based on the rate of sea-level rise, the warming was likely more rapid than what we are experiencing today, albeit the temporal match between warming air and rising sea level is not perfect. Because Florida's freshwater aquifer is perched on underlying salt water, many modern lakes held little or no water during the time of Paleoindian occupation, and modern rivers were probably no more than a "series of small limestone catchment basins or watering holes" (Milanich 1998). With lower sea level, what were then coastal Paleoindian villages are now well below sea level, some 80 km (50 mi) off the Gulf of Mexico coast (Milanich 1998).

The Page-Ladson sink is currently within the flowing Aucilla River, but at the time the Paleoindian artifacts and megafauna remains were deposited it was an isolated sinkhole with fluctuating water levels paralleling sea-level changes (Halligan et al. 2016). Spores from *Sporormiella*, the fungus unique to herbivore dung, indicate humans and megafauna coexisted in the Aucilla area for around 2,000 years until the megafauna went extinct approximately 12,600 years ago, around the same time they went extinct across most of North America. A later spike in *Sporormiella* spores from 10,400 to 10,000 years ago may indicate a reexpansion of bison into Florida at that time (Halligan et al. 2016). Not far away, in the Wacissa River, which flows into the Aucilla, divers found a skull of *Bison antiquus* with a broken stone point embedded in it (Milanich 1998). We have no clues about how Paleoindians may have altered fire regimes, but a prudent hypothesis is not much, if at all.

Archaic Culture and European Contact

Beginning around 10,000 years ago, the Paleoindian culture was replaced by the Archaic culture. With the megafauna gone, hunters were forced to concentrate on smaller game. The size of spear points dramatically decreased (Milanich 1998). By the middle Archaic period, 7,000 to 5,000 years ago, sea level stabilized at around the present level with relatively minor fluctuations (Donoghue 2011). Recall that this was the time of the most recent pine expansion in Florida (Watts and Hansen 1994), indicating regionally warm and wet but likely seasonally dry conditions (Grimm et al. 2006).

By middle to late Archaic times, Florida Indians had become more sedentary, living in villages of several acres rather than small camps. They had larger populations than before along the coasts and major waterways such as the St. Johns River. Settlements also occurred in the interior of northern Florida away from major waterways but where other large water sources such as lakes were available. An example is a middle Archaic site on the northern edge of Paynes Prairie (Alachua Savanna), next to modern-day Gainesville, which was probably a shallow lake at that time

(Milanich 1998). Shell middens from this period are common both along the coast (with marine shells) and inland (with freshwater shells).

By 4,000 years ago late Archaic Indians in Florida were making fired clay pottery, another use of fire besides cooking (Milanich 1998). The St. Johns culture, which developed in late Archaic times, was represented by the Timucua and Mayaca people of northeastern and central Florida, respectively, who had contact with the first Europeans to enter Florida (see below). These people possessed a relatively advanced technology, which included fishhooks, weirs, bows and arrows, snares, deer-hunting disguises, and the use of fire drives to surround and kill deer (Milanich 1998).

Accounts by early Spanish and French explorers indicate that most native cultures in the southeastern Coastal Plain were hunter-gatherers at the time of European contact, with some groups also practicing small-plot agriculture or sedentary horticulture. Very few practiced large-scale agriculture, apparently because soil fertility across most of the region was too poor to support it (Larson 1980; Milanich 1998).

Based on an extensive review of the narratives of Spanish and French explorers and other evidence, Larson (1980) described indigenous groups and their subsistence technology at the time of first European contact in the early to middle sixteenth century. He identified three major environmental/cultural areas on the southeastern Coastal Plain corresponding to Indian populations and their technology: (1) The Coastal Sector, which extended from Tampa Bay northward and westward to Louisiana along the Gulf of Mexico coast and from the northern Indian River Lagoon northward to southeastern Virginia on the Atlantic coast. This region extended inland a variable distance from the coast (farthest in the eastern Florida Panhandle, encompassing the Tallahassee Red Hills) and up major river valleys, especially the St. Johns, Apalachicola/Chattahoochee/Flint, and Alabama/Tombigbee. In addition to exploiting marine, estuarine, freshwater, and terrestrial resources, many Indian groups in this sector cultivated domesticated plants and practiced shifting agriculture. (2) The South Florida Sector, extending from south of Tampa Bay on the Gulf Coast and the Indian River Lagoon on the Atlantic Coast. These groups were entirely hunter-gatherers (especially fish, shellfish, marine mammals, sea turtles, and other marine and estuarine resources) and apparently made no use of cultivated plants or of fire, except for cooking. (3) The Pine Barrens Sector, which covered the remainder of the southeastern Coastal Plain outside the Coastal and South Florida Sectors—a huge area.

According to Larson (1980), the vast region constituting the Pine Barrens Sector had little aboriginal occupation at the time of the first Spanish and French explorers. Indeed, it served as a cultural barrier between the Indian groups on the coast and those in the interior of the southeastern United States (Larson 1980). Some trade occurred, however, along major river systems such as the Mississippi, Alabama/Tombigbee, and Apalachicola/Chattahoochee/Flint, based on northern cultural artifacts (including copper from Lake Superior) uncovered at Florida Indian sites (Milanich 1998). A simple explanation for the generally low human population density in this region is that "the longleaf pine forest and the floodplain areas alike offered little in the way of technologically accessible resources" for the aboriginal people of this time (Larson 1980).

Although a sparsely occupied interior southeastern Coastal Plain at the time of European contact may come as a surprise to some readers, given our new cultural myth of pre-European landscapes entirely transformed by Indian activity (Vale 1998), Larson's (1980) review of early explorers' accounts is exhaustive. He notes that "maps of the Southeast made during the years prior to the beginning of the nineteenth century . . . show Indian towns on or above the Fall Line and, in the case of the earlier maps, Indian towns on the coast; the intervening area is always shown without any occupation." No peer-reviewed literature, to my knowledge, substantially contradicts Larson's regional-scale conclusions. On the other hand, Larson (1980) probably underestimated the extent of Indian occupation away from the coast in a few areas, such as the lake district of north-central and central Florida, where abundant aquatic resources could support permanent or long-term occupation (Milanich 1998).

The Timucua and Related Groups

Within the Coastal Sector at the time of first European contact, and extending into the Pine Barrens Sector, the Timucua inhabited a large region in northeastern and north-central Florida and southern Georgia, north to the Altamaha River, west to the Aucilla River, and south possibly to Cape Canaveral. They were the most numerous people (perhaps 40,000; Milanich and Fairbanks 1980) and occupied the largest area of any culture in Florida and southern Georgia at the time (Hann 1996). The Timucua and related groups represent the most recent people of the St. Johns culture. Many different groups of Indians have been labeled "Timucua" after their shared language; for instance, the Mayaca was another St. Johns group that inhabited the area of north-central to central Florida south of Lake George, and are often considered Timucua (Milanich 1998).

Timucua groups usually were associated with water bodies and aquatic food resources, chiefly along the Atlantic coast and the St. Johns River, although they reached the Gulf Coast in the area around the mouth of the Suwannee River. Some Timucua-language groups occurred as far south as Tampa Bay (Hann 1996). Groups of Timucua and other Indian cultures located farther from major water bodies may have made greater use of fire to assist their hunting, gathering, and agriculture. The northern Timucua grew corn (maize) and some other plants, whereas the Mayaca did not (Milanich 1998). Due to the poor soils, "the St. Johns people and their Timucua descendants were not intensive farmers," and "their economic system could not support populations as large and dense as those of the full-time farmers in the interior of the Southeast" (Milanich 1998). Burning for hunting or agriculture by the Timucua and related groups was almost certainly localized, not regional.

Not as much is known about the interior groups as the coastal settlements, in part because the Spanish had far less contact with the former (Lewis 1978), but also because relatively few exclusively interior groups existed (Larson 1980). At the time of the Spanish missions in the 1600s, some Timucua groups practiced slash-and-burn agriculture, with trees girdled and left to die and the sites then burned (Miller 1998). The Spanish may have prodded the Timucua to raise more maize and other crops and rely less on fishing and hunting than they did before contact and establishment of missions.

According to some reports the Timucua prepared fields for new maize crops by burning them and the undergrowth of the surrounding "woods" during December and January. This burning facilitated a "fire hunt," known as "*hurimelas*" in the Timucua language, in which deer, rabbits, and other game were surrounded and then killed (Hann 1996). Other observers, such as Rene de Laudonniere, who commanded French troops at Fort Caroline near the mouth of the St. Johns River, reported that the Timucua kept their maize fields fallow for six months, then burned the fields in March before sowing seed (de Laudonniere 1587). Still others wrote that the Indians burned the fields in March and then again in July before planting (Braudel 1817). Hence, the seasonality of burning by the Timucua and probably other Indian groups was variable but probably mostly concentrated in the winter and early spring.

The Calusa

Within the South Florida Sector, the Calusa had a large and complex society on the Gulf of Mexico coast between Charlotte Harbor and the Florida Keys. These people, at the time of first Spanish contact, were the representatives of the Caloosahatchee culture. The Calusa relied primarily on shellfish, fish, marine mammals and turtles, and other estuarine and marine resources, rather than on terrestrial game or on agriculture or horticulture (Larson 1980). Their primary site at the time of Spanish contact was an island, Mound Key, in Estero Bay south of Fort Myers, though their rule or influence extended across much of the southern peninsula, encompassing former sites of the Glades and Belle Glade cultures (Lewis 1978; Milanich 1998).

The vast majority of complex native societies in the Americas were agricultural. The Calusa are an exception in that they raised no maize, beans, manioc, or other staple crops (Marquardt and Walker 2013); therefore, they are highly unlikely to have used fire for either agriculture or hunting. The Calusa and possibly earlier Caloosahatchee groups on Marco Island made more use of terrestrial foods (Lewis 1978), and therefore might conceivably have used fire to assist hunting or gathering; however, no accounts exist of fire use by these people.

Cabeza de Vaca, the Apalachee, and the Alachua

Alvar Núñez Cabeza de Vaca was one of four survivors of the Narváez expedition, which landed in what was probably Sarasota Bay in 1528 (Reséndez 2007; earlier authors suggest Tampa Bay). He and his companions subsequently walked northward up the Florida peninsula near the Gulf Coast, then westward across the Panhandle and ultimately to Mexico, where he reunited with his Spanish compatriots in 1536. Most of Florida was very sparsely populated at the time of Cabeza de Vaca's journey (Larson 1980; Adorno and Pautz 1999; Reséndez 2007). The narrative of Cabeza de Vaca suggests natural processes dominated human activity across most of Florida. He was impressed by the frequency and ferocity of storms—for example, writing in his journal from somewhere between the Suwannee River and the Aucilla River that the "wonderfully tall" trees were often "split from top to bottom by lightning bolts that strike in that land where there are always great storms and tempests" (Adorno and Pautz 1999).

In contrast to the small native communities that Cabeza de Vaca observed traveling up the Florida peninsula was the agriculture-based

Chiefdom of Apalachee, centered in the more fertile Red Hills region of the Florida Panhandle. Here Cabeza de Vaca reported a large agricultural society. The Apalachee represent a southern extension of the widespread and advanced Mississippian culture of that time (Milanich 1998). Apalachee was a much larger and more complex chiefdom than anything Cabeza de Vaca had encountered on the Florida peninsula, where he had found only "small villages scattered over large areas. . . . Although some of these groups cultivated maize, they were primarily hunters and gatherers" (Reséndez 2007). Recent investigations suggest maize agriculture was not practiced south of an east–west line from just north of Cape Canaveral, through Lake George on the St. Johns River, to north of Tampa Bay on the Gulf Coast (Milanich 1998).

The Apalachee population may have numbered about 25,000 at its peak (Milanich and Fairbanks 1980). Their capital at present-day Tallahassee, within Larson's (1980) Coastal Sector, contained roughly 250 homes, and they may have had more land in agriculture than modern farmers did in the 1980s, although less than in the peak of cotton agriculture prior to the Civil War (Crawford and Brueckheimer 2012). The size and density of the Apalachee population and the degree to which they modified the surrounding landscape appear to have been greater than anywhere else in Florida, and possibly anywhere within the entire southeastern Coastal Plain, at the time of European contact.

Relatively fertile soils were also located in the Hammock Belt of north-central Florida. The Alachua apparently were people of the Ocmulgee culture, who entered northern Florida from Georgia around 600 C.E. These people took advantage of the more fertile soils of the Hammock Belt in Alachua, Suwannee, Columbia, and Hamilton Counties to support maize agriculture and small chiefdoms. The various Indian groups that inhabited the mosquito-ridden Hammock Belt of north-central Florida reportedly lit small smudge fires under their beds to ward off mosquitoes while they slept (Milanich 1998).

Cabeza de Vaca's accounts of Indian use of fire are limited to a few passages about groups on the islands and mainland along the Gulf Coast, probably in Louisiana or eastern Texas, applying fire for hunting and mosquito control. To defend themselves from mosquitoes on islands, the Indians built "around the edge of the group great bonfires of rotted and wet wood that would not burn but rather make smoke" (Adorno and Pautz 1999). Inland from the immediate coast, they would "walk, with torches in hand, burning the fields and woods they encounter to drive the mosquitoes away, and also to drive out from underground lizards and other similar things in order to eat them. And they also use this to take deer, surrounding them with many bonfires" (Adorno and Pautz 1999). Accounts from Virginia describe a similar "fire-surround" method of hunting among the native groups there, as well as use of fire for signaling (d'Iberville 1968–1702, as cited by Robbins and Myers 1992).

Other Cultures and Increased Use of Fire

Besides the Apalachee, Alachua, Timucua, and Calusa, other Indian cultures in Florida and nearby around the time of European contact included the Tequesta in coastal southeastern Florida; the Jeaga and Ais on the peninsula coast north of the Tequesta; the Mayaimi around Lake Okeechobee; and the Guale on the coast of southeastern Georgia north of Timucua territory (Larson 1980). Farther beyond Florida, Indian groups in the southeastern Coastal Plain included the Choctaw, Cusabo (and subtribes including the Edisto), Orista, Secotans, Croatoans, and many others. Most groups were concentrated near the coast and relied on marine resources and small farmed plots (Larson 1980).

Extensive landscape-scale burning by Indians is well documented elsewhere in the southeastern United States, for example, by various Indian groups in the Piedmont to facilitate hunting, travel, and agriculture (Barden 1997; Juras 1997) and by the Cherokee in the southern Appalachians to facilitate chestnut gathering and deer hunting (Hudson 1976). Prior to the arrival of the Creeks and their transformation into the Seminole and Miccosukee, however, this was not the case for Florida or probably elsewhere in the lower southeastern Coastal Plain. Evidence for fire-surround hunting or other deliberate burning across large landscapes is lacking from early accounts by the Spanish and French in this region (Larson 1980; Robbins and Myers 1992). The facile notion that "anthropogenic fire is one of the few constants in Florida's human occupation" (Pyne 2016) does not stand up to scrutiny.

The increase in widespread use of fire for hunting in Florida after the arrival of the Creeks around 1750 (Milanich 1998; see below) may reflect the deerskin trade's effects on hunting pressure. This trade, which began prior to the arrival of Europeans but intensified thereafter, was most extreme north of the Coastal Plain, in the land of the Cherokee in the southern Appalachians, the Chickasaw in northern Mississippi and southwestern Tennessee, and the Creeks in portions of Georgia, Alabama, and adjacent Tennessee (Hudson 1976). Sediment cores from the Coastal Plain of Georgia and Alabama show increased use of fire by Indians in the early eighteenth century as the demand for deer hides increased (Foster and Cohen 2007).

By 1750 deer were already depleted across much of the Southeast. This could be one explanation for the migration of the Creeks southward into Florida at that time, bringing with them their practice of frequent burning. The botanical explorer William Bartram's observation that fires were set "almost every day throughout the year" (Bartram 1791) in some area or another by the Seminoles, as well as by lightning, may be an exaggeration, as Bartram was given to hyperbole (e.g., his descriptions and drawings of impossibly immense alligators attacking boats). As discussed in the following section, however, the Seminoles and white settlers alike used fire extensively in their livestock management and for other purposes.

As every good detective knows, eyewitness accounts of events are often dubious. And so "the vast majority of written and oral accounts on Indian fire use are anecdotes, fraught with uncertainty, subjective opinion, and bias" (Barrett et al. 2005). Nevertheless, secondary sources such as textbooks may be even less reliable than firsthand accounts due to their often sweeping generalizations and extrapolations from other regions. These sources regularly conclude, usually without references, that landscape-scale burning by Indians was routine in the southeastern Coastal Plain before Europeans arrived. For example, Miller (1998) states that "burning the landscape was a common practice among Florida Indians before European contact. . . . Prehistoric Florida agriculturalists practiced a slash-and-burn method, using fire not only to clear trees but also to release into the topsoil nutrients stored in vegetation."

Ecological Impacts of Indian Burning

Miller (1998) claims that "frequent burning in the southeast United States has another, more prolonged environmental effect. It is now well known that the southern pine forest is a fire-adapted vegetation community, that is, fire is necessary for its continuation." Miller implies that Indians were the source of this necessary fire, although Larson (1980) insists that "substantive support is lacking" for this customary assumption. Larson (1980) also comments, "it seems very unlikely that the fires necessary to the creation of the longleaf forest would have been started by Indian hunters rather than lightning." As reviewed earlier, the origin of pine savannas in the Coastal Plain extends back millions of years before humans entered North America.

If Indians were responsible for most landscape-scale fires prior to European settlement, we should see the signature of their burning in fire scars in the rings of old pine stumps whose early years record history before European settlement (albeit not before European contact, as no stumps that old have yet been discovered). Huffman (2006) conducted such a study in the Florida Panhandle. Lightning-season fires were the only fires recorded in the annual rings of pines at the St. Joseph Bay savanna between 1592 and 1830, before Europeans settled the area. As reviewed by Robbins and Myers (1992), reports on the seasons during which Indians burned suggest they burned at any time of year or, alternately, mostly during the dormant season (e.g., December through March; Hann 1996; de Laudonniere 1587). Indian populations were sparse in northern Florida during the 1592–1830 period, which may explain the absence of fire scars from other seasons (Huffman 2006).

Throughout the eastern United States, except the southeastern Coastal Plain, fire scars show that most fires during the late Holocene occurred during the dormant season, "outside of the thunderstorm season, but within the ever-presence of human ignitions" (Abrams and Nowacki 2015). In contrast, only three fires recorded in tree rings from the St. Joseph Bay savanna occurred during the dormant season, and all occurred after European settlement, which began in this area in the late 1820s. Huffman (2006) reasonably concludes that even if Indians were responsible for some of the pre-1830 growing-season fires, "their influence would have been overwhelmed by background climatic conditions producing very frequent growing season lightning fires."

Probably many lightning fires in the southeastern Coastal Plain were falsely attributed to Indians by naive Europeans. This mistake could result from a lack of appreciation of lightning as an ecological force in the region, given the origin of most European explorers and later EuroAmerican settlers in regions with drastically less lightning activity and little pyrogenic vegetation. Racism (i.e., viewing Indians as primitive, uncivilized, and destructive) also might have led some explorers and settlers to exaggerate Indian burning (Barrett et al. 2005). More generally, those who ascribe pre-European fires in the Coastal Plain largely to humans seem to be extrapolating evidence from elsewhere in eastern and central North America. They are ignoring the significant "north-south climate-fire gradient" (Abrams and Nowacki 2015).

The evidence for landscape-scale burning by Indians is weak for the southeastern Coastal Plain, and especially weak for peninsular Florida prior to the arrival of the Creeks. Yet the "myth of the humanized landscape" (Vale 1998) is so entrenched that evidence from many fields of inquiry for a long prehuman history of fire is often disregarded. Ignored is the extensive support for the conclusion that, unlike some other regions of North America, such as the eastern deciduous forest during the late Holocene (Stambaugh et al. 2015), the natural vegetation of the southeastern Coastal Plain is "apparently maintained by lightning fires" (Abrams and Nowacki 2015). The assertion by journalist Charles Mann (2005), in his best-selling book *1491*, that the pre-Columbian Florida peninsula and much of the Gulf Coast and Mississippi River watershed was "dominated by anthropogenic fire" is not supported by any available evidence (Noss 2013). Just as erroneous, at least for the southeastern Coastal Plain, is the incredible conclusion of Kay (2007) that "lightning-caused fires may have been largely irrelevant for at least the last 10,000 y. Instead, the dominant ecological force likely has been aboriginal burning."

Simply due to chance some landscapes would have burned less often from lightning strikes than they potentially could have. In such cases, fine fuels may have been sufficient to support occasional landscape-scale fires ignited by Indians. On islands, peninsulas, and other areas topographically protected from fire (Harper 1911), especially within the more densely populated Coastal Sector (Larson 1980), burning by Indians could have increased fire frequency above that determined by lightning. Kalisz et al. (1986) conclude that the longleaf pine–wiregrass "islands" within the extensive and less flammable scrub of Ocala National Forest are too isolated to have received sufficient lightning strikes to maintain their structure; hence, they must have been burned by Indians. Lying adjacent to the productive St. Johns River, the Ocala National Forest area probably had a denser population of native people than other areas away from the coast.

Although the southeastern Coastal Plain may not have been "fully saturated by fire" (Pinter et al. 2011) prior to human settlement, it was probably close to saturated. Across the region, Indian fires may have slightly reduced average fire-return intervals determined by lightning, but by how much we will never know. Today, with our highly altered and fragmented landscape, deliberate burning by humans is very much needed to mimic natural fire and maintain short fire-return intervals.

Were Indians capable of altering the fire regime in the southeastern Coastal Plain enough to influence the evolution of plants and animals? Reviews of fire use by aboriginal people in other regions, such as northern and western North America and Australia, show that humans are quite capable of changing fire regimes—including the frequency, seasonality, and intensity of fire—away from the previous or "natural" regime (Lewis 1982; Abrams and Nowacki 2015). The total pre-Columbian population in Florida, estimated as approximately 100,000 (Milanich and Fairbanks 1980), was about the same as the total human population during the 1800s, "when woodsburning was pervasive throughout the South" (Robbins and Myers 1992). The difference is that pre-Columbian people were concentrated along the coasts, and most noncoastal groups were hunter-gatherers or small-plot agriculturists, and did not use fire extensively.

Hiers et al. (2000), citing evidence from other regions (e.g., upper Great Lakes; Loope and Anderton 1998), speculated that indigenous people in the Coastal Plain may have changed fire regimes, including season of burns, within time frames capable of influencing natural selection. Species with short generation times can undergo substantial evolution, even speciation, over a time scale of thousands of years. This seems to have been the case for some plants and invertebrates in the Florida Keys, which apparently speciated within the approximately 6,000 years that the Keys have been separated from the mainland and from each other by higher sea level (Noss 2013). Although anthropogenic fire may have influenced the evolution of some endemic plants in the Keys (Liu and Menges 2005; see chapter 3), prior to the arrival of the Creeks (Seminole and Miccosukee) no evidence exists for use of fire by cultures in south Florida, other than for cooking. The effect of human fire use on evolution in the southeastern Coastal Plain remains an open question, but a strong influence is unlikely.

EUROPEAN AND SEMINOLE/MICCOSUKEE SETTLEMENT

Attempts at settlement of the southeastern Coastal Plain by Europeans began in the sixteenth century. The first permanent or long-lasting settlements in what is now the United States, Pensacola and St. Augustine, were both founded by the Spanish in the mid-1500s. The Roanoke colony, the "Lost Colony," was established by the English on Roanoke Island, North Carolina, in 1585, but disappeared for unknown reasons. The Jamestown colony, founded in 1607, nearly failed due to famine, disease, and fighting with Indians, but eventually it persisted and settlement spread from there (Hudson 1976). Florida was the last region to be well-settled by whites and blacks in the southeastern United States (Frost 2006). This seems odd given it was the first place to be discovered and explored by Europeans, beginning with Ponce de Leon in 1513, and had the first long-term settlements at Pensacola and St. Augustine.

As noted by Frost (2006), the Spanish effectively blocked settlement of the interior Coastal Plain and "with the exception of a handful of coastal villages . . . they never pursued immigration and settlement of the land." When Spain ceded Florida to the United States in 1821, the entire European population of Florida barely exceeded 20,000 people. By 1880 Florida still had only about 270,000 people and was considered "an outcast among outcasts within the former Confederacy" (Colburn and deHaven-Smith 2010). The southern third of the Florida peninsula was not settled by two or more people per square mile until around 1890 (Frost 2006). This is in vivid contrast to the coastal areas of Virginia and the Carolinas, which had populations denser than this before 1770 (Frost 2006).

Some early European settlers in the southeastern Coastal Plain, notably the Scotch-Irish with their Celtic roots, brought a culture of woods-burning with them from the Old Country, as opposed to the more refined and urbane English (Putz 2015). White and black settlers also learned burning traditions from the Indians and frequently "burned the woods" for similar reasons. Even today, as Putz (2003) put it, country people set fires "to improve hunting, to kill ticks, because the mower won't start, to expose snakes, and for fun." Fire scar studies from the eastern and central United States reviewed by Abrams and Nowacki (2015) show that fire frequency often remained roughly the same after European settlement as before, and in some areas it even increased, despite depopulation of Native Americans.

The land use that probably involved the most frequent burning in the southeastern Coastal Plain, particularly Florida, was cattle grazing, both by whites and by Indians. As noted earlier, the ancestors of the Seminole and Miccosukee tribes were Creeks, largely from Georgia and Alabama. The Lower Creeks entered Florida in the mid to late eighteenth century to gain independence from the Upper Creeks. By this time, the ranches and missions of the Spanish and native tribes had been abandoned, and the Lower Creeks colonized these lands and herded the remaining cattle (Milanich 1998). By the 1760s these Creeks were becoming known as the Seminoles, a derivation of the Spanish word "*cimarrones*," meaning runaways. At one Seminole village, Cuscowilla, near Paynes Prairie south of Gainesville and visited by William Bartram in the 1770s, the Seminoles kept cattle descended from Spanish herds. The chief of the Cuscowilla Seminoles was known as Cowkeeper (Milanich 1998). Frequent burning was applied by these people to provide fresh green forage for cattle. This was critical because many of the common grasses, especially wiregrass

(*Aristida beyrichiniana*), of Florida and most of the southeastern Coastal Plain are quite unpalatable except for a brief period after burning (Robbins and Myers 1992).

Another major colonization of Florida by Creeks, in this case Upper Creeks, occurred in the early nineteenth century, following Andrew Jackson's defeat of Upper Creek warriors in 1814 (Milanich 1998). The Miccosukee, formally recognized as a tribe in 1962, are descendants of another Lower Creek group, the Mikasuki. Conflicts between the Seminoles and white settlers over agricultural and grazing lands, among other issues, led to a series of three Seminole Wars between 1817 and 1858. Many Seminoles and Miccosukee were relocated to Oklahoma, whereas others were progressively pushed southward to the Everglades–Big Cypress region (Milanich 1998). The Florida panther (*Puma concolor*) suffered a similar fate. For both these Indians and the panther, the inaccessibility of the Everglades–Big Cypress region offered the last refuge in Florida.

The Seminoles and Miccosukee in Florida regularly burned the landscape for hunting as well as for their cattle, as they previously did in their native Creek territory to the north. Three decades before Bartram visited Florida, Stork (1769, cited in Robbins and Myers 1992) described Indian use of fire here: "the hunting parties of the Creek Indians, who are dispersed through the whole province, continually set the grass on fire, for the conveniency of hunting." Burning of the landscape by Seminoles, Miccosukee, white and black settlers, and recent rural folks continued well into the twentieth century. It continues today in some places. The statement below, made by an administrative forester (Eldredge 1911, cited in Scott et al. 2014), provides a cogent description of the purpose, frequency, and seasonality of deliberate burning by white (and probably black) settlers in Florida in the early twentieth century:

> The popular sentiment of the residents within the forests, in common with nearly all the people of the South, is unqualifiedly in favour of the annual burning over of the pineries. The homesteader and the cattleman burn the woods to keep down the blackjack undergrowth and to better the cattle range. The turpentine operator burns over his woods annually, after raking around the boxed trees. . . . The camp hunters, of whom there is a large number during the fall and winter months, set out fires in order to drive out game from the thickets. . . . The turpentine operator burns his woods and all other neighboring woods during the winter months, generally

> in December, January, or February. The cattleman sets fires during March, April and May to such areas as the turpentine operator has left unburnt. During the summer there are almost daily severe thunder-storms, and many forest fires are started by lightning. . . . It is only by chance that any area of unenclosed land escapes burning at least once in two years.

Burning by settlers in the South, usually outside the natural lightning fire season, was done not only for resource management but also as "an attempt to transform fire . . . from an uncontrollable and unpredictable force of nature to one controlled by man" (Robbins and Myers 1992). Or as Pyne (2016) put it, people "found ways to keep the right kind of fire on the ground, because otherwise fire will fill that vacuum with the wrong kind." What is the "right" or "wrong" kind of fire depends on one's perspective. Ecologists and conservationists may think that fires ignited by lightning, or controlled burns designed to mimic lightning fire, are right, because many native species benefit most from such burns. Land managers driven by other values, such as forestry, may think that easily and comfortably applied dormant-season fire is right. Humans have become all too adept at controlling and transforming nature.

Fire scars in tree rings tell a story of the effects of settlement on fire regimes. Such dendropyrochronology studies had not been conducted in the southeastern Coastal Plain until recently, but we now have records from several sites in the region. These fire-scar records reveal very frequent fire both before and after settlement by EuroAmericans, largely within the one-to-three-year fire-return intervals summarized by Frost (1995, 2000, 2006) for the most fire-exposed portions of landscapes.

In the West Gulf Coastal Plain of central Louisiana, fire frequency increased markedly during the period 1791–1880, coinciding with a rapid influx of white settlers in response to the Spanish governor's opening of the region to immigration prior to the Louisiana Purchase (Stambaugh et al. 2011). In the Florida Panhandle, dormant-season fires first appear in the tree-ring record between 1831 and 1848, and late growing-season fires increase in frequency. EuroAmerican settlers moved into that portion of the Panhandle in the late 1820s. Shortly thereafter they initiated open-range grazing of cattle and burned pine savannas for winter forage production (Huffman 2006).

At Avon Park Air Force Range in south-central Florida, Huffman and Platt (2014) related changes in fire-return intervals and season of burns in tree-ring records to distinct historic periods. They divided the period sampled into a Pre-extractive Era (1787–1919), during which humans did not greatly alter the landscape, and an Extractive Era (1920–2005), which corresponds to widespread logging of old-growth pines and changes in grazing practices. The Pre-extractive Era was further subdivided into the Seminole Period (1780–1839), the Seminole War Period (1840–1859), and the Open Range/Homestead Period (1860–1919). The Extractive Era was subdivided into the Extractive Period (1920–1939) and the Bombing Range Period (1940–2005).

Fires remained frequent (mostly one-to-two-year fire-return intervals) throughout the entire Pre-extractive Era (Huffman and Platt 2014). Open-range cattle grazing was the dominant land use during this period of more than a century, when "sparse populations of Seminoles and subsequently open range cattlemen and settlers took advantage of the large, open, grassy prairies and flatwoods of the Kissimmee River region to raise cattle" (Huffman and Platt 2014). Both the Seminoles and the whites burned this landscape for the benefit of their stock.

An interesting finding of Huffman and Platt (2014) is that fires remained frequent during periods when people were probably absent from the landscape. Prior to 1823 this region was very sparsely inhabited (Larson 1980); whatever native groups that may have been present had been decimated by European diseases after Spanish contact. The Seminoles came into this region of Florida after 1823, driven south from north Florida (citations in Huffman and Platt 2014). The landscape was again virtually uninhabited during portions of the Seminole War Period (1840–1859). That fire frequency remained high throughout these periods is testimony to the abundance of lightning.

FIRE EXCLUSION

Fire frequency across the southeastern Coastal Plain remained high until active fire suppression campaigns began in the 1920s. Beginning then, burning by settlers was strongly discouraged by forestry agencies and organizations, who feared loss of commercially valuable timber. This was partly a legitimate concern because massive clear-cutting of pine savannas, combined with the ravages of the naval stores industry (tar, pitch, rosin, and turpentine), had virtually eliminated mature stands of longleaf pine. Swarms of feral hogs, first introduced to the Coastal Plain by Hernando de Soto in 1539, ate longleaf pine seedlings and reportedly reduced pine regeneration across much of the southeastern Coastal Plain (Frost

2006). Consistent annual burning further hindered pine regeneration, and the logging slash left behind after clear-cutting, along with dead and weakened trees after terpentining (Frost 2006), fueled large and uncontrollable fires. Large fires occurred in the 1920s and 1930s across much of Florida and the southeastern Coastal Plain. In 1935, the Big Scrub Fire in the Ocala National Forest burned some 14,175 ha (35,000 ac) in just four hours (Pyne 2016). Predictably, the forestry agencies overreacted to these large fires and paid no attention to the ecology of the ecosystems they were affecting with their simpleminded fire-exclusion policy.

The U.S. Forest Service also showed an astounding lack of respect for local customs and no awareness of the adaptive link that often exists between local cultures and the ecosystems in which they are embedded. A major propaganda campaign was launched by the U.S. Forest Service and the American Forestry Association (AFA) to spread an evangelical message of fire prevention across the rural South. This was an all-out effort. As described by Rooney (1993):

> A fleet of special trucks—equipped with generators (many of the hamlets visited lacked electricity) and motion-picture projectors, and manned by articulate young southern foresters—headed for the woods in September 1928. Between then and June 1931, the Dixie Crusaders, as they came to be known, preached the gospel of fire prevention to three million people in Florida, Georgia, Mississippi, and South Carolina. Highlight of the whistlestops was a series of movies produced for the Crusaders to show. Two of them were created by AFA itself, and directed by Erle Kauffman, then editor of American Forests. Kauffman even wrote the script for "The Burner," and got McCormick, the project's director, to star as the villain/hero, Burnin' Bill. The Southern Forestry Education Project was at the time one of the most intensive educational outreaches ever undertaken.

Ultimately the fire-prevention caravans of the Dixie Crusaders traveled 300,000 miles through the Deep South. Their motion pictures were alluring because few people in the region had access to movies at this time. Despite this elaborate effort, however, woodsburning in the rural South continued (Pyne 2016).

The failure of the Dixie Crusaders to stop woodsburning did not convince the Forest Service to abandon its effort. Particularly appalling is a paper published by John P. Shea, a psychologist with the U.S. Forest Service, titled "Our Pappies Burned the Woods" (Shea 1940). One might wonder why the Forest Service would hire a psychologist to investigate southern woodsburners. The reason was that "approximately one-half of the forest fires occurring yearly in the United States are recorded in the eleven southern states" (Shea 1940), and the Forest Service wanted to figure out what kind of cultural pathology would cause people to set these destructive blazes. As Shea (1940) put it, they wanted "to find the 'inner-most' reason why inhabitants of the forest lands of the South cling persistently to the custom of burning the woods," and hoped by doing so, they might find "a point of vaccination that with an improved educational serum would reach the germs of the woods-burning desires." Shea went out of his way to ridicule rural southerners and portray their burning practices as uncivilized, ignorant, and reckless.

Shea (1940) paints a picture of cultural depravity among rural southerners by noting a general lack of musical instruments, books, periodicals, art, and craftsmanship in the dwellings of the impoverished people he visited (which were in the southern Blue Ridge, a population he curiously considered representative of country people across the Deep South). Shea (1940) recommended a strategy of law enforcement, education, and cultural indoctrination to stop the burning. He concluded: "Southern woods-burning is a human problem and should be tackled in the scientific and human way."

Shea's (1940) article, with its antifire and culturally prejudiced overtones, is representative of the attitude of the forestry establishment in the United States during the first two-thirds of the twentieth century. This attitude frustrated and often obstructed the pioneers of the science of fire ecology, men such as H. H. Chapman, Roland M. Harper, Herbert L. Stoddard, and Edward V. Komarek Sr. Ironically, however, it helped vitalize the development of this new science as a rebellion against the status quo.

The propaganda machine against fire intensified during World War II, when many able-bodied men were away at war and unavailable to fight fires at home. In August 1942, the animated motion picture *Bambi* was released. The movie portrayed both deer hunters and forest fires quite unfavorably. Walt Disney loaned the Bambi character to the government for one year to star in a public service campaign for fire prevention. In 1944, a more potent and permanent character was created for the U.S. Forest Service by the Advertising Agency: Smokey Bear. The name was inspired by "Smokey" Joe Martin, a New York City Fire Department hero. The slogan attached to Smokey that proved most enduring—"Remember . . . only

YOU can prevent forest fires"—was coined in 1947. "Forest fires" was amended to "wildfires" in 2001 to encompass habitats beyond forests. Although the U.S. Forest Service has improved its outlook on fire as an ecological process, Smokey Bear continues to present a misleading message to Americans.

The primary influence of humans on fire regimes in the southeastern Coastal Plain over the past century has been to reduce fire activity. Yet, paradoxically, individuals and agencies in Florida and adjacent states led the way to dampen this trend and sometimes even to reverse it. It is ironic that the region where the fire-exclusion doctrine was most adamantly promoted also became the national and global leader in controlled burning. Just as the Dixie Crusaders concluded their propaganda campaign in the early 1930s, St. Marks National Wildlife Refuge, south of Tallahassee, became the first unit in the entire national refuge system to initiate controlled burning (Pyne 2016).

In 1943, shortly after the article by Shea (1940) was published and one year before Smokey Bear hit the scene, the Florida national forests became the first national forests in the country authorized to conduct controlled burns. In 1958, less than a decade after its designation, Everglades National Park became the first national park to apply prescribed fire (Pyne 2016). As discussed in the following section, the originators of the science of fire ecology worked mostly in Florida and adjacent states, and came together in 1962 at Tall Timbers Research Station to formally launch this new field.

Changes in Understanding of Fire and the Development of Fire Ecology

A comprehensive history of the science of fire ecology could fill volumes, a task well beyond my scope. Instead, in this section I discuss some key figures and events in the history of fire ecology in the southeastern Coastal Plain, especially in Florida and southern Georgia, where the field was born. The development of fire ecology was, at its heart, a grassroots effort to combat the misinformation perpetrated by federal agencies and professional foresters, who typically portrayed all fire as destructive. In contrast to this prevailing attitude, and many decades before the term "fire ecology" was coined, a handful of astute naturalists noticed the dependence of some of the iconic species of the South on regular fire. They described how many of the characteristic and desirable plants and animals of the region, such as longleaf pine and northern bobwhite quail, flourished under conditions of frequent fire and declined in its absence. These protofire ecologists were in the minority among ecologists, and they were actively scorned by most foresters.

Early field botanists and other naturalists traveling through the southeastern Coastal Plain showed variable knowledge and mixed feelings about fire. Coming mostly from Europe or the northeastern United States, where lightning fire was relatively rare, they lacked appreciation of fire's ecological and evolutionary roles. Fire was frightening to these folks, and only a few recognized that it could have beneficial effects.

Most leading botanists and ecologists of the nineteenth century opposed controlled burning. For example, Charles Sprague Sargent, the first director of Harvard University's Arnold Arboretum and the leading authority on the trees of North America during his career, wrote a report on forests to the U.S. Census Office in 1884, in which he opposed the common practice of woodsburning by ranchers and farmers in the South. Sargent stated that such burning in longleaf pine communities "often caused serious destruction of timber," and "less valuable species now occupy the ground once covered with forests of the long-leaved pine, through which annual fires have been allowed to run to improve the scanty pasturage they afford. Stockmen have been benefited at the expense of the permanency of the forest" (Sargent 1884).

An unfortunate example of bias against fire can be found in the writings of John Kunkel Small, an outstanding field botanist and author of the monumental *Flora of the Southeastern United States*. Small particularly loved the hammocks and forested wetlands of Florida, where human-set fires combined with artificial drainage could be destructive—for example, by generating severe peat fires. As he put it emphatically: "DRAINAGE and FIRE! The two processes are tending to eliminate all native life from the State" (Small 1929). Yet Small extended his hatred of fire to entirely natural fire in pyrogenic communities. Although he asks ecological questions such as "Why are pinelands? Why are black-jack ridges? Why is scrub?" (Small 1929), he fails to recognize the obvious answer: fire.

Observing the extensive dry prairies of the Kissimmee River region, one of the most fire-prone of all Florida's landscapes (Bridges 2006a), Small (1929) notes that "extensive fires had swept not only the prairie, but many of the palm oases, leaving them in a deplorable state from which it will take a long time for restoration to their normal condition." Referring to the same region, Small (1929) comments, "the prairies and pine woods have been burned over so much that their former growth of animal foods has disappeared, and as a consequence the native animals have disappeared." One wonders how Small could have been such an astute botanical observer, yet was unable to appreciate the fire-dependence of the plants and animals he observed.

Another superb botanist and ecologist with a conflicted understanding of fire was B. W. Wells. In his book *The Natural Gardens of North Carolina*, Wells (1932) had a section titled "Fire Helps Make the Savannah." This section, showing cognizance of the ecological role of fire, is followed a few pages later by a section, "The Sandhill Villain," in which Wells (1932) identified fire as "the villain, or ogre, who periodically makes his visitations in all of the sandy regions. . . . Government aided by science has not conquered him. Like the ruthless gangster he defies society successfully. He is seldom caught, for his allies are too strong. His name is Fire."

Shortly thereafter, as if returning to ecological sensibility, Wells (1932) describes how low-moisture fuels, lack of water-holding leaf litter, and sandy, droughty soils create conditions that foster frequent fires and the perpetuation of the "curious sandhill vegetation" (Wells 1932). In his scholarly papers, Wells portrayed fire as a factor that reset succession toward a hardwood climax community, the prevailing view of the time. Adding to the ambiguity over whether Wells thought fire was "the villain, or ogre" (Wells 1932) or an important ecological factor, Wells and Shunk (1931) correctly observed that the longleaf pine–wiregrass community and much other vegetation of the Coastal Plain are fire-dependent: "[T]he present native communities of plants throughout the coastal plain represent plant aggregations surviving under frequent fires. So frequent and so certain everywhere is fire that the Aristida communities represent what might be called 'fire subclimaxes.'"

Perhaps the earliest published scientific statement of the positive role of fire in the ecology of the southeastern Coastal Plain is from the eminent British geologist Sir Charles Lyell, known as a major influence on Charles Darwin, including the key idea of small, gradual changes producing large effects over long spans of time. In 1849, on his second trip to the United States, Lyell described the salubrious effects of fire on longleaf pines near Tuscaloosa, Alabama: "These hills were covered with longleaved pines, and the large proportion they bear to hardwoods is said to have been increased by the Indian practice of burning the grass; the bark of the oaks and other kinds of hardwoods being more combustible, and more easily injured by fire, than that of the fir tribe. Everywhere the seedlings of the longleaved pine were coming up in such numbers that one might have supposed the ground to have been sown with them" (Lyell 1849, cited in Chapman 1932).

Ellen Call Long is a largely unrecognized pioneer of southern fire ecology. She did not conduct research on fire, but she was an incredibly perceptive observer of the effects of fire, and, conversely, fire exclusion, on vegetation. Long was the daughter of Florida governor Richard Keith Call and a prominent woman in Tallahassee society. She was a member of the American Forestry Congress, and at its meeting in Atlanta in 1888, she presented a paper that described fire as essential to the growth of longleaf pine. Titled "Notes of Some of the Forest Features of Florida," this was one of the earliest published papers to challenge the accepted forestry practice of fire suppression. Roland Harper later described Long's paper glowingly as "heretical" (Shores 2008), and it was cited approvingly by other early leaders of fire ecology, including H. H. Chapman and Herbert Stoddard.

Long (1889) told the foresters, "'Forest fires' is a term of very different significance throughout the Southern pine belt from the meaning attached to it in the North and West. There they are destroyers of forests, pure and simple. . . . But there is sound reason for believing that the annual burning of the wooded regions of the South is the prime cause and preserver of the grand forests of *P. palustris* to be found there; that, but for the effect of these burnings, the pine forests would never have been, and but for the continued annual wood firing that prevails so generally throughout the South the Maritime Pine Belt would soon disappear and give place to a jungle of hardwood and deciduous trees."

Invoking Darwinian natural selection, Long (1889) stated, "In the persistent application of the law of the survival of the fittest *Pinus palustris* alone has been able to contend with the condition of fire as it annually occurs over the grassy surface of the Southern forest." She described how, in contrast to other tree species, the stem and growing buds of longleaf pine are protected from fire "except when subjected to much more intense heat than generally results from the slowly burning wire-grass covering the pine forest." Without fire, Long noted, longleaf pine stands quickly convert to hardwoods: "Localities protected for a short time from fire rapidly become covered with oak, hickory, magnolia, dogwood, etc., and become in time 'Hammock' lands." She sternly warned the foresters about the dangers to the timber industry posed by the official policy of fire exclusion: "The total abolition of forest fires in the South would mean the annihilation of her grand lumbering pineries."

Herman H. Chapman of the Yale School of Forestry was mentioned earlier (chapter 1) for his then-novel idea that the concept of climax vegetation should include types such as longleaf pine communities, for which "fire at frequent but not necessarily annual intervals is as dependable a factor of site as is climate or soil" (Chapman 1932). Chapman had promoted the use of controlled burning to stimulate southern pine regeneration since around 1908 (Shores 2008). In a 1912 article in *American Forests* (the magazine of the American Forestry Association, the same organization that launched the Dixie Crusaders in 1928), Chapman argued that trying to keep fire out of longleaf pine stands would result in their destruction. He pointed out that southern and northern pines differ profoundly in their relationships to fire, and that "the three southern pines are all remarkably fire resistant and the longleaf pine has adapted its whole structure and growth as a seedling to the primary object of surviving ground fires" (Chapman 1912). He wrote to a readership largely hostile to his views: "A promiscuous enforcement of forest fire laws, borrowed whole from the Northern States, and utterly unsuited to the South, will never result in anything but dissatisfaction and contempt on the part of practical men for forestry" (Chapman 1912).

Chapman became a leading advocate of burning southern pine stands. He may properly be considered the father of controlled burning. In 1952, Chapman summarized his views on fire in a paper on "the place of fire in the ecology of pines." He explicitly recognized fire in the evolutionary environment: "While man is responsible for a many-fold increase in the occurrence of forest fires, lightning as a natural cause has operated with a frequency that places fire among the determining ecological factors that influenced development, modification, and survival of species, especially of pines" (Chapman 1952).

Two fellow Ph.D. students of John Kunkel Small at Columbia University in the late 1890s and early 1900s were Roland McMillan Harper and Henry Allen Gleason. Gleason was rediscovered by ecologists in the late twentieth century for his pioneering work demonstrating the "individualistic" nature of plant communities, where each species is distributed primarily according to its unique set of environmental tolerances (Gleason 1917, 1926). At the time Gleason published these ideas, they were not popular among ecologists, who were more enamored with the climax theory of Frederic Clements (Clements 1916). Disillusioned, Gleason left ecology in the 1930s and focused on plant taxonomy, a field in which he produced one of the major floras of the northeastern United States with

Arthur Cronquist, as well as a 1964 book, *The Natural Geography of Plants* (Gleason and Cronquist 1964).

Roland Harper is a lesser known figure to ecologists, at least outside of the southeastern Coastal Plain, but he was arguably farther ahead of his time on the topic of fire ecology than anyone. His compelling descriptions of the role of fire in ecosystems, especially in Florida and Alabama, were not equaled in several decades, if ever. Although his work was dismissed by Frederic Clements as "descriptive ecology" (Shores 2008), Clements (1916) nevertheless cited Harper, among others, in a statement in his book: "The role of lightning in causing fire in vegetation has come to be recognized as very important."

Harper described the ubiquity of fire in southern landscapes and the adaptation of the dominant species to frequent fire. In an excerpt (deleted by the editor!) from a popular article in the *Florida Review* in 1910, Harper explained that "the pines have been accustomed to fire since long before the dawn of history and do not seem to mind it much, and the palmetto when its leaves are burned soon puts out a new crop" (Shores 2008). A year later, in one of his most insightful papers, Harper (1911) suggested that fire is so common in the landscapes of Florida and the southeastern Coastal Plain that hammocks are restricted to topographically protected sites such as islands and peninsulas, usually out of the reach of fire. He noted that fire alone explains this distribution, as "the only apparent difference between the soil of the hammocks and that of the neighboring pine land was that due to the vegetation itself, namely, a small amount of humus (which is necessarily present in all hammocks and other shady upland forests) on the peninsulas" (Harper 1911).

In an unfragmented landscape, fire spreads widely: "Although fires may not be started by lightning on any one square mile oftener than once in several decades, a fire once started in the grassy carpet of an unbroken pine forest might easily spread over several square miles, so that every acre of such forest if not protected in some way would be likely to be burned over every few years" (Harper 1911). But when topographic barriers to the spread of fire exist, other types of natural communities are able to develop, enriching the diversity of species and communities across the landscape. Harper (1911) pointed out that hammocks, once formed, typically become highly nonflammable and are capable of maintaining themselves against fire. This phenomenon we now recognize as alternative stable states, where two or more vegetative states are possible under the same environmental conditions: "The relation between fire exemption and climax vegetation is reciprocal, for when the hardwoods are once well established the herbaceous vegetation under them is very sparse, and the humus is usually too damp or too thoroughly decomposed to burn readily" (Harper 1911).

In a particularly prescient statement, Harper (1911) recognized that in the developed landscape, "numerous highways, clearings, etc. . . . serve as barriers to fire," a problem ecologists now recognize as habitat fragmentation. One of the consequences of fragmentation is passive fire exclusion, where artificial barriers are able to stop fire just as surely as active fire suppression (Noss 2013). It would be difficult to find in the contemporary fire ecology literature an understanding of the relationships among fire, vegetation, and landscape morphology more sophisticated and comprehensive than what Harper demonstrated in 1911. Like Chapman, Harper was able to publish, in 1913, a pro-fire paper in *American Forests* (then called *American Forestry*), which was excerpted by *Literary Digest* with the title "A Defense of Forest Fires" (Shore 2008).

These early southern fire ecologists were well aware of each other's work and occasionally banded together to confront the forestry establishment. They knew fire must be recognized as a natural and beneficial feature of southeastern pine savannas, not just for the sake of the flora, fauna, and aesthetics, which they deeply appreciated, but also for production of wood products and other economic purposes. Most of these ecologists had experienced scorn and condemnation from their colleagues who lacked an appreciation for fire's role. Opposition came not only from foresters but also from botanists, including Harper's old schoolmate John Kunkel Small.

The man who contributed the most to our understanding of the importance of fire to wildlife, and who also encountered tremendous opposition from the forestry establishment, was Herbert L. Stoddard Sr. Stoddard was different in many ways from his new comrades in fire ecology. Chapman and Harper had earned Ph.D.s at top universities, whereas Stoddard had only an eighth grade education. He was essentially self-taught, which even in those days was uncommon for such an accomplished naturalist.

Stoddard spent his formative boyhood years where I now live, just outside Chuluota, Florida, and he credits his early outdoor experiences there with shaping his worldview and his affection for the landscapes of the South. At a 1929 meeting of the American Forestry Association in Jacksonville, Florida, Stoddard presented his research on the deleterious impacts of fire suppression on northern bobwhite quail. He received the same kind of unfriendly resistance as Harper had recently experienced in Alabama (Shores 2008). Stoddard later wrote in his memoirs, "I have never attended a meeting with a more pervasively hostile atmosphere. The Chairman, who obviously feared that I might contaminate my hearers, actually cautioned them not to take too seriously what I might say. I fear I took a perverse pleasure in reading the chapter to them, though I had a feeling of futility at the solid wall of opposition with which it was received" (Stoddard 1969). One sympathetic listener in Jacksonville, however, was S. W. Greene of the U.S. Department of Agriculture's McNeill Experiment Station in Mississippi. Greene subsequently described the relationship between longleaf pine and fire in a controversial article in *American Forests*, "The Forest That Fire Made" (Greene 1931).

As summarized by Way (2006), Herbert Stoddard "came to the longleaf pine–grassland forests of south Georgia in 1924 to study the bobwhite quail, and stayed to develop a method of land management that stressed ecological habitat over the dominant production-oriented model." Stoddard is now considered, along with his close friend Aldo Leopold, one of the fathers of wildlife management. Nevertheless, the forestry establishment vigorously attempted to prevent his book *The Bobwhite Quail* (Stoddard 1931) from being published and to quash his career.

Stoddard's research on the life history of northern bobwhite in the Red Hills of southern Georgia and adjacent Florida stands out as one of the most significant contributions of what is now viewed as old-fashioned natural history to the conservation of an iconic species and the ecosystem upon which it depends. The work was conducted through the Bureau of the Biological Survey and the Cooperative Quail Investigation, the latter funded by local landowners. These landowners were mostly wealthy northerners who spent winters on their quail-hunting plantations in the Red Hills. Stoddard's exhaustive research showed beyond doubt that quail benefited from frequent burning of their habitat. Years of fire exclusion from many lands was a major reason for the decline of this immensely popular game bird. As summarized by Stoddard (1969):

> [Q]uail do not thrive in the rough. Not only do the seeds that are their main food decline as the rough builds up but also the few seeds that are produced fall into the tangle of grasses and duff, where they are largely unavailable to the weak-scratching quail. Quail are conditioned to pick seeds off the surface of mineral soil, not to dig them out of a layer of duff. The preserves of the region had long been

> protected from open-range grazing. Because of this, and the earlier regular and frequent burning of the pinelands and the old fields, the perennial legumes and grasses that quail depend upon largely for food had been abundant. Fire exclusion had rapidly changed this situation for the worse.

This well-supported link between quail abundance and fire did not impress mainstream foresters, who tried to suppress Stoddards's findings. Stoddard (1969) noted that "before publication, manuscripts by employees of the various agencies had to go 'through channels' for approval. . . . Since the United States Forest Service was one of the largest and most powerful bureaus in the Department of Agriculture, it exercised enormous influence upon the reports of other agencies. My chapter on fire in *The Bobwhite Quail* immediately ran into trouble when it arrived at the Forest Service, in spite of the fact that I had made my recommendations as mild as my conscience would permit."

Ultimately Stoddard's fire chapter was rejected and rewritten several times as Stoddard tried to make it "acceptable without weakening it to a point where it would be entirely counter to the findings. Finally I passed the word that if it was not cleared in its diluted form I would resign and write a book that would really burn them up" (Stoddard 1969). Because Stoddard had the financial support of some wealthy and politically powerful landowners, the Forest Service took Stoddard's threat seriously and finally backed down. The chapter was cleared. Over the years, the Forest Service softened its antifire stance. But as Stoddard wryly noted, "In a gesture that might be termed face-saving, the substituted the word 'prescribed burning' for 'controlled burning' and insisted that experts must 'prescribe' the practice."

Despite rejection and ridicule by the vast majority of foresters, as well as being ignored by mainstream ecologists, Harper, Chapman, Stoddard, Greene, and like-minded southern naturalists redoubled their efforts to increase professional and public appreciation of fire as a critical component of southeastern ecosystems and to overturn the official policy of fire suppression. Edward V. Komarek Sr. dubbed this group of iconoclasts the "Dixie Pioneers" as a contrast to the Dixie Crusaders. They were ultimately joined by longtime southern foresters such as Austin Cary and Inman Eldridge (Way 2006).

Edward V. Komarek Sr. went on to become one of the most outspoken and influential of all fire ecologists. A student of the eminent animal ecologist W. C. Allee at the University of Chicago, Komarek had worked at the Chicago Academy of Sciences and collected specimens for the academy in the southern Appalachians and the Florida Everglades. Ed and his brother, Roy, visited the Sherwood Plantation in Thomasville, Georgia, in 1933, where they met Herb Stoddard, who was working there on his quail research. Stoddard was much impressed by Ed Komarek and hired him as his assistant for the quail studies (Stoddard 1969). Another protégé of Stoddard was Leon Neel, who in 1950 also made a visit to Sherwood and enthused Stoddard enough that he was hired as his assistant. The two men went on to create a wildlife-oriented style of forestry, most applicable to longleaf pine, which makes abundant use of controlled burning. This forestry philosophy and method is now known as the Stoddard-Neel Approach (Neel et al. 2010).

An enormously positive outcome of the fire wars fought by the pioneers of fire ecology was the formal creation of this field and the ultimate acceptance of controlled burning as a legitimate land management practice. The pivotal event that marked this revolution was the founding, in 1962, of Tall Timbers Research Station by Stoddard, Komarek, and Henry L. Beadel. Located just north of Tallahassee, the station quickly became the leading center for fire ecology research. The history of Tall Timbers is an engaging story that has been covered by several authors. I refer the reader to the comprehensive book *The Legacy of a Red Hills Hunting Plantation: Tall Timbers Research Station and Land Conservancy* (Crawford and Brueckheimer 2012).

As the first executive secretary of Tall Timbers, Ed Komarek organized the inaugural Fire Ecology Conference at Florida State University in 1962 to bring attention to this "most neglected ecological subject." It was at this conference that for the first time the words "fire" and "ecology" were combined as "fire ecology." This was the title of one of Ed Komarek's talks at the conference, as well as the title of the entire meeting. Komarek (1962) defined fire ecology as "the study of fire as it affects the environment and the interrelationships of plants and animals therein." I suspect Komarek and his colleagues used the term "fire ecology" for a considerable time before he presented his paper at this conference.

Although early Tall Timbers Fire Ecology Conferences concentrated on the southeastern Coastal Plain, over the years they increasingly encompassed other regions and ecosystem types globally. The conferences were annual at first, but after 1974 have been held at an irregular frequency. After its highly successful beginnings, Tall Timbers languished for a number of years, from which it recently rebounded (Pyne 2016). Active research on the ecological effects of fire continues at Tall Timbers, at Archbold Biological Station on the southern Lake Wales Ridge, at Kennedy Space Center, at the Jones Center in Georgia, and at several other locations across Florida and the Coastal Plain.

With the dawn of fire ecology, an appreciation of the natural history of lightning (Komarek 1964) and its ecological effects was awakened, and the more thoughtful fire managers began to consider "whether to relinquish some of our hard-won control over fire by returning to a more natural fire regime" (Robbins and Myers 1992). Building on the early history of fire ecology in the southeastern Coastal Plain, as summarized above, the field blossomed worldwide, but with its center still arguably right where it all began. I need not review here the recent history of fire ecology in this region, because it is evident in the scholarly literature cited throughout this book.

3

Adaptation to the Fire Environment of Florida and the Southeastern Coastal Plain

> It is assumed that through natural selection primarily, over long periods of time, plants and animals have developed "adaptations" that allow them to live where fire is a factor in the environment.
>
> Komarek (1962)

Species from a regional species pool come together and form a local community when they have environmental requirements and tolerances in common. The assembly of communities involves a complex interplay of stochastic and deterministic processes, as well as a mix of abiotic and biotic constraints. As reviewed in chapter 1, fire is a largely deterministic process in the southeastern Coastal Plain. Although precisely where and when a lightning bolt will strike is largely stochastic, bolts are certain to strike many places on a regular basis and ignite vegetative fuels. The seasonal dry-wet climate, the high incidence of lightning strikes during a period of time when vegetation is relatively dry, and the high flammability of many plants assure that fires occur regularly. Thanks to our long tradition of controlled burning in this region, human ignitions are also to a considerable degree deterministic.

Fire is among the primary constraints or filters that restrict membership of species in Florida's fire-prone communities. If you can't handle frequent fire, you are excluded. If you somehow sneak in, you won't last long. Those species that persist in fire-prone ecosystems possess adaptive strategies and traits that allow them to thrive with fire.

Macroevolutionary and microevolutionary research are both necessary to understand fully the effects of fire on species' traits. Studies in macroevolution seek to determine the origin of traits and the long-term evolutionary history of clades: groups of organisms consisting of common ancestors and all their lineal descendants. Microevolutionary research addresses genetic-based divergence of traits and local adaptation in populations of species from different selective environments (Pausas 2015a). To show that a trait is shaped by natural selection, one must not only demonstrate the trait is variable and enhances fitness. The trait also must be heritable, and showing that requires microevolutionary research.

Fire-adaptive traits are often visible, sometimes strikingly so such as the thick bark of many pines and oaks, but others are revealed only by detailed study. Proving beyond doubt that thick bark or underground storage organs, for example, evolved in response to fire as opposed to some other environmental factor, such as drought or herbivory, may not be possible. Nevertheless, the most parsimonious explanation for multiple species possessing such traits in a fire-prone ecosystem is adaptation to fire. For the purposes of ecological restoration and management, the intricacies of evolutionary history may not matter much, providing we know what type of fire regime benefits native species and communities at the present time. On the other hand, consideration of evolutionary history may provide insights about the relationships between fire regimes and life-history traits that call for adjustment in fire management practices. Adjustments in management to simulate more accurately the specific fire regimes under which native species evolved is expected to better attain conservation goals.

In this chapter I first provide a brief review of how species assemble into communities, with an emphasis on fire as a powerful environmental filter that restricts membership to species that have evolved ways to avoid, tolerate, exploit, or facilitate fire. I then review components of fire regimes that were introduced in chapter 1: frequency, seasonality, intensity/severity, extent, and heterogeneity. For each of these components I discuss how species in the southeastern Coastal Plain respond to variation in the regime—for example, changes in fire frequency or seasonality. Finally, I review fire-adaptive strategies and traits of plants and animals and provide examples of species in the region that possess such traits.

Community Assembly

The theory of niche-based community assembly has abundant scientific support for species-rich communities (Kraft et al. 2008). The assembly process is hierarchical, beginning with dispersal of species to a site and followed by species sorting through environmental and biotic filters. Given the presence of fire on Earth for hundreds of millions of years and its ability to alter habitat conditions dramatically, we should expect fire to be a strong environmental filter and a force of biological evolution.

For fire-prone ecosystems, recent research has confirmed the expectation from niche theory that biological communities are not random assemblages of species that just happened to arrive on a site. Most communities are not determined solely by biotic interactions, either. Rather, the species assemblage one finds in a fire-prone ecosystem consists of species that arrived there and then were able to pass through the environmental filter of fire. Fire-sensitive species were filtered out by virtue of not possessing traits that permit persistence in a fiery environment.

Of the niche-based processes that determine community membership, environmental (abiotic) filtering and local biotic interactions, especially competition, have received the most attention from ecologists. Contrasting predictions can be made from the environmental filtering vs. competition models. If environmental filtering is the dominant process in community assembly, we would expect species in the community to be more closely related or more similar in appearance (phenotype) than expected by chance. For plants, functional traits such as height, seed mass, leaf size and shape, leaf nitrogen concentration, relative biomass of belowground organs, bark thickness, pollination mode, and seed dispersal mode are expected to be convergent across species. The trait values possessed by co-occurring species in an environmentally filtered community should be a limited subset of all possible trait values (the morphospace) in the regional species pool.

Pausas and Verdú (2008) found such convergence in a comparison of high-fire-frequency and low-fire-frequency communities in the western Mediterranean Basin. High-fire communities were clustered both phenotypically and phylogenetically, with vacant zones in the morphospace, and were composed of related species with fire-resistant traits. Moreover, the observed pattern of environmental filtering did not match communities simulated under a neutral evolutionary model, suggesting a limited role of stochastic processes (Pausas and Verdú 2008).

The oaks (*Quercus*) are a genus of trees (including stunted, largely underground trees) that are especially diverse in the southeastern United States. Weakley (2015) lists 43 native species of oaks in his *Flora of the Southern and Mid-Atlantic States*, and many of these are characteristic of fire-prone ecosystems. Seventeen species of oaks studied in north-central Florida by Cavender-Bares et al. (2004a, 2004b) are phylogenetically overdispersed. This means oak species that occur together in local communities are not closely related members of a single clade, but rather usually belong to separate clades, specifically the white oak, red oak, and live oak clades. This overdispersion of close relatives stands in contrast to some other communities shaped by frequent fire (e.g., Pausas and Verdú 2008), but is not unexpected in local communities, where high similarity in niches among close relatives might make coexistence challenging due to competition.

Niche conservatism, a phenomenon in which closely related species are similar in appearance or ecology because traits within species remain relatively static over time, is a widespread pattern in nature (Wiens et al. 2010). More distantly related oak species coexist in local communities in Florida more than expected by chance due to convergent evolution of phenotypic traits that allow them to pass together through environmental filters, in particular the filters of soil moisture, nutrient availability, and fire regime. On a landscape scale, all 17 oak species studied by Cavender-Bares et al. (2004a, 2004b) coexist because they partition their niches along these important environmental gradients, with oak species clustering into three major community types: hammock, sandhill, and scrub. These three communities have distinct fire regimes, namely rare fire for hammocks, frequent surface fire for sandhill, and less frequent stand-replacing fire for scrub.

An experimental manipulation of seed arrival and ecological filters in herbaceous communities in longleaf pine (*Pinus palustris*) savannas in Louisiana (Myers and Harms 2011) evaluated the two competing general models of community assembly, the niche-based model vs. the dispersal assembly model. Under the dispersal model community membership is determined by the size of the source species pool and chance events in species colonization and demography. As predicted in the southeastern Coastal Plain with its large species pool containing many rare species, dispersal (seed rain) was important in determining species richness. Nevertheless, seed rain interacted with the environmental filters of fire intensity and soil moisture to determine local community membership. High-intensity fire decreased richness of resident species, but increased recruitment and richness of immigrating species, probably by moderating other filters such as competition and resource limitation (Myers and Harms 2011).

Local heterogeneity in fire intensity, produced by variability in prefire fine fuel loads (e.g., grass thatch and pine duff), appears to promote coexistence of dominant and rare species and to increase recruitment opportunities for arriving seeds. In the long term, "variation in local fire intensity may help maintain diversity by enhancing individual performance (growth, survival, and fecundity) of rare recruits in sites with low densities of competitors . . . as well as by promoting coexistence of species with different postfire regeneration traits" (Myers and Harms 2011).

Seldom considered explicitly in community assembly theory and research is how characteristics of the biotic community and individual species might alter the expression of environmental filters. Various aspects of a fire regime, such as frequency, seasonality, and intensity/severity, do not operate independently of vegetation. Fire is not exogenous to the community, and it is not purely abiotic. It is an interactive and coevolving biotic-abiotic phenomenon. Some of the species that compose fire-prone vegetation possess traits, such as high flammability, that promote fire (i.e., they are pyrogenic) and can increase its local intensity, often to the detriment of their more fire-sensitive neighbors. Jennifer Fill, Bill Platt, and colleagues developed a vegetation-fire feedback model in which differences in pyrogenicity and plant-fire feedbacks among species determine community structure. In this model "pine savannas are conceptualized as persistent, nonequilibrium communities maintained by endogenous, coevolutionary vegetation-fire feedbacks" (Fill et al. 2015a).

Fire shapes life on Earth at several levels of biological organization. At the genetic level, fire favors genes that create fire-adaptive traits. At the levels of individual organisms and populations, the fire regime permits entry only to those individuals possessing certain combinations of functional traits that allow them to tolerate or promote fire, exploit the postfire environment, and persist and reproduce through subsequent fires. At the community level, environmental filtering, often in combination with more stochastic processes such as dispersal, as well as biotic filtering processes such as competition, results in communities of phenotypically similar and sometimes closely related species that represent a nonrandom subset of the potential morphospace. At the ecosystem level, vegetation-fire feedbacks demonstrate the limited realism of any model that considers the abiotic environment separately from the biotic environment. Fire regimes result from fire and vegetation functioning together as an integrated whole. At the landscape level, where species assemblages are arrayed across environmental gradients determined by elevation, topographic position, soil and moisture conditions, and other variables, different fire regimes are associated with different types of communities or patches.

According to alternative stable-state theory, the composition and structure of communities are stabilized by webs of feedbacks. If some of the stabilizing positive feedbacks are weakened, the community can undergo an abrupt state or regime shift (Bowman et al. 2015). Due to vegetation-fire feedbacks and other mechanisms of homeostasis, fire-dependent communities can be seen as self-organizing systems resistant to change unless disrupted by human activities such as fire exclusion, extreme soil disturbance, or paving over. These are the kinds of activities that disrupt critical feedbacks and can cause state/regime shifts because they lie outside of the range of variability to which species in the community have been exposed during their evolutionary histories. When a community shifts to an alternative stable state—for example, from pine savanna to hardwood forest—it develops a whole new set of feedback webs, which function to maintain it in that state and make restoration to any previous state difficult.

Adaptation vs. Exaptation

> Distinguishing between adaptations and exaptations is a high bar that is difficult to demonstrate as there is not a single dichotomy between paths leading to adaptations and exaptations.
>
> Keeley et al. (2011)

Plants, animals, and other organisms exposed to fire over long stretches of their evolutionary histories possess functional traits that allow them to cope with fire or even to benefit from fire at the expense of potential competitors that are more fire-sensitive. Otherwise, these species would not have survived for long in fiery environments. Despite this compelling logic, demonstrating that particular traits arose in response to fire, as opposed to other environmental factors, developmental constraints, or pure chance, has proven challenging.

Epistemologically, it is difficult to prove any trait is truly adaptive. Famed evolutionary biologist George Williams (1966) explained that "evolutionary adaptation is a special and onerous concept that should not be used unnecessarily, and an effect (a fitness-increasing use to which a trait is put) should not be called a function (a designed fitness-increasing use) unless it is clearly produced by design and not by chance." Researchers are still battling out in the scientific literature the question of fire adaptations vs. "exaptations" (Gould and Vrba 1982), traits that evolved under some other selective pressure or perhaps just by chance, and were later co-opted as quasi-adaptations to fire.

Much of the argument seems academic—and it is. Although environmental factors other than fire, such as drought and herbivory, can select for plant traits that subsequently serve a species well in a fire-prone environment, the presence of many such traits spread across diverse phylogenetic lineages suggests that fire is a dominant selective agent driving the evolution of these traits. This is not just a matter of plant evolution. Animals possess many traits, especially behavioral, that protect them from fire and allow them to prosper in fire-prone environments.

Skeptics of fire adaptation insist that many traits of organisms that confer survival and reproduction in fire-prone environments may have evolved for entirely different purposes. In a commentary about assumed fire adaptations in Mediterranean ecosystems, Bradshaw et al. (2011) drew attention to an ostensible weakness of the adaptation assumption: "Exapted traits can be as effective as adapted traits in enhancing fitness in fire-prone environments, but ever-increasing fire frequencies might overwhelm the potential fire protection afforded by traits that have evolved in response to other, quite different, environmental factors." For instance, if a trait evolved in response to herbivore browsing but also proved valuable for surviving infrequent fire, it might not still be beneficial if fire frequency were increased, which has happened in many Mediterranean-climate regions around the world (Bradshaw et al. 2011).

In California, invasion of chaparral by highly flammable, nonnative annual grasses has increased fire frequency in some areas beyond the ability of obligate-seeding native shrubs to cope. These shrubs cannot reach maturity and reproduce within the narrow fire-return interval imposed by the annual grasses and the mostly anthropogenic ignitions (Regan et al. 2010). Nitrogen pollution from smog compounds the problem. Nitrogen fertilization increases biomass of nonnative annual grasses, whose fine fuels increase fire risk in arid and semiarid ecosystems, especially in years with higher precipitation (Talluto and Suding 2008; Rao et al. 2010). Why can't presumably fire-adapted obligate-seeding shrubs adjust to this increase in fire frequency?

The fallacy of this argument is it assumes that if a species is fire-adapted, then it should respond favorably to all fire. In reality, species adapt through evolution to particular fire regimes and not necessarily to others (Keeley et al. 2011). The importance of fire in the evolution of traits can be shown by the independent evolution of fire-adaptive traits multiple times in unrelated taxa and in many different ecosystems. In any case, for those interested in the conservation of fire-dependent species and ecosystems, the present function of a trait matters more than some hypothetical original function (Keeley et al. 2011). Thick bark protects a tree's cambium from the heat of a fire, even if bark thickness was originally influenced as much or more by herbivory from large animals or some other selective pressure.

Parameters and Components of Fire Regimes in the Southeastern Coastal Plain

The climate of the lower Coastal Plain produces a fiery evolutionary environment. The components of a fire regime were introduced in chapter 1 (e.g., Fig. 1.3; Table 1.1). Chapter 2 reviewed evidence, albeit much of it indirect, that fire has been a prominent component of the evolutionary environment of the southeastern Coastal Plain for millions of years. What we do not know, unfortunately, is how stable the components of fire regimes have been through this vast stretch of time. For example, was fire more frequent or more intense during some periods of time than others? The charcoal record and other potential evidence are insufficient to answer such questions.

Another question is whether the primary lightning-fire season in Florida and the southeastern Coastal Plain always occurred predominantly during the transition from the dry to the wet season. We can assume it has, since this fire season is characteristic of savannas globally and it is when relatively dry and combustible vegetation is exposed to lightning; however, we lack direct evidence. The timing of the dry and wet seasons likely varied some over time, as it still does today, though it is probably safe to assume that the spring dry/summer wet seasonal pattern is ancient in this region.

The observed compatibility of native species (especially old taxa) to natural or seminatural fire regimes today strongly suggests that modern natural fire regimes in the southeastern Coastal Plain have a long history. The products of this ancient ecological/evolutionary history have not been erased by recent human alteration of fire regimes. They may be erased in the long run, and that would inevitably result in a loss of biodiversity.

Evolution does not produce perfect adaptation to a set of environmental conditions. Species associated with a particular natural community today evolved within a somewhat different environmental context with an at least partially different set of interacting species. That previous community may no longer exist anywhere. In some cases, a species may have been forced out of a community by a superior competitor and persists today in a marginal habitat less than optimal for meeting its needs, but just favorable enough to support a sparse population.

Such a scenario brings to mind Hutchinson's (1957, 1978) concept of the fundamental and realized niche. The fundamental niche represents the entire multidimensional environmental space potentially occupied by a species in the absence of competition, predation, disease, parasitism, or other biotic interactions. Niche space can exist independently of a species, but a niche is an attribute of a population or species in relation to its abiotic and biotic environment (Colwell and Rangel 2009). The realized niche represents the portion of the fundamental niche in which a species finds itself, and often helps create, at any given time in the presence of biotic interactions such as predation and competition, among other constraints.

With environmental change and biological evolution ongoing, and with communities continually changing in membership, both the fundamental and realized niches of species change over time. The match of a species to its abiotic and biotic environment will never be perfect, and the optimal fire regime for a species will usually differ somewhat from the modal fire regime of any defined natural community. Different species occupying the same fire-prone community can be assumed to have somewhat different optimal fire regimes; they have unique fire niches.

For a set of species with varying requirements and vulnerabilities with respect to fire to coexist in the same community usually requires some spatial or temporal variability in the fire regime. For example, plant species that cannot reproduce successfully with annual or biennial burning usually must have some unburned or lightly burned patches on a site burned at that frequency. Because most fires are spatially heterogeneous, this is not usually a problem unless managers make a concerted effort to achieve a "clean burn" across a site.

The idea that fire can shape species and entire ecosystems in a positive and creative sense is relatively new to the sciences of ecology and evolution. Any aspect of a fire regime could theoretically influence the evolution of plant and animal traits and the assembly of communities. When the components of fire regimes change, species composition and richness can be predicted to change as well.

Below I focus on several components of fire regimes that have been shown, or are theoretically expected, to have major influences on community structure and ecosystem processes: frequency, seasonality, intensity/severity, spatial extent or spread, and heterogeneity. Each of these components, alone or in combination, has the power to influence trait evolution in plants and animals. More important for conservation, careful consideration of these components of a fire regime within an evolutionary context, along with measurement of the responses of species to alternative regimes, will help us continually improve restoration and management practices.

FIRE FREQUENCY

Fire Frequency in Florida and the Southeastern Coastal Plain

Fire occurs often in the southeastern Coastal Plain under natural conditions, except where vegetation is protected topographically from fire or is too wet or otherwise unable to burn. One reason for frequent natural fire is that an ignition source is abundant. Florida has the greatest number of thunderstorm days per year and the highest density of cloud-to-ground lightning strikes of any sizable region of North America. Virtually the entire southeastern Coastal Plain is a frequent-fire region due to its high incidence of lightning (Fig. 1.6), suitable physical and chemical conditions (Fig. 1.5; Guyette et al. 2012), and rapid growth of flammable fuels.

Frost (1995, 1998, 2006, and in Noss 2013) provided maps of estimated presettlement fire-return intervals for the southeastern United States, based on multiple sources of information. The intervals apply to "the most fire-exposed parts of the landscape" (Frost 2006), which comprise generally flat uplands as well as many lowlands, including some wetlands. These "fire compartments" allow fire to spread from ignition points across often broad areas. A fire compartment is a contiguous, internally unfragmented and undissected landscape patch that is capable of burning across its entire extent in a single fire event. As Frost (2000) explains, fire compartments are "elements of the landscape with continuous fuel and no natural firebreaks, such that ignition in one part of the element would be likely to burn the whole."

Considerable evidence supports the suggestion by Frost (2000, 2006) that, all else being equal, natural fire frequency is proportional to the size of the fire compartment. The lower Coastal Plain is the flattest part of the southeastern United States. The extremely flat and undissected dry prairie (subtropical hyperseasonal grassland) landscape of south-central

Florida may have the highest intrinsic fire frequency in the region because the flow of fire across these open grasslands is uninterrupted, even compared to many pine savannas in the same region (Bridges 2006a).

The combination of generally large landscape patches (up to 1,000 km^2 or more; Ware et al. 1993; Frost 2000), flammable vegetation, and the highest incidence of lightning strikes in the nation virtually assured high fire frequency in the southeastern Coastal Plain prior to EuroAmerican settlement. Exposed landscape positions across Florida and the Atlantic Coastal Plain had fire-return intervals mostly in the one-to-three-year range, compared to four to six years on more dissected terraces away from the immediate coast in the Gulf Coastal Plain, and longer in "wet swamps" and other fire-protected locations in the region (Frost 2000, 2006).

Frost's estimates of very frequent fire contrast with some earlier accounts of somewhat longer fire-return intervals in the region. Christensen (2000) summarized that "the natural interval probably was 3–10 years" in pine savannas, in part because "three to four years are required for sufficient accumulation of dry fuel to carry a surface fire after which the probability of fire is determined by the availability of ignition sources" (Christensen 2000). Documentation now exists for fine fuel accumulation sufficient for combustion within one to two years, and sometimes less, in southeastern pine savannas and similar communities. Although natural fire-return intervals for pine savannas may span a 1–10-year range, recent studies show intervals concentrated at the lower end of that range, particularly in peninsular Florida.

Fire Scars

Frost's (2000, 2006) summary of fire-frequency patterns has been validated by studies of fire scars in the rings of old, dead longleaf pines (stumps and snags) in the southeastern Coastal Plain (Henderson 2006; Huffman 2006; Huffman and Platt 2014) and in the West Gulf Coastal Plain (Stambaugh et al. 2011). The fire-return intervals documented in these studies are largely but not entirely within the one-to-three-year range determined by Frost (2000, 2006), both before and for some time after EuroAmerican settlement. Scars from surface fires in pine savannas provide a conservative estimate of fire frequency, because many low-intensity fires create no obvious scars (Stambaugh et al. 2011). In addition, the earlier years in tree-ring records are underrepresented because so few stumps remain from the few very old trees that grew prior to EuroAmerican settlement (Huffman and Platt 2014). Nevertheless, techniques exist

Figure 3.1. Researcher Jean Huffman (Louisiana State University; *left*) and fire manager Steve Morrison (formerly with The Nature Conservancy) examine a section from a pine stump, which has been cut and polished in preparation for dating fire scars. Photo by Reed Noss at Tiger Creek Preserve, Florida.

for cutting sections from the bottoms of old stumps and snags to maximize the probability of finding buried scars from fire events (Huffman and Rother 2017; Fig. 3.1).

Old longleaf pine stumps are now exceedingly rare. Beginning in the early twentieth century, the naval stores industry began pulling stumps from the ground and cooking them to extract turpentine, rosin, and pine oil. It is a tragedy that so few old stumps of longleaf pines remain, as they are critical not only to document fire history but also as wildlife habitat and as organs for fire storage. Means (2006) states that "probably the most overlooked refuge for longleaf pine savanna vertebrates is the subterranean base (butt, stump) of dead longleaf pine trees and their associated rotting roots," which include a taproot up to 5 m deep. The heartwood of these roots resists decomposition for many years, as the surrounding sapwood decays and provides an ideal burrowing substrate for animals.

Across a landscape, even within the same general vegetation type, fire frequency is variable due to effects of topography, weather, fire history, fuel loads, fuel connectivity, chance, and other factors. Variability, however, appears to be higher in some parts of the Coastal Plain than in others. For example, although Stambaugh et al. (2011) found a mean fire-return interval of 2.2 years across a 254-year period (1650–1905) in Louisiana, variability was high, with fire intervals ranging from 0.5 to 12 years over this period. In contrast, Huffman (2006) documented a 3.2-year mean fire-return interval between 1679 and 1868 in the Florida Panhandle, with low variability; 92 percent of all fires occurred at <5-year intervals. With an increasing sample size of sectioned stumps, quantified fire-return intervals here are getting shorter (Jean Huffman, personal communication).

Observational Evidence of Effects of Disappearing Fire

Given the limited number of fire history studies, historical trends in fire-return intervals have been poorly documented quantitatively in the southeastern Coastal Plain. Anecdotal accounts abound, however. Every field ecologist in the region can point to dozens of sites that show strong evidence of a marked decline in fire over the last century. In many cases, fire has virtually disappeared from what were formerly annually or biennially burned landscapes. The evidence, though indirect, is persuasive. It is based on reading the landscape.

I took the students in my Ecosystems of Florida class to Mills Creek Woodlands, near my home in Chuluota, Florida. This is a place Herbert Stoddard explored as a child (Stoddard 1969), which makes it more poignant to me. I ask students to decipher the "original" vegetation and history of the site from the vegetation they observe there now. They usually require a few hints, but then the history becomes palpable. Much of the site is now a mature mesic hammock (hardwood forest) with the largest hardwood trees (mostly live oak [*Quercus virginiana*], sand laurel oak [*Quercus hemisphaerica*], water oak [*Quercus nigra*], and sweetgum [*Liquidambar styraciflua*]) on the order of a century old. Some cabbage palms (*Sabal palmetto*), southern magnolia (*Magnolia grandiflora*), and pignut hickory (*Carya glabra*) are also present.

Emerging well above the hardwood canopy at Mills Creek are scattered old longleaf pines, probably more than 150 years old. Here and there throughout this hammock, usually under persistent canopy gaps, are patches of wiregrass (*Aristida beyrichiana*). These wiregrass plants, which usually require fire during the growing season to flower and produce viable seed, probably haven't reproduced sexually here in 100 years. And yet they hang on. Also scattered throughout the hammock are old turkey oaks (*Quercus laevis*), a characteristic tree of sandhill, not mesic hammock.

These cues suggest a series of events since EuroAmerican settlement of this landscape. The presettlement vegetation was sandhill (also known as high pine), with longleaf pine and wiregrass dominant in the overstory and ground layer, respectively, and with turkey oaks in the ground layer (when kept low by fire) or sometimes midstory. The early white settlers were mostly cattlemen, as we know in part from the memoirs of Herb Stoddard, who spent seven years of his youth (1893–1900) in Chuluota (Stoddard 1969). As reviewed in chapter 2, cattlemen burned the woods often; what lightning didn't ignite, they did. During these open-range times cattle were mostly on the move, often following fire across the landscape; overgrazing to the extent of reducing fire frequency was probably rare.

In the early twentieth century, after Stoddard moved back north with his family, active fire suppression began. Also, the largest longleaf pines were logged. Younger trees were left, which today are the remnant pines we see emerging above the canopy of the hammock. With fewer fallen pine needles to fuel fires, coupled with direct fire suppression and increasing firebreaks across the landscape, fire-sensitive mesic hardwood trees invaded the Mills Creek Woodlands. After approximately 100 years of growth, these trees now dominate the site, which has converted to mesic hammock. The site is now surrounded by improved pasture and exurban low-density development (with higher-density development on the way). A few fires likely crept into the hammock from its edges over the last century, but probably not many, and no evidence of recent fire is visible.

A wholesale type conversion, from sandhill to hammock, has occurred at Mills Creek Woodlands, primarily due to the elimination of fire. This story is repeated across much of the southeastern Coastal Plain. The hammocks that replaced frequent-fire communities are alternative stable states. They resist fire by creating a more humid microclimate, shading out the flammable grasses, and producing litter that is much less flammable than the previous pyrogenic community. Only under the most extreme weather conditions will these hammocks burn, and even then not thoroughly. More important, the hammock alternative state is less species-rich, especially of Coastal Plain endemics, than the pine savanna community it replaces.

A quantitative study of the effects of long-term fire exclusion in pine savannas was conducted by Gilliam and Platt (1999) in the Sandhills of North Carolina, on the edge of the Coastal Plain bordering the Piedmont. Mesic and xeric sites within the Boyd Tract had last burned about 80 years previously. Hardwoods invaded both sites, but with different species composition and densities. On the mesic site, longleaf pine recruitment diminished quickly after fire exclusion, but on the xeric site the pines recruited for approximately 60 years. Regeneration of pines was minimal at both sites by the time of the study. As predicted, both xeric and mesic sites showed an age-class distribution of longleaf pines sharply different from that of the virgin stand at the Wade Tract in southern Georgia, studied intensively by Platt et al. (1988a). Hardwood invasion of the Boyd Tract occurred when fire exclusion allowed hardwoods to occupy the large gaps between pines that are characteristic of longleaf pine savannas naturally disturbed by fire and wind (Gilliam and Platt 1999).

Active and Passive Fire Exclusion

That fires no longer burn as frequently as they once did in the southeastern Coastal Plain is a consequence of two kinds of fire exclusion: active and passive. Active fire exclusion includes direct efforts to fight fires and protect sites from fire; what fires do occur are extinguished as quickly as possible. Efforts to dissuade rural people from woodsburning, as epitomized by the Dixie Crusaders and Smokey Bear (see chapter 2), also constitute active fire exclusion. Passive fire exclusion is less direct, but historically may have affected more total area than active fire exclusion in this region. The primary cause of passive fire exclusion is fragmentation of the landscape by agricultural fields, roads, urban areas, and other anthropogenic barriers to the spread of fire. Fire compartments have shrunk. Today, however, active fire exclusion (suppressing wildfires) is dominant; most wildfires are rapidly put out.

Both active and passive fire exclusion result in often dramatic changes in ecosystems, as illustrated by the Mills Creek Woodlands example given above. These changes are most pronounced in communities with characteristically short fire-return intervals, such as pine savanna and dry prairie. Across 232 longleaf pine sites studied by Brudvig et al. (2014), fire suppression was the primary cause of community degradation, based on progressive deviation from high-quality reference sites in understory species composition and diversity, litter depth, and other parameters.

Although the development of a largely fire-proof hammock—the alternative stable state—is a common end result of fire exclusion, other outcomes are possible. The build-up of fuels in long-unburned communities—anything from pine duff, grass thatch, and other fine fuels up to shrubs and coarse woody debris such as down logs, snags, and dead lower branches—often produces a "tinderbox" situation, where a fire, when it inevitably occurs, will be much hotter than fires characteristic of the community. Unlike natural low-severity surface fires, fires in long-unburned sites can result in superabundant smoke and high mortality of mature pines and other trees. The height of flames and crown scorch is much greater than in frequently burned sites, sometimes spreading completely through the tree canopy. The intensity of heat on the living cambium of trees can be extreme. Flow of water upward through the xylem also can be interrupted, and fine roots (which typically penetrate through duff) can be killed (Varner et al. 2009 and references therein).

The Importance of Fire Frequency to Biodiversity in the Southeastern Coastal Plain

How frequently a site burns in the southeastern Coastal Plain is often the principal determinant of its ecological condition and biodiversity. This is especially true for natural communities such as pine savannas and other grasslands, which are characterized by stand-maintaining surface fires. In these communities, burning much of a site as frequently as fuels permit often, though not always, produces the optimal results, especially in terms of herbaceous species richness.

Fire Frequency and Plants

A review of fire-frequency effects in longleaf pine flatwoods of South Carolina and northeast Florida confirmed that the "most frequent fire hypothesis" best accounts for the high species richness of native herbaceous plants, as opposed to the intermediate disturbance hypothesis or other hypotheses regarding the influence of fire frequency on species richness (Glitzenstein et al. 2003). This study demonstrated the positive influence of annual or biennial fires for most plants native to pine savannas in this region. Nevertheless, some herbaceous species performed better with longer fire-return intervals, which suggests that some temporal variability or spatial heterogeneity in burn frequencies is desirable. In wet pine savannas in southern Mississippi, a few plant species tolerated or benefited from fire-free intervals of up to eight or nine years (Hinman and Brewer 2007).

Shrubs generally increase in cover and stature with longer intervals between fires. Saw palmetto, which often becomes dense and tall enough in pine flatwoods to competitively exclude herbs and reduce the capacity of a site to burn frequently, had significantly lower cover in annually burned plots (Glitzenstein et al. 2003). Although light interception from a midstory of trees and shrubs is a major factor reducing herbaceous species richness in long-unburned sites, other factors may be at play. In a study of xeric longleaf pine communities in the Florida Panhandle, forest floor development due to duff and other litter accumulation after fire suppression was the primary cause of reduced herbaceous species richness (Hiers et al. 2007). Most vulnerable are plants, such as grasses, that grow from a basal meristem, as they are not able to penetrate the thick litter and grow upward. These are dominant or foundation species in longleaf pine communities, so their reduction has ecosystem-wide impacts. Given these results, Hiers et al. (2007) recommend, at least for xeric longleaf pine stands, that managers should "refine their application of fire and fire surrogates to focus on forest floor reduction," rather than emphasizing reduction of the midstory canopy.

Other studies underscore the ecological value of very frequent fire in pine savannas and the negative consequences of reducing fire frequency. Based on the results of their study in the Green Swamp Preserve in North Carolina, one of the most species-rich savannas in the world on a fine scale (Fig. 3.2), Palmquist et al. (2014) conclude "nearly annual fire is necessary for the maintenance of high plant species richness. . . . [E]ven a modest reduction in fire frequency can have dramatic negative impacts."

Highest species richness with low fire-return intervals is typical of tropical and subtropical savannas but not of some temperate grasslands. A comparison of relationships of plant species richness to fire frequency in a South African savanna and a tallgrass prairie in Kansas showed conflicting responses. The frequency-richness relationship in South Africa is essentially equivalent to that in the southeastern Coastal Plain: increasing richness as fire frequency increases. Notably, grasslands in both regions are dominated by nonrhizomatous C_4 bunchgrasses. In contrast, in Kansas species richness declines with more frequent fire, primarily due to increasing dominance by the rhizomatous C_4 grass, big bluestem (*Andropogon gerardii*) (Kirkman et al. 2014). The frequency-richness relationship

Figure 3.2. Some of the highest fine-scale species richness in the world—up to 50 species in 1 m^2 plots—has been documented in the Big Island Savanna in the Green Swamp Preserve in North Carolina (Walker and Peet 1984). Species richness here is highest with annual fire and has been declining recently with reductions in fire frequency (Palmquist et al. 2014). Photo by Reed Noss.

in eastern tallgrass prairie of northeastern Illinois, however, is more like that in tropical savannas and the southeastern Coastal Plain, with increasing fire frequency leading to higher richness of summer forbs and increased compositional stability of late-successional vegetation (Bowles and Jones 2013).

In pine sandhill, a community drier and generally less productive than flatwoods, somewhat longer fire-return intervals might be expected because more time is required for sufficient fine fuels to accumulate. Beckage and Stout (2000) sampled six sandhill sites in central Florida that had burned from one to six times in the previous 16 years. They found no relationship between fire frequency within this range and plant species richness, diversity, or flowering stem density. They suggested their results support a "saturation model," where species richness increases with fire frequency, but only up to some point determined by competition from the woody midstory or overstory (but see Hiers et al. 2007 for an alternative mechanism). They acknowledged that too low a fire frequency would allow dense stands of fire-intolerant tree species to establish and eventually cause declines in understory species richness (Beckage and Stout 2000).

More recent research shows benefits of more frequent fire. Whereas earlier studies at Tall Timbers Research Station supported a saturation model, with a threshold at six-to-seven-year burn intervals, a longer-term study there shows saturation of canopy cover at three-year intervals. Species richness increased along a spectrum from three- to one-year intervals, with maximum herbaceous dominance achieved at one-to-two-year intervals (Glitzenstein et al. 2012).

A paradox remains concerning the apparent benefits of very frequent fire in longleaf pine savannas. On one hand, numerous studies show fire as frequent as annually or biennially maximizes herbaceous plant species richness. But on the other hand, seedling and juvenile longleaf pines are often killed by fire. Although the adult pines are highly fire-resistant, very young seedlings and the juvenile "bolting" stage, during which the pines transition from the extended seedling (grass) stage to the sapling stage, are not. This is especially true in flatwoods, where the stresses of waterlogged soils and possibly intense competition from dense groundcover vegetation may make pines more vulnerable to fire. Juvenile pines are frequently killed during the bolting stage under such conditions with high fuel loading, until they reach a fire-resistant height of a meter or more. Glitzenstein et al. (1995) found declining densities of longleaf pines in all of their flatwoods (but not sandhill) plots at St. Marks National Wildlife Refuge in the Florida Panhandle under an annual or biennial fire regime. They worried about pines being lost from the plots if the experiment had continued.

Glitzenstein et al. (1995) concluded, "lower fire frequencies or, at the least, occasional extended fire-free intervals would be necessary for recruitment of longleaf pines in the flatwoods." Those fire-free intervals need not be long; three years would generally suffice, and longer intervals would produce higher fuel loads that kill bolting pines when burned. Another way that juvenile pines might survive with annual or biennial burning is through increased spatial heterogeneity in burning. Natural fires are usually patchy in intensity (Komarek 1965), whereas controlled burns are often more homogeneous (Ryan et al. 2013; see chapter 5). Unburned or less intensely burned patches (which occur in the gaps between adult pines, where most recruitment takes place; Platt et al. 1988) may be critical to survival of juvenile longleaf pines and other plants that are sensitive to very high fire frequency or intensity in flatwoods.

Fire Frequency and Animals

Animals as well as plants in the southeastern Coastal Plain have adapted to frequent fire to the extent that many species depend on it. In a Florida sandhill, the density of grasshoppers on plants was significantly higher in one-year, two-year, and five-year burned plots than on a seven-year or unburned plot (Kerstyn and Stiling 1999). Although fires kill grasshoppers, many species of which are flightless, the authors reasoned that "differences in grasshopper densities could be due to a higher density of forbs and the occurrence of healthier forbs in the more frequently burned plots" (Kerstyn and Stiling 1999).

Other animal species, however, including some of conservation concern, seem to respond most favorably to fire-return intervals slightly longer than what is often recommended on the basis of plant species' responses, groundcover richness, or vegetation structure. Maintaining viable populations of all animal species native to a natural community requires some variability in fire regimes within and among sites. Variability in fire regimes ("pyrodiversity") is beneficial only up to a point, however. That point is difficult to specify, but it should be determined on the basis of responses of individual species (with emphasis on imperiled taxa and those of high ecological importance) as well as overall species richness to pyrodiversity, and ultimately by the evolutionary histories of those species (see chapter 5).

Many amphibians in the southeastern Coastal Plain require frequent fire. They provide instructive examples of adaptation to fire because fire affects both the terrestrial habitats in which many species spend their adult lives as well as the wetland and aquatic habitats in which they lay their eggs, develop as embryos and larvae, and undergo metamorphosis into juveniles. Fire can cause direct mortality of juvenile and adult amphibians in terrestrial environments, but the indirect effects of fire on aquatic environments appear to be more important at a population level. For example, fire releases nutrients into aquatic habitats through a variety of mechanisms (see references in Noss and Rothermel 2015). Perhaps vitally important to some amphibians is the release of alkaline cations (Ca^{2+}, Mg^{2+}, K^{+}, Na^{+}) from burned vegetation into water, which increases pH. Because the embryos and larvae of some amphibian species cannot tolerate acidic water, frequent fire may increase survival by raising pH; indeed, some amphibians may be fire-dependent from this mechanism alone.

Noss and Rothermel (2015) conducted an experimental study of the effects of time-since-fire on larval growth, development, and survival of oak toads (*Anaxyrus quercicus*) in seasonal ponds. These ponds (depression marshes) were embedded in a scrub matrix and had been burned four months, 3–4 years, and 11 years in the past. Mean survival was significantly higher in the most recently burned ponds and was positively associated with pH, which was highest in those same ponds. Time-since-fire was somewhat confounded with fire frequency in this study, but it is not unlikely that both influence oak toad recruitment.

The dependence of many amphibians in the pine savannas of the southeastern Coastal Plain on frequent fire is so well accepted among herpetologists in the region that a paper by Schurbon and Fauth (2003) questioning this relationship created quite a stir. They monitored amphibian assemblages in temporary ponds in South Carolina 0, 1, 3, 5, and 12 years after burns, and found species richness increased with time-since-fire. Salamanders were rarely encountered at sites burned within two years. From these and other results, Schurbon and Fauth (2003) suggested that "decreasing the frequency of prescribed burns from the current 2–3 years to 3–7 years will better maintain diverse amphibian and plant assemblages." They also controversially recommended against growing-season burns to "avoid repeatedly interrupting amphibian breeding." Several responses poured into the editorial office of *Conservation Biology*, where the article was published, pointing out potential flaws in the methodology and the interpretation of results (see Means [2006] for a summary of the responses).

A limitation of the Schurbon and Fauth (2003) study is they paid little attention to the landscape context of their ponds and did not separate longleaf pine specialists from generalist amphibians. Robertson and Ostertag (2004) point out that Schurbon and Fauth (2003) "use their results to describe the ecosystem as one that simply accumulates amphibian species with time since burn. However, their data more strongly suggest a pattern of community change in which dominance by an initial suite of species gives way to a new amphibian community composition," composed in large part by forest species. Species associated with forest are expected to be poorly adapted to frequent fire but are also less vulnerable to extinction in the modern landscape than longleaf pine specialists.

Klaus and Noss (2016) studied pond-breeding amphibians within the same landscape (Francis Marion National Forest) as Schurbon and Fauth (2003) and combined data from the two studies to evaluate overall patterns in amphibian response to fire history. They confirmed that generalist and longleaf pine specialist species differed in their relationship to vegetation and time-since-fire. Some generalists were not found in recently burned ponds, and some specialists were absent from long unburned sites. They suggested longleaf pine specialists be given preference in land management, in part because fire refugia are available at multiple spatial scales for generalists sensitive to frequent fire (Klaus and Noss 2016). To maintain specialists, they recommend burning on one-to-three-year intervals and in the growing season "for maximum control of woody encroachment . . . and to maximize the probability that wetlands burn."

Other vertebrates associated with longleaf pine savannas include nine species of snakes and three species of lizards that Means (2006) considers longleaf pine specialists. All of these species prosper in the presence of frequent fire, and most appear to have been associated with pine savannas in the Coastal Plain since at least the early Pleistocene. Two of the snakes most closely associated with the longleaf pine ecosystem, eastern indigo snake (*Drymarchon couperi*) and eastern diamondback rattlesnake (*Crotalus adamanteus*), are the largest nonvenomous and venomous snakes, respectively, native to North America. The catastrophic loss of pine savannas across the southeastern Coastal Plain, closely linked to declining fire frequency, is the principal cause of the decline of this rattlesnake (Waldron et al. 2008; Fill et al. 2015b) and likely contributed to the decline of the indigo snake as well.

The gopher tortoise (*Gopherus polyphemus*), which feeds mostly on herbaceous vegetation, is also primarily associated with pine savannas, as well as other pyrogenic communities such as scrub. This southeastern representative of what is now a mostly southwestern United States and Mexican genus is the last in a long line of tortoises associated with savanna and scrub habitats in the Coastal Plain, extending back to the Miocene (see chapter 2). The gopher tortoise is significant ecologically because its deep burrows serve as temporary refuges or primary homes for dozens of other vertebrates as well as invertebrates, making it one of the most ecologically pivotal species in the southeastern United States (Noss 2013). Reintroduction of frequent fire to Archbold Biological Station in south-central Florida led to an increase in the tortoise population, with number of fires the best predictor of active tortoise burrow density (Ashton et al. 2008). A study in the Florida Panhandle shows burrow abandonment by tortoises is tied to hardwood encroachment associated with fire-return intervals that exceed three years (Legleu 2012).

Grassland/savanna birds in the southeastern Coastal Plain are fire birds, and virtually all have been declining at least in part due to fire exclusion. The red-cockaded woodpecker (*Leuconotopicus* [*Picoides*] *borealis*), a federally listed Endangered species, may be the most iconic species of the longleaf pine ecosystem besides longleaf pine or wiregrass. This bird is closely associated with open pine savannas (especially longleaf pine in the lower Coastal Plain) with minimal hardwood midstory. It achieves greatest demographic success with fire as frequent as fuels will permit, usually a one-to-two-year fire-return interval (James et al. 1997; Ramirez and Ober 2014). As soon as a significant midstory starts to develop as a result of longer fire intervals, the woodpeckers typically abandon the site. The beneficial effects of fire are not limited to midstory reduction, however; they extend to maintenance of a natural food web with sufficient availability of calcium obtained from the woodpeckers' insect prey, largely ants (James et al. 1997).

The most frequent fire-dependent vertebrate in all of North America may be the Florida grasshopper sparrow (*Ammodramus savannarum floridanus*), a federal Endangered species much closer to extinction than the red-cockaded woodpecker. This highly specialized sparrow is endemic to a small area of south-central Florida, where it is completely dependent on well-burned Florida dry prairie (subtropical hyperseasonal grassland). It is on the verge of extinction due to historic habitat loss, fire mismanagement, perhaps increased predation on nestlings, possibly disease, and other factors not well understood due to perennial underfunding of field research.

The highest densities and reproductive success for Florida grasshopper sparrows appear to occur in the first breeding season after fire, with zero reproduction after more than 24 months following fire (Pranty and Tucker 2006; Noss et al. 2008, and references therein). Our research team observed marked sparrows shifting their breeding territories into areas burned just a few weeks earlier at Kissimmee Prairie Preserve. This sparrow requires grassland with minimal shrub/palmetto cover and abundant bare ground, conditions maintained by fires at one-to-two-year intervals. Habitat suitability can shift radically from year to year, from season to season, and even within seasons, due to variability in fire history and hydroperiod (i.e., flooding of nest sites). At any given time, relatively little of the landscape is suitable habitat, and next year—or next month—it may be distributed differently.

Other grassland/savanna birds in the southeastern Coastal Plain that depend on frequent fire for breeding habitat include northern bobwhite quail (*Colinus virginianus*) and Bachman's sparrow (*Peucaea aestivalis*). The northern bobwhite was already noted with respect to Herbert Stoddard's demonstration of its fire dependence in this region (chapter 2). With optimal fire-return intervals of two years or less for nesting habitat, the bobwhite is a serious contender with the Florida grasshopper sparrow for the bird most dependent on very frequent fire (Cox and Widener 2008). Bachman's sparrow, which like most grassland birds has been declining throughout its range, achieves highest densities and reproductive success within the first three years following fire. After more than three years without fire, populations decline (Tucker et al. 2004, 2006).

Wintering birds of savanna/grassland habitats in the southeastern Coastal Plain also require frequently burned habitats. This is noteworthy because the species richness, density, and biomass of birds in these habitats in winter far exceed those in summer. As summarized by Korosy et al. (2013) for wintering sparrows in Florida dry prairie, "Infrequent fire results in encroachment of woody shrubs and trees that shade out pyrogenic bunchgrasses, increased herbaceous density at ground level, and reduction or elimination of bare ground areas, which in turn impair movement, foraging efficiency, and predator detection by ground-dwelling sparrows."

The Special Case of Florida Scrub

The contribution of frequent fire to biodiversity is best established in this region for pine savannas and other grass-dominated ecosystems. Nevertheless, fire frequency is also a critical consideration for Florida scrub and other communities that have longer fire-return intervals and more intense burns than grasslands. Scrub occurs almost entirely within Florida on ancient ridges as well as on more recently deposited coastal dunes. It differs from the highly flammable grassland/savanna communities by having less combustible vegetation, a higher heat of ignition, and characteristically longer fire-return intervals than most grasslands/savannas.

The characteristic range of fire frequency is also much broader in scrub. The natural fire-return interval for scrub has been summarized as 10–100 years (Myers 1990). This broad span is misleading if one assumes scrub sites have natural fire-return intervals more or less evenly spread across this 90-year spectrum. Some coastal scrubs in the Florida Panhandle may have burned on approximate 100-year intervals or longer even before human disruption of fire regimes, but these are special cases.

Recent research on the life-history requirements of scrub endemics on the Florida peninsula suggests many species require relatively frequent fire. Natural fire-return intervals are clustered toward the lower end of the 10–100-year range and are generally shorter in peninsular scrub than in the Panhandle, where fire can be "locally nonexistent" in coastal scrub (Drewa et al. 2008). As an indication of this difference, many sand pines (*Pinus clausa*) in the Panhandle possess cones that open and release seeds in the absence of fire, whereas some 70 percent of the cones of sand pine on interior ridges in the peninsula are serotinous and open only when exposed to the intense heat of fire (Eric Menges, personal communication).

No species drives fire management in scrub more than the charismatic Florida scrub-jay (*Aphelocoma coerulescens*), in those scrub sites where it still persists. The only full species of bird endemic to Florida, the Florida scrub-jay is highly disjunct from its congeners in the West and is federally listed as Threatened. This bird is a cooperative breeder specialized on oak scrub and secondarily scrubby flatwoods, which are basically oak scrub with a scattered canopy of longleaf, slash (*Pinus elliottii*), or south Florida slash pines (*P. densa*).

Fire at approximately 5–20-year intervals is required to maintain a mosaic of oak scrub with an appropriate height distribution to meet the needs of Florida scrub-jays. Scrub of medium height (1.2–1.7 m) with abundant patches of bare sand is optimal for nesting, foraging, and survival (Breininger and Carter 2003). On the Lake Wales Ridge, scrub-jays abandon territories when oaks reach a height greater than approximately 3 m (Fitzpatrick et al. 1991). The lower fire-return interval limit of five years is based on the time required for resprouting oak stems to bear acorns (Ostertag and Menges 1994), as acorns are an essential component of the scrub-jay's diet.

In the presettlement landscape, lightning fires would have created a dynamic shifting mosaic of recently burned scrub and patches in various stages of recovery from fire, thus assuring that oak scrub of the necessary height for nesting by scrub-jays remained continuously available in the landscape. With fragmentation of the landscape and disruption of fire regimes, managers must now laboriously divide sites into burn units and burn them in a sequence and timing that approximate the regime required by this imperiled bird.

Multiple habitat states (regeneration patches representing different heights and times since fire) must be maintained within jay territories, which can be challenging. As noted by Breininger et al. (2014), "Complexity can be reduced by subdividing landscapes into units (e.g. territories) relevant to demographic processes (e.g. recruitment, survival) and classifying units by habitat states that have enough important habitat features (sandy openings and medium height) and not too many detrimental features (tall shrubs, trees)." State-transition models can then be used to predict the proportion of source patches (where reproduction exceeds mortality) and sink patches (where mortality exceeds reproduction) in response to fire history (Breininger et al. 2014).

A managed fire regime that suits scrub-jays might not be ideal for other animals of the Florida scrub, however, which adds complexity back into management. Three of the lizards that co-occur in scrub—the federally Threatened Florida sand skink (*Neoseps* [*Plestiodon*] *reynoldsi*), Florida scrub lizard (*Sceloporus woodi*), and six-lined racerunner (*Aspidoscelis sexlineata*)—occupy different microhabitats and might be expected to respond differently to fire. Schrey et al. (2011a) found that the Florida scrub lizard has highest genetic variation at more recently burned sites, whereas increasing time-since-fire leads to higher genetic diversity in the Florida sand skink and the six-lined racerunner. The sand skink also shows increased variance in genetic diversity in locations with shorter time-since-fire (Schrey et al. 2011b).

The response of the racerunner is counterintuitive because this lizard uses open habitat and avoids densely vegetated areas; thus, it should be favored by frequent fire. The racerunner regularly disperses greater distances than the patches burned during prescribed fires, however, so its metapopulation genetic processes probably operate across a greater spatial extent than investigated in this study. Because responses of genetic diversity to fire are species-specific, the authors recommend maintenance of habitat diversity "through a mosaic of burn frequencies" (Schrey et al. 2011a).

Continued study of the sand skink found effective population size greatest in areas not burned for more than 10 years (Schrey et al. 2016). With decreasing fire-return intervals, effective population size declined, as did abundance. This finding is consistent with the life history of the sand skink. These animals are fossorial and prey on insects in the surface leaf litter, which is diminished by fire. Although these results might imply fire is harmful to sand skinks, such an interpretation would be erroneous. The authors acknowledge that this response of sand skinks to time-since-fire is compatible with the natural fire regime of Florida scrub. The skinks fare well with a 10–14-year fire-return interval (Schrey et al. 2016), which is well within the natural as well as the managed fire regime for much

Florida scrub, and is generally compatible with management for scrub-jays after all.

Plants in scrub respond best to fire at different frequencies, depending on the species. This variability in response is not surprising because the characteristic frequency of fire in scrub spans a considerably broader range than in pine savannas or other grasslands. Considering five species of resprouting woody shrubs, Ostertag and Menges (1994) examined effects of time-since-fire (ranging from 0 to 64 years) on reproductive effort. Four of their study species had the greatest level of reproductive effort (fruit biomass/aboveground biomass) within five years after fire, whereas the fifth species peaked in reproductive effort seven years postfire. The authors suggest "the frequency of scrub fires may have been too unpredictable to select for reproductive allocation patterns precisely reflecting particular fire-return intervals. Early peaks in postfire reproductive effort may be a bet-hedging strategy to allow for greater chances of seedling establishment and survival."

From an evolutionary perspective, fire-return intervals select for specific aging properties in a population as well as alternative strategies for recolonizing sites after fire (Olivieri et al. 2016). In scrub, obligate-seeding plants killed by fire must have enough time for their seedlings from the soil seed bank to grow and mature before the next fire, so that they can replenish the seed bank. With repeated fires at intervals too short for plant maturation, the seed bank eventually will be depleted and the population will go extinct. Compared to oak scrub, rosemary scrub, dominated by the allelopathic Florida rosemary (*Ceratiola ericoides*), has very sparse vegetation and correspondingly longer fire-return intervals, on the order of 10–40 years. The lower limit represents the age at which rosemary first produces seed and the upper limit the age at which the shrub senesces and seed production declines (Johnson 1982). The allelopathy of *Ceratiola* produces bare sand gaps around adult plants, which effectively reduce fire frequency immediately around the plants, allowing them to grow old enough to flower and set seed.

Hypericum cumulicola is a rare endemic plant of rosemary scrub studied over many years by Pedro Quintana-Ascencio. Survival, growth, fecundity, and establishment for this species all decline markedly with time-since-fire. Germination and growth are stimulated by fire because fire topkills dominant shrubs that compete for light, nutrients, and water; enriches soils by releasing nutrients; removes ground lichens and surface organic matter; creates unoccupied gaps with higher levels of soil water; and destroys allelopathic chemicals that affect seed germination and survival (Quintana-Ascencio et al. 1998). An innovative fire-explicit population viability analysis of *Hypericum cumulicola* determined that fire-return intervals greater than 50 years significantly increased extinction risk. Twenty-five years or longer without fire results in loss of aboveground individuals (Quintana-Ascencio et al. 2003).

Other scrub plants have responses to fire frequency that differ from *H. cumulicola*. Contrasting responses to changes in shrub and lichen cover may explain some of these differences. *Eryngium cuneifolium* is similar to *H. cumulicola* in that both species form persistent soil seed banks, on which they depend for recruitment after fire. *E. cuneifolium* is even more dependent on frequent fire, however, with populations declining markedly within less than 15 years postfire (Menges and Kimmich 1996). In contrast, *Polygonella basiramia* relies on dispersal and immediate germination of seeds to colonize recently burned patches (Quintana-Ascencio and Menges 2000). *P. basiramia* is not as sensitive as *E. cuneifolium* or *H. cumulicola* to time-since-fire, although it requires open sand with minimal cover of ground lichens. Ground lichens in scrub reach their highest abundance with much longer fire intervals.

Divergent responses of co-occurring plant species to fire intervals support the idea that coexistence is favored by a "storage effect" when at least one life-history stage can survive periods of poor recruitment, and when fluctuations in recruitment rates among species allow competitively inferior species to sometimes prosper (Warner and Chesson 1985; Quintana-Ascencio et al. 2003). Conflicting species-specific responses to fire regimes within the same natural community are instructive to scientists and managers alike because they tell us fire regimes are naturally variable in time and space, and that fire management must maintain this spatio-temporal heterogeneity in order to preserve biodiversity.

FIRE SEASONALITY

> Much of our thinking in relation to the ecological effects of fire has been based on man-caused fires, even though several investigators have pointed out that the longleaf pine must have evolved under a regime of summer fires. . . . Ecologists must clearly understand *that it was summer fires in the past that have had the most effect in "molding" the landscape of North America.*
>
> Komarek (1965)

The seasonal timing of fire is ecologically important in many ways, but it has been altered by human activities, including fire management from aboriginal times to the present. Across North America the number of fires and the acreage burned by season differ between lightning fires and fires set by humans (Keeley et al. 2009). In the southeastern Coastal Plain, the season of burning has generally shifted away from the season determined by lightning.

Komarek (1965) explained the ecological role of growing-season fires and why they are preferable to dormant-season fires. For example, he noted that "summer fires have a much more profound effect on the control of woody growths than winter fires." Still, the literature on seasonality of fire is confusing in part because terminology has not been consistent. As noted by Hermann et al. (1998), "the phrases warm and cool seasons, summer and winter, growing and dormant season, and lightning and non-lightning season are related, but not synonymous terms; dates associated with each phrase may shift depending on the latitude of a site." In this chapter I generally use the terminology applied by the authors of the studies cited, but the reader should recognize that overlap among growing season, lightning season, and summer, for example, is imperfect.

In 1992, Louise Robbins and Ronald Myers published *Seasonal Effects of Prescribed Burning in Florida: A Review* as a miscellaneous publication of the Tall Timbers Research Station. Robbins and Myers (1992) acknowledged that "nothing is known about the effect of fire season on a number of species or communities, and the data we do have are, in many cases, incomplete or unreliable." A flurry of new studies of fire seasonality and its effects have been conducted and published since then, so this statement is not as valid today as it was a quarter-century ago. Another, geographically broader review of fire season effects (Knapp et al. 2009) summarized the more recent literature. Nevertheless, detailed studies of the demographic responses of most plant and animal species to fire during different seasons are still lacking.

Despite the need for additional research to answer complex questions about fire seasonality, evidence reviewed below shows many to most of the dominant, characteristic, and rare native plants studied in southeastern ecosystems respond most favorably, in terms of flowering and producing viable seed, to fire during the dry-wet transition season of late spring through early summer. Our native animals in fire-prone communities also generally respond best to lightning-season fires. Relatively few species are favored by fires during other times of year. These findings suggest fire seasonality has not varied significantly over the evolutionary histories of species here. Below, I describe the lightning-fire season in the southeastern

Coastal Plain and then review literature showing (or not showing) different responses of species to different seasons of burns.

The Lightning-Fire Season in Florida and the Southeastern Coastal Plain

Because rain usually accompanies thunderstorms, the months with the most thunderstorm activity are not necessarily the months with the most lightning-ignited fires. Komarek (1964) showed that the annual peaks of thunderstorm days and lightning fires in Florida are out of synchrony: lightning fires are most common in late spring and early summer (especially May and June), whereas the peak in thunderstorms occurs later, in July and August (Fig. 3.3). A study of lightning ignitions over 16 years in east-central Florida showed ignitions strongly concentrated in July. However, fires were nearly five times larger, on average, in early summer (1 June–15 July) than late summer (16 July–31 August) (Duncan et al. 2010).

In the early part of the thunderstorm season in Florida, vegetative fuels are still dry from the spring drought, which in central and south Florida can last from fall (October–November) through spring (April–May) or sometimes into June, July, or even later in severe drought years. The moisture content of live vegetation is lowest during drought (Robbins and Myers 1992). Therefore, lightning strikes that contact vegetation during late spring and early summer are more likely to cause large fires than when fuels have higher moisture content later in the wet season. Dry lightning (strikes occurring in the absence of substantial rain) is relatively common in the southeastern Coastal Plain (Komarek 1965; Abrams and Nowacki 2015). Although north Florida and areas northward and westward have an additional wet season in the winter resulting from cold fronts that rarely penetrate into southern Florida, lightning fire is still most common there during the late spring and early summer.

The main lightning-fire season in the southeastern Coastal Plain is often called the "transition season" because it marks the transition from the dry to wet seasons. In addition to having relatively dry fuels, the transition season also typically has low relative humidity and intense sunlight (Slocum et al. 2010). Huffman (2006) and Platt et al. (2006) compared the annual mean number of successive days with <5 mm rainfall with lightning strike density for the Florida Panhandle (St. Joseph Bay area, with rainfall data from nearby Apalachicola) and southwestern Florida (Myakka River State Park), respectively (Fig. 3.4). The peak lightning-fire season at both sites is clearly indicated as the period, peaking in May,

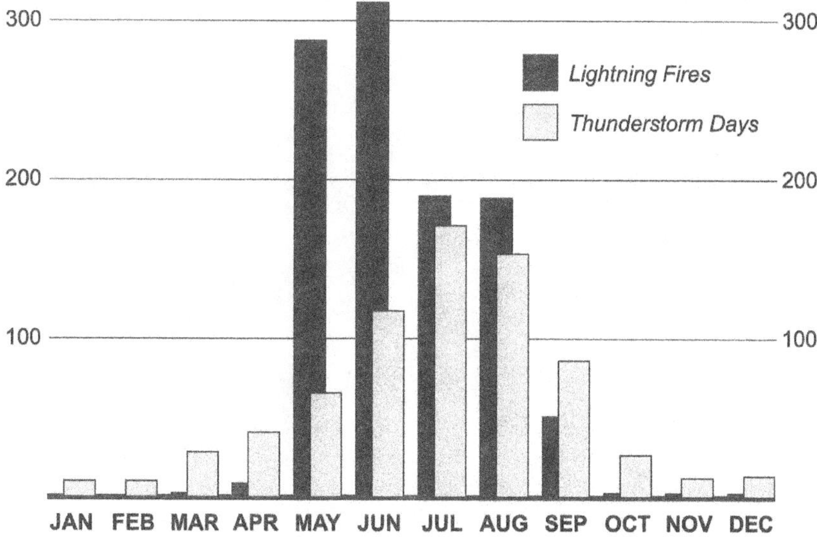

Figure 3.3. Annual distribution of thunderstorm days (gray bars) and lightning fires (black bars) for the years 1962 and 1963 combined. Whereas thunderstorm days peak in late summer, lightning fires peak in late spring, when lightning strikes vegetation still dry from the spring or (in central and south Florida) winter–spring dry season. Adapted from Komarek (1964). From *Forgotten Grasslands of the South*, by Reed F. Noss. Copyright © 2013 Island Press. Reproduced by permission of Island Press, Washington, DC.

when many cumulative days without significant rainfall coincide with an increase in cloud-to-ground lightning strikes.

This peak lightning-fire season is also the time when groundwater levels are usually the lowest of the year, allowing more fire to pass through wetlands and even to burn into the organic soil or peat. Jean Huffman (personal communication) has photographs of wintering white pelicans (*Pelecanus erythrorhynchos*) using depression marshes in dry prairie at Myakka River State Park whose central open-water zones had been restored by transition-season fires burning out the shrubs and organic soils (see Huffman and Blanchard 1991).

Especially in low, flat landscapes, such as subtropical hyperseasonal grasslands, transition-season fires followed by flooding favor grasses over woody plants. Graminoid plants, especially C_4 grasses, which are relatively tolerant of high water levels, can usually grow tall enough within a few weeks after fire to survive the flooding that comes with the heart of the wet season. Woody species, on the other hand, are less able to tolerate this double-whammy of fire followed by prolonged flooding (Platt et al. 2006).

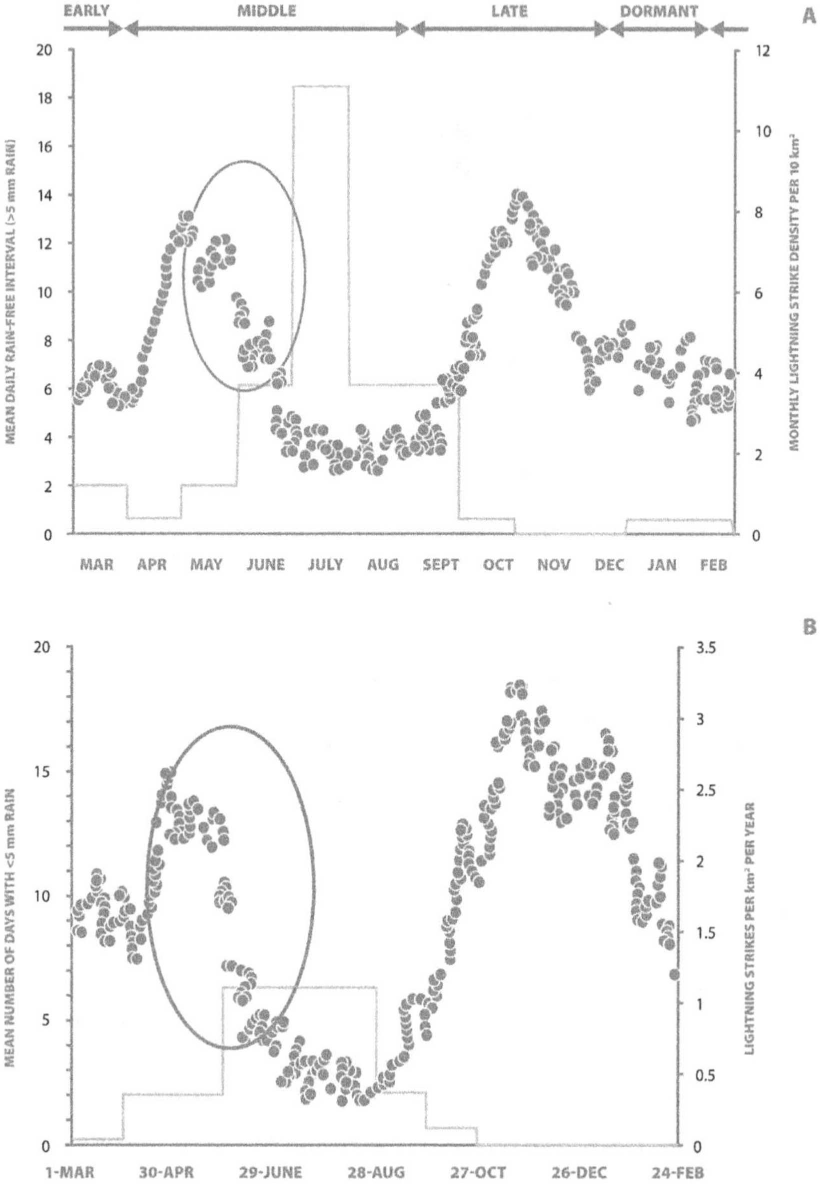

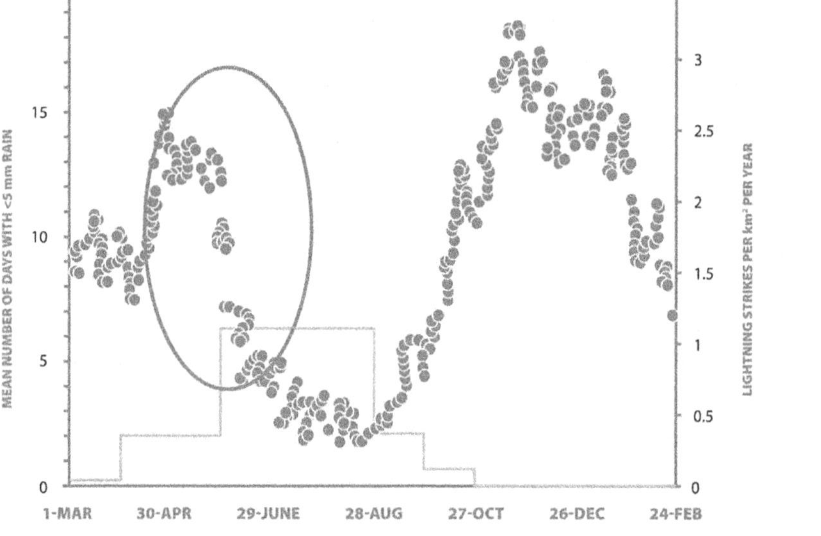

Figure 3.4. (*A*) Monthly lightning strike density per 10 km^2 (vertical bars) and mean number of days since last recorded precipitation >5 mm (dots) in Apalachicola, Florida. Lightning: 1986–1995; Rain-free periods: 1931–1993. Circled area indicates when long rain-free intervals and an abundance of lightning strikes coincide and lightning fires are most likely to occur. "Early," "middle," "late," and "dormant" refer to the growing season of pines. Adapted from Huffman (2006). (*B*) Rain-free days vs. thunderstorms at Myakka River State Park, Florida. Black circles: mean number of successive days with <5 mm rainfall recorded at Sarasota, Florida, for each day of the year during the period from 1944 to 1997. The histogram is the mean number of cloud–ground lightning strikes per km^2 from 1944 to 1997. Adapted from Platt et al. (2006).

The frequency of lightning fires by season is, in itself, not particularly meaningful ecologically. As pointed out by Robbins and Myers (1992), "the important statistic is not occurrence but rather area burned." Using data on the yearly distribution on lightning- and human-caused fires in Florida's national forests from 1968 through 1987, they showed that regardless of the source of ignition, the greatest area burned occurs in May (Fig. 3.5). The prominence of transition-season fires in Florida was further illuminated by a study in the Everglades, which found 53 percent of the total area burned by lightning fires was from fires starting within one week of the onset of the wet season, with an additional 36 percent of the total area burned in fires starting 7–21 days after this onset (Slocum et al. 2007).

Although the existence of a prominent lightning-fire season in Florida has been accepted by most fire ecologists for half a century (since Komarek 1964), the weather conditions that promote susceptibility of vegetation to fire had not been investigated in detail until recently. Slocum et al. (2010) characterized the lightning-fire season by transforming daily rainfall data over a 58-year period (1950–2007) at Avon Park Air Force Range into "cumulative rainfall anomalies," which allow the dry and wet seasons to be more precisely defined. Although the average wet season lasted from 21 May to 1 October and the average dry season from 2 October to 20 May, both the onset dates and seasonal durations were highly variable. The ENSO cycle produced major statistical outliers: large amounts of rainfall in El Niño years, a huge amount during the "Super El Niño" of 1997, and longer droughts during La Niña years. Despite substantial interannual variability, the wet season in peninsular Florida is distinct in having on average twice as much rain as the dry season despite being half as long. As for the impacts of weather on wildfire, Slocum et al. (2010) found that more area burned after dry seasons that were longer and had less rainfall, and when rainfall during the previous year's wet season was inconsistent.

Platt et al. (2015) conducted a discriminant function analysis of 13 years of daily weather data from Avon Park Air Force Range, combined with an excellent fire history record. In the model that explained the most variation, a lightning-fire season characterized by drought, intense solar radiation, low humidity, and warm air temperature was clearly delineated (Fig. 3.6). In discriminant function analysis, points in a two-dimensional graph that represent similar conditions are clumped close together. In Figure 3.6, each dot represents daily weather over the 13-year period. The dots cluster nicely into a dry season, a wet season, and the transitional

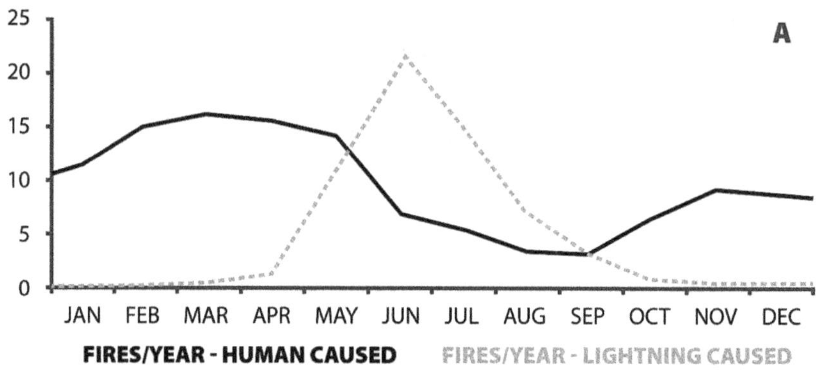

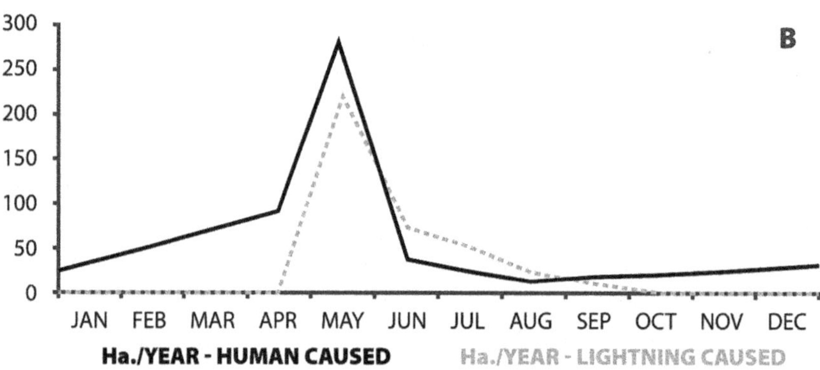

Figure 3.5. (*A*) Yearly distribution of lightning-ignited and human-ignited wildfires in the national forests of Florida, 1968–1987, from U.S. Forest Service fire records. Although human fires occur year-round, most are ignited during the dormant season (winter/spring). Lightning fires, on the other hand, occur almost year-round, but with virtually none in winter and the vast majority in late spring and summer. (*B*) Area burned (ha/yr) by human-ignited and lightning-ignited fires over the same period. Note that the area burned by both lightning and human fires peaks in May, the height of the spring drought. Adapted from data in Robbins and Myers (1992), who note "only fires that were observed are recorded. Many small lightning fires, particularly during the peak lightning period, may go unnoticed or unreported if they do not create a fire-control problem."

period between the dry and wet seasons. The transition season is the lightning-fire season, when the most and the largest lightning fires occur (Fig. 3.6A). This is the season when landscape-scale fires would normally burn across both uplands and wetlands (Platt et al. 2015). Human-caused fires at Avon Park overlap with the lightning-fire season, but imperfectly. Military fires, set by bombs and other training activities, are concentrated

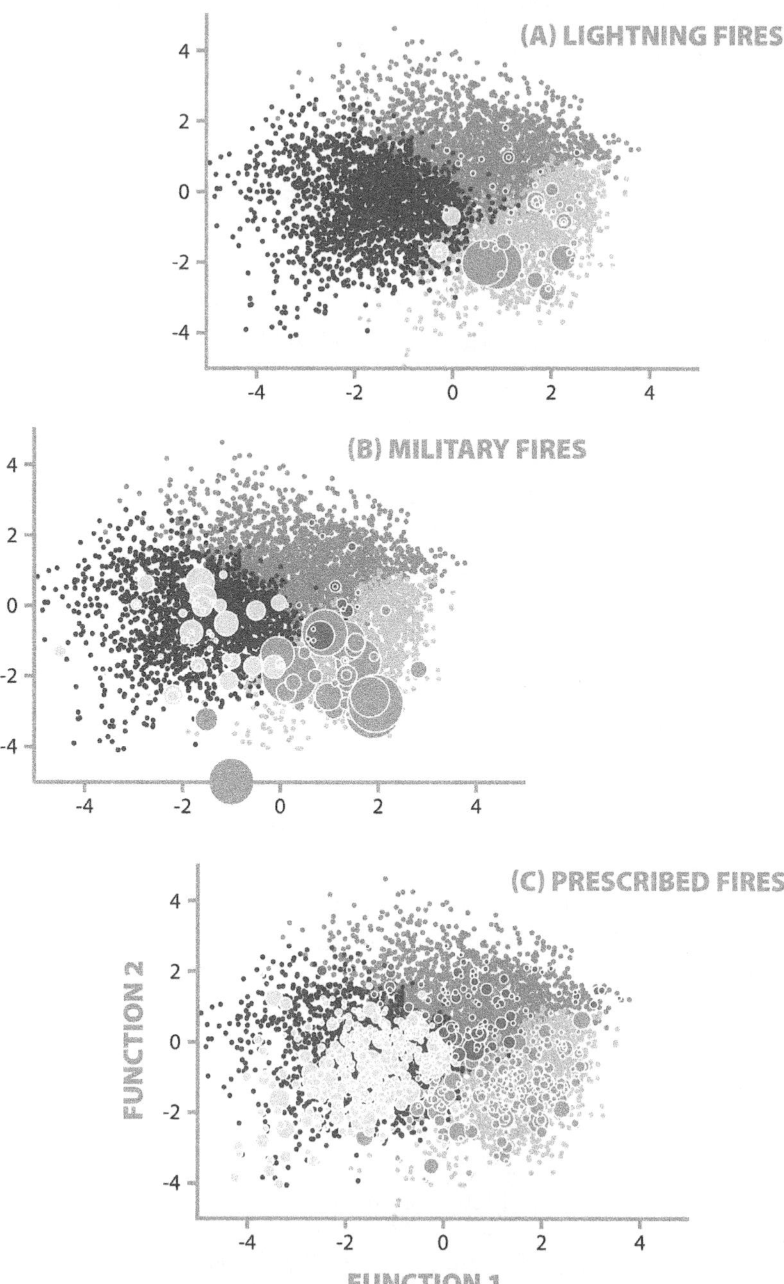

Figure 3.6. Discriminant function analysis of 13 years of daily weather and fire history data from Avon Park Air Force Range, south-central Florida. Weather-days are represented by small dots in all three diagrams: black dots for the dry season, light gray dots for the lightning-fire season, and dark gray dots for the wet season. Fires are indicated by the variably sized dots ("bubbles"), with larger bubbles indicating larger fires. (*A*) In the model that explained the most variation, the lightning-fire season (light gray dots) is characterized by drought, intense solar radiation, low humidity, and warm air temperature. (*B*) Military fires also are concentrated in the lightning-fire season, but many also occur in the dry season. (*C*) Prescribed fires are concentrated in the dry and lightning-fire seasons, with some in the wet season. Adapted from Platt et al. (2015).

in the transition season (Fig. 3.6B), but many also occur in the winter. In contrast, prescribed fires over this 13-year period were concentrated in the winter dry season as well as in the lightning-fire season (Fig. 3.6C).

Fire weather conditions elsewhere in the southeastern Coastal Plain have not been studied as thoroughly as in the Florida peninsula. Given the reduced wet season–dry season contrast northward from central Florida, it is likely the three seasons defined by Platt et al. (2015) are not as distinct elsewhere. Nevertheless, the dendropyrochronology study by Huffman (2006) in the Florida Panhandle, a region with a typical period of winter rain, showed fires prior to European settlement were concentrated in the lightning-fire (transition) season. Lightning-season fires were the only fires recorded in the annual rings of pines at the St. Joseph Bay savanna between 1592 and 1830. Only three fires in the fire-scar record there occurred during the dormant season (winter), and all were post-1830, after European settlement (Huffman 2006). The weather variables that promote fire in Florida—drought, intense sunlight, warm air temperature, and low humidity—should promote fire anywhere, given an ignition source.

Prior to active and passive fire exclusion, anecdotal reports suggest fires often traveled for many miles across the landscape, sometimes igniting tens to hundreds of thousands of acres. They burned until they encountered a major natural firebreak, heavy rain fell, or the wind shifted or died. A fire ignited in May might burn into June, July, or longer, dying back and smoldering during rainy periods, only to flare up again during dry periods, especially if prodded by high winds. These were likely not complete burns, however, but rather mosaics of burned and unburned patches. Nevertheless, fire season was probably more variable and complex than in our fragmented and intensively managed landscapes today. This historic complexity supports the argument for a seasonally variable controlled burning regime, up to a point (see chapter 5).

Responses of Species and Communities to Fire Season in Florida and the Southeastern Coastal Plain

Virtually all ecologists and land managers now accept that frequent fire is necessary to maintain the biodiversity of many natural communities of Florida and the southeastern Coastal Plain. They also recognize the utility of frequent controlled burns to reduce the risk of unwanted wildfire, whether caused by lightning or intentionally or accidentally by humans. More controversial is the significance of fire season. Do controlled burns during the natural lightning-fire season have more beneficial ecological effects than burns during other seasons? Fire ecologists and managers still argue vigorously over this question. Nevertheless, sufficient data are now available to suggest that season of burning is important, and that the lightning-fire season is usually the optimal time to burn from the standpoint of biodiversity conservation. Practical concerns may dictate burning in other seasons, however, a topic I address in chapter 5.

I propose an evolutionary fire season hypothesis: if fire is an important agent of selection, and if fires occurring during the evolution of species in a region were concentrated in a particular season, then species should have evolved mechanisms of growth and reproduction timed to respond to seasonal cues. Therefore, burning primarily during the evolutionary fire season should provide conservation and restoration benefits, including higher fitness of fire-adapted species, additive to those achieved by burning at the evolutionary fire frequency. The key word here is "additive." Researchers and managers mostly agree that burning as frequently as lightning fires occurred prior to European settlement is a worthy objective. The crux of the argument is whether burning during the evolutionary fire season provides any additional benefits beyond burning frequently.

Resprouting

A frequently observed effect of changing the season of fire is the resprouting response of hardwood shrubs and trees. Early research (e.g., Hodgkins 1958; Ferguson 1961; Brender and Cooper 1968) demonstrated that top-kill or complete-kill of hardwoods is higher after spring or summer fires than after winter fires. A classic investigation of the effects of fire seasonality and frequency on hardwood resprouting was the long-term study in the Santee Experimental Forest in the lower Coastal Plain of South Carolina (Waldrop et al. 1987, 1992). Study plots were set up in 1946 in a 42-year-old stand of loblolly pine (*Pinus taeda*) that had naturally regenerated after logging. Three replicate 0.1-ha plots were established in each of five treatments: (1) periodic winter burning, (2) periodic summer burning, (3) annual winter burning, (4) annual summer burning, and (5) unburned control. The winter burns were applied as close as possible to 1 December, and the summer burns as close as possible to 1 June. Periodic burns were conducted when 25 percent of the understory hardwood stems reached 2.5 cm in diameter, which resulted in fire-return intervals of three to seven years.

The results after 30 years (Waldrop et al. 1987; Fig. 3.7) and 43 years

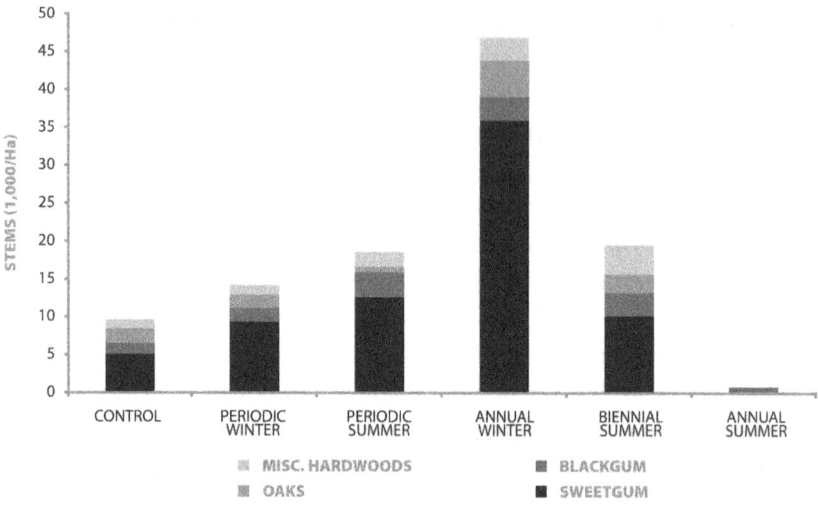

Figure 3.7. Number of understory hardwood stems <2.5 cm (1 in) in diameter after 30 years of controlled burns in loblolly pine stands in the Santee Fire Plots, South Carolina. Note particularly the difference between the annual winter and annual summer burned plots. Whereas annual winter burning led to a proliferation of small hardwood stems after topkill by fire, annual summer fires strongly reduced the number of hardwood stems. Adapted from Robbins and Myers (1992) based on information in Waldrop et al. (1987).

(Waldrop et al. 1992; Fig. 3.8) were stunning. Annual burning in the winter dramatically increased the number of small hardwood (especially sweetgum; *Liquidambar styraciflua*) stems resprouting after topkill by fire after 30 years. Not only are abundant hardwood stems abhorred by foresters who are interested in pine production, they are also considered a major problem by ecologists interested in restoration of the grassy herbaceous layer in pine savannas. Annual summer burning resulted in the fewest resprouting hardwood stems, whereas other treatments had intermediate effects (Fig. 3.7)

The Santee Fire Plots study was terminated in 1989 after Hurricane Hugo caused severe damage to the overstory and midstory that year. The full 43 years of data from the study were analyzed and reported by Waldrop et al. (1992). Hardwood tree and shrub stems less than 1.5 m tall were more abundant in plots burned periodically in winter or summer, or annually in winter, than in unburned plots. Only annual summer burns significantly controlled hardwood resprouting. In fact, these fires virtually eliminated understory woody vegetation (Fig. 3.8A). Regarding

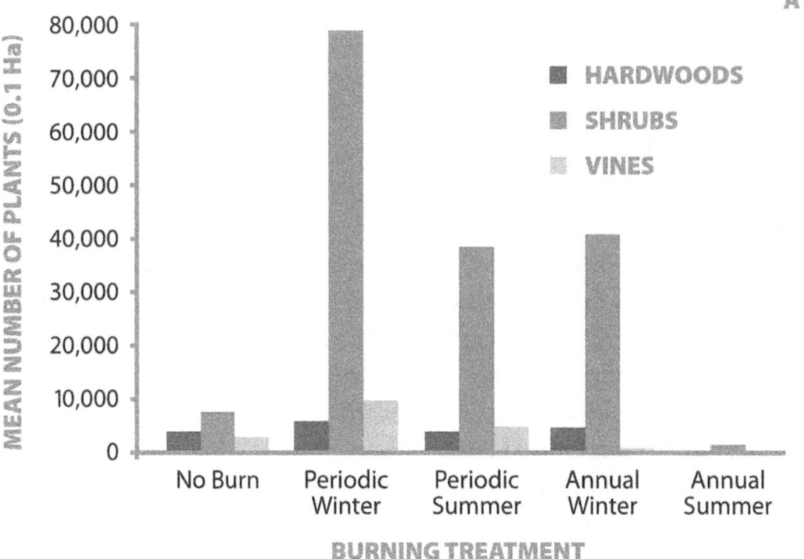

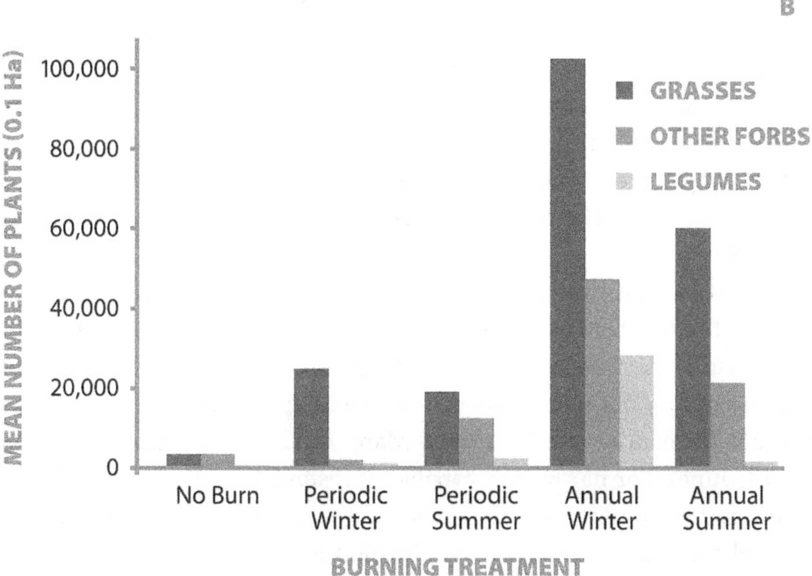

Figure 3.8. Response of mean number of stems <1.5 m tall per 0.1 ha for (*A*) woody plants (hardwood trees, shrubs, and vines) and (*B*) herbaceous plants (grasses, legumes, and other forbs) to four burn treatments and a no-burn control over a 43-year period in the Santee Fire Plots, South Carolina. Adapted from Waldrop et al. (1992).

effects of treatments on herbaceous vegetation, the no-burn controls and the periodically burned plots (winter or summer) had sparse groundcover, whereas plots burned annually in either summer or winter had understories dominated by grasses and forbs, which "resembled the grassland communities described by Komarek (1974)" (Waldrop et al. 1992).

Surprisingly, annual winter burns produced the best responses of grasses, legumes, and other forbs, even better than the annual summer-burned plots (Fig. 3.8B). Even this 43-year study, however, may have been too short to allow firm conclusions about the effects of summer vs. winter burning on herbaceous vegetation. The authors caution that the high number of resprouted hardwood and shrub stems in the annual winter-burned plots "would soon regain dominance of the understory and replace the grassland community if burning were delayed a few years" (Waldrop et al. 1992). Another caveat to the Santee studies is that this part of South Carolina is an area devoid of wiregrass (i.e., a hiatus between the ranges of *Aristida stricta* and *A. beyrichiana*), and therefore is atypical of most of the southeastern Coastal Plain (Hermann et al. 1998).

In a study of fire effects on resprouting shrubs in upslope longleaf pine savannas and downslope seepages in Louisiana and Florida, shrubs in both site types and geographic locations resprouted more vigorously following dormant-season fires than growing-season fires (Drewa et al. 2002). Shrub responses were not related to fire temperature (intensity), which suggests resprouting may be more dependent on the physiological status of the shrubs at the time of the burn than on the characteristics of particular fires.

Robertson and Hmielowski (2014), in southern Georgia, similarly addressed the question of effects of fire season on resprouting hardwoods. Their study site was largely old-field pine-grassland but with remnant patches of longleaf pine–wiregrass. Fire was applied to separate plots in the late dormant season (February–March) and the early growing season (April–June). For hardwoods capable of resprouting after fire, topkill is followed by mobilization of carbohydrates stored in roots and underground storage organs. Such mobilization is also necessary for aboveground growth in spring, which depletes carbohydrate stores that could be used for resprouting after fire. Topkill during the period of time when carbohydrate reserves are lowest—during the early growing season—best reduces the growth rate of resprouting stems after fire (Robertson and Hmielowski 2014).

A sequence of many fires over a period of years is often needed to knock back perennial resprouting plants that have become overly dense due to infrequent fire. In the study by Drewa et al. (2002), however, even repeated growing-season fires did not reduce established shrubs. This result led the authors to conclude that "long-term shifts in characteristics of fire regimes, even in fire-frequented habitats, may produce effects that are not reversible in the short term (<10 yr)."

Postfire Mortality of Pines and Hardwoods

Research on postfire mortality of pines and hardwoods in the Coastal Plain has confirmed the importance of season of burn. The St. Marks studies, directed by William (Bill) Platt over more than a decade at St. Marks National Wildlife Refuge in the Florida Panhandle, addressed several topics in fire ecology. In one of the papers from this group, Glitzenstein et al. (1995) investigated effects of fire season, frequency, and intensity on tree population dynamics in longleaf pine sandhills and flatwoods over an eight-year period. In contrast to the Santee study, which compared only January vs. June fires, this study compared controlled burns during eight months of the year.

Longleaf pine population dynamics were not affected directly by season of burn at St. Marks (Glitzenstein et al. 1995), which confirmed an earlier study (Streng et al. 1993). This result stands in contrast to some earlier studies and observations showing greater mortality of pines after growing-season burns. For example, Boyer (1987) found 8.0 percent mortality of longleaf pines following summer burning, compared to 4.3 percent with winter or spring burning; this difference, though small, was statistically significant. Glitzenstein et al. (1995) opined that Boyer's study "continues to provide the only truly reliable evidence for the possible existence of an ambient-temperature effect on longleaf pine."

Fire affects the various life stages of longleaf pine differently. Fires during the growing season provide a suitable seedbed for establishment of seedlings in fall and winter (Hermann et al. 1998). Growing-season fires applied frequently do not usually harm adult longleaf pines, and they appear to enhance growth of the extended seedling or "grass stage," which lasts from 2 to 15 years, sometimes longer (Brockway et al. 2006). Seedlings and pregrass-stage juveniles less than one or two years old with a root collar diameter thinner than about 1.3 cm are often killed by fire (Brockway et al. 2006). Juveniles are fire resistant by approximately two years after germination (Grace and Platt 1995), which argues for at least occasional fire intervals longer than two years or for heterogeneous burns that leave unburned or lightly burned refugial patches for seedling and juvenile pines. Juvenile survivorship is higher within patches of low adult pine density and low needle accumulation, where fire temperatures are lower (Grace and Platt 1995).

The rapid stem elongation or "bolting" stage of longleaf pine, during which the trees can grow 30–90 cm in a single season, is also vulnerable to death from fire. Once the saplings have reached approximately 1 m in height, they are usually above the flame zone in a frequently burned site, have thick bark, and are resistant to moderate-intensity fire (Brockway et al. 2006). Lightning-season fires, however, are often more intense than winter fires in longleaf pine communities, although intensity is dependent on fuel loads and fuel moisture, among other factors such as temperature and wind. Especially if excessive duff has built up due to infrequent fire or if fire occurs late in the growing season or in the fall, pine mortality can be high (Robbins and Myers 1992; Hermann et al. 1998).

Three deciduous oak species (*Quercus laevis*, *Q. margaretta*, and *Q. incana*) in the Glitzenstein et al. (1995) study were most vulnerable to fire during the early growing season (April and May). Both topkill and complete-kill of oaks peaked after these spring fires, accompanied by a decline in oak recruitment, density, and basal area and a shift toward increasing dominance by longleaf pine in the stands. Oak trees in sandhill plots experienced higher rates of topkill and declines in density with a biennial fire regime than an annual regime, presumably because biennial burning allowed a greater accumulation of fine fuels, which produced more intense fires (Glitzenstein et al. 1995).

Fire Season Effects on Plant Flowering and Demography

Another well-demonstrated effect of fire season is enhanced flowering of many grasses and forbs following burns during the lightning-fire season, a result expected if fire exerts selective pressure on plant phenology. This effect can be expressed as number of flowers, proportion of plants flowering, or the dates and duration of flowering (Robbins and Myers 1992). The effects of fire season on plant phenology can be obscured by various factors, however—for instance, time lags in the response of species to fire, regional variation in fire seasonality, or changes in species' distributions—such that the present range of a species may not correspond to the region and lightning-fire season where it evolved (Beckage et al. 2005).

Wiregrass (*Aristida stricta* and *A. beyrichiana*) is the plant best known as finely tuned to the lightning-fire season. Biologists since the mid-twentieth century have noted this effect, as well as exceptions, although questions have not been fully resolved about the viability of wiregrass seeds after fires in different seasons (Robbins and Myers 1992). The viability of wiregrass is important because it is a keystone or foundation species due to its high flammability and characteristically high density and biomass, allowing it to carry surface fire quickly through pine savannas and other grasslands and to promote high herbaceous species richness (Clewell 1989; Hardin and White 1989; Noss 1989; Fill et al. 2012, 2016).

Likely every field biologist and land manager in the southeastern Coastal Plain has observed the striking difference in response of wiregrass burned in the growing season (particularly the early lightning-fire season) vs. the dormant season. I have seen several sites where two sides of a road were burned at different seasons and show this effect (Fig. 3.9). The mechanism behind the enhanced flowering of wiregrass after growing-season burns was illuminated by Fill et al. (2012), who showed the response of wiregrass to environmental variability represents a trade-off between allocating meristems (tissues that give rise to various plant organs and keep the plant growing) to the competing life-history strategies of vegetative growth vs. sexual reproduction. Fires in May or June stimulated greater production of inflorescences than did fires in early spring (March–April) or in late summer (August). Fill et al. (2012) concluded that the historical fire regime selected for optimal timing of resource allocation to flowering. Nevertheless, wiregrass does occasionally bloom and produce viable seed outside of the growing season. As explained by Hermann et al. (1998), "ambient temperature interacts with season of burn in a complex fashion so that late dormant-season burns on unusually warm days may mimic the outcome of later fires by stimulating flowering."

How exactly might wiregrass and other plants benefit from timing their sexual reproduction to follow fire during the lightning-fire season? Fill et al. (2012) summarize the reasons suggested in the literature, which include increased light availability, removal of litter to provide suitable microsites for germination, reduced competition, and increased nutrient availability. They also suggest a novel but not mutually exclusive explanation:

> We propose that wiregrass reproduction is closely tied to the phenology of the natural fire regime because the occurrence of

Figure 3.9. Photo of the Wade Tract, near Thomasville, Georgia, illustrating effects of fire season. The area to the left of the road was burned 5 April 2009 (in the late dormant season), whereas the area on the right side of the road was burned just 19 days later on 24 April 2009 (in the early growing season). The taller herbaceous growth on the right is mostly flowering wiregrass (*Aristida beyrichiana*); no flowering of wiregrass is evident on the left. Photo courtesy of Jim Cox, Tall Timbers Research Station.

> lightning-ignited fires was more predictable than that of anthropogenic (i.e., dormant-season) fires. If fire were a stochastic process in this ecosystem, it is unlikely that such strong selection would . . . be so readily apparent or persistent in wiregrass over centuries. We also propose that synchronizing reproduction after predictable fires would increase survival and continued dominance.

Besides wiregrass, many other plants in the southeastern Coastal Plain have been observed flowering most vigorously after lightning-season fires. The Florida endemic cutthroat grass (*Coleataenia abscissa*) flowers profusely when burned between mid-April and mid-August, but rarely if burned at any other time (Myers and Boettcher 1987). Flowering of three grasses of wet prairies in south Florida, *Muhlenbergia capillaris* (now *M. sericea*), *Paspalum monostachyum*, and *Schizachyrium rhizomatum*, was stimulated by growing-season fire, and was significantly greater for all species following fire in May as opposed to fire in January. Flowering decreased substantially for all three species during the second and third years postfire (Main and Barry 2002).

Platt et al. (1988b) found an increase in the number of flowering stems, especially among composites, after growing-season burns at St. Marks, as well as decreased flowering duration per species, increased synchronization of flowering within species, increased dominance of fall flowering forbs, and delayed peak fall flowering. Although increased synchronization of flowering might be expected to increase pollination rates and outcrossing, Robbins and Myers (1992) point out that a greater number of ramets (genetically identical stems) per genet (genetic individual) could increase the rate of self-pollination. Therefore, higher flower density could reduce outcrossing and genetic diversity if pollinators move short distances.

Also in the St. Marks studies, Streng et al. (1993) found a large positive effect of growing-season fires in flatwoods on flowering and presumably seed production of the dominant grasses (*Aristida beyrichiana*, *Andropogon* spp., *Sporobolus floridanus*, *Ctenium aromaticum*) and one forb, the grass-leaved goldenaster (*Pityopsis graminifolia*), as compared to responses after winter fires. Similar positive responses to growing-season fire were found in their sandhills plots for *A. beyrichiana*, *Andropogon* spp., *Schizachyrium scoparium*, and *Pityopsis aspera*. Several other grasses and forbs, however, were not influenced by season of burn (Streng et al. 1993). In the nearby Apalachicola National Forest, blazing stars (*Liatris* spp.), goldenrods (*Solidago* spp.), and goldenasters (*Pityopsis* spp.) showed enhanced flowering after May or June fires, whereas the annual senna (*Seymeria cassiodes*), which is parasitic on roots of pines and some other trees, did not flower after growing-season burns (Hermann et al. 1998).

Brewer and Platt (1994) examined the response of *Pityopsis graminifolia* at St. Marks in greater detail. Floral induction and fecundity were greater following May and August fires than following January fires. Seedling emergence was highest in May-burned plots, intermediate in August-burned plots, and lowest in January-burned plots. This result is precisely what would be predicted if May is the peak of the evolutionary fire season in Florida.

Curiously, though, even after more than a decade of study, the enhanced sexual reproduction of many species at St. Marks after growing-season fires was not accompanied by changes in species composition. With a couple possible exceptions, the species that flowered more profusely after growing-season fire did not become more abundant in the plots burned during that season. Although the reason for this finding is obscure, it may reflect a general rarity of opportunities for seedling establishment in pine savannas due to domination of the community by long-lived perennials that are seldom killed by fire (Streng et al. 1993). Therefore, changes in species composition related to season of fire might not become apparent until after decades of burning different plots in different seasons. Stronger effects of fire season on species composition would be expected on restoration sites, which often have more space available for seedling establishment.

In longleaf pine savannas at Eglin Air Force Base in the Florida Panhandle, the two dominant grass species, *Sporobolus junceus* and *Schizachyrium scoparium* var. *stoloniferum*, had the best flowering responses when burned in April, May, or July (Shepherd et al. 2012). *Aristida purpurascens* had more flowering culms per plant after May or July burns. The best germination of *A. purpurascens* and *Aristida ternarius* was achieved with July burns, and of *S. junceus* with May burns. In contrast, *Aristida mohrii* had the fewest flowering culms per plant and the fewest seeds per flowering culm after July burns. Seeds of this species germinated only after February or April burns. Although this study generally confirmed the superiority of lightning-season fire for grasses in pine savannas, no single burn month optimized responses among species. The authors recommended variability in season of controlled burning for maintenance of diverse groundcover, but with an emphasis on growing-season fires to restore dominant matrix grasses on degraded sites (Shepherd et al. 2012).

Hiers et al. (2000) found a range of reproductive responses of native legumes to season of burn on the J. W. Jones Ecological Research Center in southwestern Georgia. Only 2 of 12 legume species showed significantly increased reproduction following lightning-season fires, whereas 3 species showed significant decreases in reproduction following lightning-season fires as opposed to fires in late winter or early spring. For a subset of legume species studied more intensively, season of burn had no effect on pollination activity (Hiers et al. 2000). Perhaps something about the reproductive strategies of legumes makes them more variable across species in their responses to fire seasonality.

To summarize, most of the herbaceous plants studied with respect to season of burn in the southeastern Coastal Plain show a better response to growing-season fire than to dormant-season fire. Some species show no or variable responses to season of burn, but lagged effects cannot be ruled out in these cases. A few plants, such as *Aristida mohrii* noted above and the sedge *Scleria ciliata*, are stimulated more to flower by dormant-season fires (Robbins and Myers 1992). An evolutionary explanation for this type of response has not been offered, but it could relate to unusual mechanisms of pollination, seed dispersal, or seedling establishment.

An intriguing case, and possibly an exception that proves the rule, is the Big Pine partridge pea (*Chamaecrista keyensis*). This legume, endemic to pine rockland in the Lower Florida Keys, shows mixed fitness responses to summer and winter fires. A field demographic study showed winter fires promote greater survival in the first year after fire. This result was not unexpected, as the summer fires were hotter than the winter fires (Liu and Menges 2005). After three years, however, effects were reversed, with plants in summer-burned plots showing greatest survival. Summer fires also promoted greatest vegetative growth.

Both winter and summer fires stimulated fruit production of *Chamaecrista keyensis* compared to unburned plots, but winter fires produced the greatest recruitment of seedlings (Liu and Menges 2005). Early summer (June) fires produced much higher vital rates than the late summer (August–September) fires usually applied by managers in this area to better control burns because of higher fuel moisture later in the summer. The authors reasoned that June fires occurred "when flowering had just started, prior to most fruiting, and when plants had not yet invested much of their resources into growth and reproduction that year" (Liu and Menges 2005). Therefore, fewer negative consequences would occur from consumption of aboveground biomass early in the summer than later in the season. Similarly, coontie (*Zamia floridana*) shows superior reproduction and recruitment after early-summer burns in pine rockland in the Everglades (Negrón-Ortiz and Gorchov 2000). Many plants flower the same year if burned early in the growing season, often very soon after fire, but flower in the subsequent year if burned later in the growing season.

A population viability analysis based on multiple matrix models and demographic data showed highest finite population growth of *Chamaecrista keyensis* during the first two years postfire, as well as a lower risk of extinction or population decline with winter fires on five-to-seven-year intervals (Liu et al. 2005). From these results and those reported by Liu and Menges (2005) the authors concluded that *Chamaecrista keyensis* may have evolved under a regime that included both early summer lightning fires and anthropogenic winter fires. This interpretation may be plausible because the Florida Keys are geologically young, separated from each other and the mainland for only around 6,000 years. Endemic taxa, especially on the Lower Keys, presumably evolved over this short period (Noss 2013), and humans were probably present at the time of divergence of these taxa. A problem with this interpretation is that the Indian groups most likely present in the Keys were Calusa, and no records exist of Calusa using fire for anything but cooking (see chapter 2).

Another south Florida endemic, pineland clustervine (*Jacquemontia curtisii*), also shows a mixed response to fire season. Spier and Snyder (1998) found mortality of adult *Jacquemontia* twice as high after June than January burns, even though in this case fire temperatures were higher in the winter. Plants burned in June produced more than three times the number of flowers than plants in unburned plots, but January burns resulted in five times more seedlings germinating from the seed bank than June burns. These authors similarly recommended diversity in the season and severity of controlled burns. *J. curtisii* is restricted to the mainland of south Florida, which is also a geologically young landscape. Again, however, no records exist of Indians in this region burning the landscape, so an evolutionary explanation for the favorability of winter fire is elusive.

Fire Season Effects on Animals

Animals also respond to seasonality of fire, but have not been studied as intensively as plants with respect to this variable. Many invertebrates are vulnerable to mortality from fire but also depend on fire to the extent that the plants they require are fire-dependent. Hall and Schweitzer (1992, cited in Hermann et al. 1998) suggest that many arthropods in southeastern pine savannas are not harmed by growing-season fire because they have a mobile (i.e., winged) phase during this season, and hence can escape the flames and heat. They recommended fire years be alternated among sites and that fires be patchy to reduce the probability of local extinctions.

Animals closely associated with embedded wetlands in savanna/grassland landscapes in the southeastern Coastal Plain are highly sensitive to the vegetation structure of these communities, which in turn is affected by seasonality of fire. Not only is frequent fire required to maintain the herbaceous vegetation that dominates depression marshes and similar natural communities, but fire must occur during a time of year when water levels are low and the wetland will burn through. As recommended by Kirkman (1995), managers "should consider episodic fire associated with prolonged drought periods as a desirable component in the preservation of herb-dominated depression wetlands." Typically, the main annual drought coincides with the lightning-fire season. Controlled burning is conducted mostly during the winter, which often produces a shift to an alternative, hardwood-dominated stable state that resists fire. After such a state shift has occurred, only very high-intensity fire during extreme drought might push the community back across the transition threshold to herbaceous wetland (Martin and Kirkman 2009). Because such extreme fires are usually forbidden (i.e., by burn bans enforced by forestry agencies) in today's fire management culture, intervention in the form of hardwood removal is often required to restore such degraded wetlands.

As discussed earlier with respect to fire frequency, many amphibians in the southeastern Coastal Plain are fire-dependent. Those that breed in ephemeral wetlands are also sensitive to the seasonality of fire. Two species of flatwoods salamander are confined to a portion of our study region, where they are closely associated with longleaf pine savannas and embedded depression marshes (seasonal ponds). The frosted flatwoods salamander (*Ambystoma cingulatum*) is distributed east of the Apalachicola/Flint River system in northern Florida, adjacent southern Georgia, and about half of the lower Coastal Plain of South Carolina; it is federally listed as Threatened. The reticulated flatwoods salamander (*Ambystoma bishopi*) inhabits the western Florida Panhandle and a small part of adjacent Georgia west of the Apalachicola/Flint River system; it is probably extirpated from Alabama and is federally listed as Endangered. The river apparently served as a barrier to gene flow and promoted divergence of these sister species. Their life histories appear to be essentially identical.

Recall that the northern portion of our study region has two wet seasons: one in summer and another in winter. Ephemeral wetlands in this area typically fill with water in winter due to a combination of rain and low evapotranspiration. They then dry out in the spring and early summer (encompassing the lightning-fire season), fill again during the summer, then dry out again in the early fall (September/October). Adult flatwoods salamanders, which spend most of the year underground, particularly in crayfish burrows, migrate to their natal wetlands in late fall during rains associated with passage of cold fronts. They mate within dry pond basins or shallow water, then the female lays eggs in dry to moist vegetation or by the entrances of crayfish burrows. By December most adults are migrating out of the pond basins (Palis 1997). Eggs hatch after inundation by rising water levels from winter rains. After a larval period of about three months, metamorphs leave the breeding ponds in March–April (Means 1972; Palis 1996, 1997; Bishop and Haas 2005).

In addition to the serious threat of conversion of natural pine savannas to bedded pine plantations (Means et al. 1996), flatwoods salamanders are extremely vulnerable to the prevailing practice of winter burning. Winter fires not only cause mortality of migrating adults; more important, they change the structure of the breeding ponds by allowing shrubs to invade and reduce the herbaceous vegetation (Kirkman 1995). Herbaceous cover is a primary predictor of occupancy of depression wetlands by larval flatwoods salamanders (Gorman et al. 2009).

Woody vegetation proliferates in wetlands with winter burning because these fires seldom burn through the wetlands. Any woody plants that are topkilled will resprout more vigorously after winter than summer fire. In addition, managers often plow fuel breaks (fire lines) through the ecotones around wetlands to prevent possible muck fires. Fire lines alter hydrology, prevent most fires during any season from penetrating wetlands, interfere with the movements of adults and metamorphs, potentially disrupt the oviposition behavior of females, and often become death traps for larval amphibians (Bishop and Haas 2005; see chapter 5). Invading woody vegetation in these wetlands increases evapotranspiration, leading to premature drying of wetlands and loss of egg and larval habitat (Bishop and Haas 2005; Gorman et al. 2009).

These findings for flatwoods salamanders are generally repeated for other amphibians associated with ephemeral wetlands in pine savanna landscapes, such as the critically endangered dusky gopher frog (*Rana* [*Lithobates*] *sevosa*) (Thurgate and Pechmann 2007). In the case of the related gopher frog (*R. capito*), individuals emigrating from breeding ponds select habitat characterized by an open canopy, few hardwoods, small amounts of leaf litter, and abundant wiregrass (Roznik and Johnson 2009). Fire during the drought of the lightning-fire season appears essential to maintain native pond-breeding amphibian communities within these landscapes.

Effects of burn season on birds in savanna/grasslands of the southeastern Coastal Plain has been argued for decades. Robbins and Myers (1992) provided vividly contrasting quotations from two of the originators of the science of fire ecology:

> Nothing . . . can be more destructive to ground nesting game birds than summer fires which destroy nests and young, together with

> growing food supply and cover, and all conservationists should combine against them. (Stoddard 1935)

> Summer fires . . . occur not only during the . . . reproductive stages of grasses and forbs but also of the animal inhabitants. These animals have had to evolve under such conditions, too, so that natural selection over a long period of time has "fitted" them to survive under a summer, lightning-caused fire regime. (Komarek 1965)

Evidently Herb Stoddard finally became convinced by the data and arguments of his friend and colleague Ed Komarek, because his later writings reflect no such disdain of lightning-season fire. In his memoirs Stoddard writes lovingly of the ancient pine savanna landscape maintained by natural fire: "The pinelands were quite open, with short grass and some palmetto ground cover, kept that way from time immemorial by frequent grass fires" (Stoddard 1969). Nevertheless, even since the 1960s many ornithologists and game bird managers have argued against controlled burning during the avian nesting season. Recently, the balance of opinion among ornithologists has shifted toward the benefits of growing-season fire in southeastern ecosystems, with the exception of some special cases of highly endangered birds on the brink of extinction. The shift in thinking occurred primarily because ornithologists came to recognize the superiority of fires during the early growing season for controlling woody invasion of grass-dominated groundcover, which savanna/grassland birds require both for nesting and in winter.

Winter habitat quality is crucial for many birds in the southeastern Coastal Plain. Because species richness and density of grassland birds in this region are highest in the winter, intra- and interspecific competition for food potentially could be more intense then than during the breeding season, even considering the increased demands of feeding young. Fall or winter fires reduce seed and invertebrate food sources for wintering birds, which would enhance competition. Vegetation recovers slowly from fire during the dormant season, so a winter burn may annihilate habitat and food supplies for the remainder of the season. Henslow's sparrow (*Ammodramus henslowii*), a declining grassland bird, winters in the southeastern Coastal Plain. It shows highest winter survival and densities in longleaf pine savannas burned during the previous growing season (Thatcher et al. 2006; Johnson et al. 2009).

Cox and Widener (2008) asked whether lightning-season fire is "friend or foe of breeding birds." They answered: friend (usually). The challenge is that fires between late February and August coincide with the breeding season for most savanna/grassland birds in our region. These birds nest primarily on the ground (exceptions being mostly woodpeckers, nuthatches, bluebirds, and other cavity-nesters), and their nests are constructed largely of highly combustible dried grasses and other grasslike plants. When nests contain eggs or nestlings, or when recent fledglings with limited mobility are in the area, mostly from March through midsummer, mortality from fire could be high. Nevertheless, studies generally do not show evidence of such mortality affecting grassland birds at the population level. This may be because other sources of nest mortality, especially predators, overwhelm the effects of fire. Moreover, many birds nest very soon after lightning-season fires.

Fire in May enhances vegetation structure for the northern bobwhite for up to six months longer than winter fire. At Tall Timbers Research Station and vicinity, the nesting season for bobwhite peaks in June. Burns in mid-May, when less than 10 percent of hens are incubating, provide the habitat benefits of lightning-season fire without threatening many nests. If a nest is destroyed, the quail will renest unless the burn occurs very late in the growing season (Cox and Windener 2008).

For Bachman's sparrow in the Florida Panhandle and adjacent Alabama, season of last burn did not affect population density (Tucker et al. 2004). This study did not consider density immediately after fire, however. No doubt fires in late April and early May could result in mortality of nestlings or fledglings of Bachman's sparrows during the first nesting cycle. These sparrows, like many birds, usually renest after nest destruction depending on the timing (Tucker et al. 2004). Florida grasshopper sparrows will extend their breeding season and renest after fire in dry prairies in June, but perhaps not in July when less of the summer season remains (Shriver et al. 1996, 1999). This species has been documented to renest several times after nest loss. Birds that nest on bare ground may do so almost immediately after fire. Several nests of common nighthawk (*Chordeiles minor*) were found at Tall Timbers Research Station only three days after a controlled burn in June (Cox and Windener 2008).

Savanna/grassland birds usually prosper with lightning-season fire because fire in this season better enhances their habitat quality, especially for the first one to three years after a burn. The increased reproductive output that occurs in those first years after lightning-season fire offsets mortality associated with these burns (Cox and Widener 2008). The biomass of prey for red-cockaded woodpeckers increases more with the positive herbaceous response and the reduction of hardwood midstory with lightning-season fire than with winter fire (Collins et al. 2002; Hess and James 1998).

For the brown-headed nuthatch (*Sitta pusilla*), which commences nesting in early March in north Florida and a month or more earlier in central Florida, winter or early spring fires can destroy their cavity nests, which are usually close to the ground. Lightning-season fires avoid this conflict, and also have the advantage of not forcing renesting of this pine savanna specialist into the warm season, when predatory snakes are active (Cox and Widener 2008). For the loggerhead shrike (*Lanius ludovicianus*) in longleaf pine savannas, lightning-season fires increase densities of its preferred insect prey, especially grasshoppers, by up to 90 percent (Cox and Widener 2008).

Still, burning an entire site every year during the peak nesting/fledgling period could potentially threaten rare savanna/grassland birds. For species on the verge of extinction, such as the Florida grasshopper sparrow, shifting burning outside the nesting season in occupied habitat may be a prudent emergency tactic. The Florida Grasshopper Sparrow Working Group (unpublished) recommends avoidance of burning in April through July if territorial males occupy a burn unit. Instead, managers should plan ahead and burn sites that contained territorial males the previous year during the following February through March. Because lightning-season fires provide the highest-quality habitat for this species, however, fire during this season should be used to create new habitat in areas no longer supporting territorial males. For all birds in these habitats, the basic question is whether improvements in habitat quality generated by lightning-season fire outweigh the nest losses experienced with fire during this season. In most cases, the benefits of lightning-season fire appear to exceed the costs.

FIRE INTENSITY AND FIRE SEVERITY

"Fire intensity" and "fire severity" are two terms often used interchangeably, but they are different things. Sometimes they are not even positively correlated. Fire intensity represents the energy released from the combustion of organic matter, and can be measured as reaction intensity, fireline intensity, temperature, heating duration, or radiant energy (Keeley 2009). Temperature, often considered equivalent to intensity, may be the least useful metric. The flame from a single burning pine needle and a massive crown fire may both reach a maximum of approximately 1,000°C (Van Wagner and Methven 1978). What matters much more ecologically, and to firefighters, is the rate of heat released by a moving fire. Fireline intensity is the amount of heat transfer per unit length of fireline or fire front (Byram 1959). This is considered by many fire scientists the best measure of fire intensity.

Fireline intensity may vary by more than a thousandfold, from approximately 15 to at least 100,000 kW/m, with levels above 4000 kW/m usually considered "high intensity" (Alexander 1982). Most crown fires fall within a range of 10,000–30,000 kW/m, whereas the low-intensity surface fires characteristic of savannas and grasslands usually do not exceed 550 kW/m (Alexander 1982). A surface fire with a fireline intensity of around 10 kW/m is barely able to sustain itself (Scott et al. 2014). Fireline intensity can be calculated for any part of the perimeter of the fire. The head is the fastest spreading part of the fire and has the highest fireline intensity, whereas the back spreads slowest and has the smallest flames and lowest intensity; flanks are usually intermediate (Scott et al. 2014).

The difficulty of measuring fireline intensity in the field led researchers and managers to investigate the empirical relationship between fireline intensity and flame length, with the latter relatively easy to estimate (Byram 1959). Wildland firefighters often use observed flame length to judge how to attack a fire or to recognize at what point control is likely to be ineffective. Fireline intensity is useful for comparing fires and for calculating lethal scorch height as well as the expected temperatures at certain heights above surface fires. Empirical relationships between fireline intensity, flame length, and crown scorch are not universal, but depend on details of fuel structure and other variables (Alexander and Cruz 2012).

In ecosystems such as savannas and grasslands, the modeled relationship between flame length and fireline intensity may be unreliable (Keeley 2009). Moreover, other measures of fire intensity are more appropriate for other important effects of fire. For instance, smoldering combustion is better measured by temperature at the soil surface or by the duration of heating than by fireline intensity (Keeley 2009). Smoldering fire is a case where fire intensity and fire severity are often not correlated or are negatively correlated.

In contrast to fire intensity, fire severity and the related term "burn severity" have to do with the effects of a fire on vegetation, or, more precisely, on aboveground and belowground organic matter. Fire severity is often estimated by field observations of biomass loss after a wildfire.

Fire severity estimates, however, do not capture ecosystem responses such as vegetation regeneration, recolonization of a site by animals, or soil erosion. As pointed out by Keeley (2009), "although some ecosystem responses are correlated with measures of fire or burn severity, many important ecosystem processes have either not been demonstrated to be predicted by severity indices or have been shown in some vegetation types to be unrelated to severity." Ecosystem responses are arguably the most important consequences of fire for ecologists and managers to understand, but they cannot be encapsulated by simple metrics of intensity or severity. More detailed studies of ecosystem responses should be pursued whenever feasible.

Fire intensity and severity are influenced by the amount and condition of the live and dead vegetation (fuels) and a number of meteorological variables, including air temperature, relative humidity, and wind speed and direction. The lightning-fire season delineated by Platt et al. (2015) is characterized by drought, intense solar radiation, low humidity, and warm air temperatures. Fires during this season are not only naturally more frequent and larger but also are typically more intense and severe. Exceptions occur, as they always will in fire ecology and ecology generally. A hot, windy day in winter may produce a fire more intense than a humid, still day during the lightning-fire season. Eric Menges (unpublished data) at Archbold Biological Station has found higher intensities (burn temperatures) of scrub fires in winter, on average, than in summer. Residence times of fires in the two seasons are similar.

The biomass and distribution of fuel that has built up since the last fire and the moisture content of that fuel are key predictors of fire intensity and severity (Platt et al. 1991). All else being equal, the longer the time since fire, the greater the fire intensity and severity, especially in savannas and grasslands. Pine savannas may be especially vulnerable to abnormally high-intensity fire after a period of fire exclusion, due to the build-up of highly flammable pine duff. Glitzenstein et al. (1995) found higher rates of topkill and density declines for oaks in sandhills with a biennial fire regime than with an annual regime, presumably due to the greater accumulation of fine fuels with biennial burning. In flatwoods, longleaf pines showed greater crown scorch and lower growth rates in biennially burned plots than in annually burned plots. A few large longleaf pines in the flatwoods died during this study. Mortality in this case was not related to heat flux near the ground surface, but rather to fireline intensity and the resulting high degree of crown scorch (Glitzenstein et al. 1995).

Are fires today more or less intense than before anthropogenic alteration of fire regimes? Systematic studies to address this question on a broad scale are lacking. Most fire ecologists with whom I've discussed this issue observe that fire intensity is not as great under typical controlled burning conditions today as under natural lightning-fire conditions, at least for sites frequently burned. This is because fire managers generally prefer to burn during easier and presumably safer conditions, either cooler (and often wetter, depending on latitude) winter or early spring weather or in the late growing season, when groundwater and fuel moisture levels are higher. Either set of conditions will lead to fires less intense than most lightning fires. On the other hand, on fire-excluded sites with high fuel loads, fire intensities are generally higher than the natural condition.

The current trend in some parts of Florida and elsewhere to burn later in the summer is worrisome to many fire ecologists. Research in the Everglades shows that, as the summer progresses and fuel moisture and water levels rise, fires become less intense and patchier (Slocum et al. 2003). This pattern is likely generalizable across the southeastern Coastal Plain. Although heterogeneity in burns is important ecologically (see below), so too are intense landscape-scale fires that burn through wetlands and other patches that function as firebreaks in seasons outside the peak lightning-fire season. On the other hand, too intense a fire due to high fuel loads can destroy soil seed banks and kill adult trees, such as longleaf pines, which are normally fire-resistant.

FIRE SIZE/EXTENT AND SPREAD

Nearly a century ago Roland Harper (1927) offered some well-founded speculation on the extent of fires prior to European settlement in "high pine land" (i.e., sandhill) on the Lake Wales Ridge of south-central Florida. He also made a comparison to fire on the modern landscape:

> Fire is a frequent and important part of the environment. In prehistoric times the fires must have been started mostly by lightning, and although that may not have happened very often on any one square mile, a fire once started might run for many miles, so that every spot in the high pine land must have been burned over something like once in two years, on the average. With the multiplication of roads, orange groves, etc., any one fire cannot spread far now as formerly.

Changes in the extent of individual fires or series of fires over long periods of time have not been quantified in the southeastern Coastal Plain.

Multiple observations, however, show that the area affected by individual fire events has declined dramatically since EuroAmerican settlement. The causes of this decline include: improved "fire protection" from forestry agencies; superior fire-fighting capability—that is, successfully extinguishing wildfires before they spread very far; and passive fire exclusion resulting from landscape fragmentation. In addition to the human land uses Harper mentioned (roads, orange groves), now urban and suburban areas are important barriers to the natural flow of fire.

At a finer spatial extent, fuel breaks (fire lines; not the same as fireline) are intentionally constructed. These usually plowed or disked strips of mineral soil create barriers to the spread of fires. The ecological consequences of reduced fire size and spread are not well studied. One obvious consequence is that it now takes many small fires to accomplish the habitat restoration or management objectives that one large fire would have accomplished in the past. Anecdotal observations from many regions also suggest that large lightning fires are more internally heterogeneous than small fires of any origin. Large fires typically leave an adequate but not excessive supply of refugia for fire-sensitive species (see below).

FIRE HETEROGENEITY OR PATCHINESS (AND THE IMPORTANCE OF FIRE REFUGIA)

Individual species associated with a particular natural community often show divergent responses to time-since-fire and other attributes of fire regimes. That these species have been able to coexist for presumably tens of thousands to millions of years in the southeastern Coastal Plain shows unequivocally that fire regimes have been spatially and temporally heterogeneous. That is, different patches in the landscape burned at different frequencies and intensities within any given span of time. Frequencies also varied over time in response to changes in climate, the herbivore community, and other factors. This recognition of beneficial patchiness should not be exaggerated, however. In pine savannas and other grasslands, fire regimes appear to have been remarkably uniform in many landscapes. Fire history reconstructions from fire scars in tree rings suggest that variability in fire frequency often was low (Huffman 2006; Huffman and Platt 2014).

Even so, variation in microtopography, hydroperiod, fuel density, fuel moisture, species-specific flammability, fuel continuity and connectivity (e.g., fuel disruption by grazing herbivores, down logs, gopher tortoise burrow aprons, pocket gopher excavation mounds), and other factors of usually fine spatial grain assure that almost all natural fires leave at least small patches of unburned vegetation intact. These fire refugia or "microrefugia" allow some individuals of plant and animal species that are fire-sensitive or which achieve higher reproductive success with longer fire intervals to persist in the community. In the less flammable scrub vegetation, as reviewed above, fire frequencies are quite variable and patchy, so species with widely varying responses are able to coexist on a landscape scale.

The spatiotemporal heterogeneity of fire regimes and the availability of fire refugia have been poorly researched in the Coastal Plain. Usually we simply observe the patchiness in vegetation and infer that the fire regime must be heterogeneous. An explicit study of patchiness and intensity of fires was conducted in the Long Pine Key area of Everglades National Park, which is the southwestern extension of the Miami Rock Ridge (Slocum et al. 2003). This area is notable for having limestone at or very near the ground surface and a complex microtopography created by solution holes. Plots within several types of natural communities were burned at various times within the lightning season. Fires in higher elevation communities, especially pine rocklands, were more intense but slightly less patchy than in lower-elevation communities, with 91 percent of 10 m^2 subplots burned in high pine areas and 89 percent in low pine areas.

More complete burns are expected at higher elevations due to the lower moisture content of fine fuels. Intermixed shrub/palm patches within the pine rocklands were 80 percent burned. These patches are typically associated with water-filled solution holes and may serve as microrefugia for fire-sensitive species. Within short-hydroperiod prairies 68 percent of subplots burned, compared to only 19 percent within long-hydroperiod prairies. Thus, patchiness of fire increases as elevation declines and length of hydroperiod increases in this landscape.

Fire frequency also affected patchiness in this study, with frequently burned subplots being more patchy (73 percent burned) than infrequently burned subplots (82 percent burned). Timing of fire was important, too, with burns in the late lightning season being less intense but more heterogeneous than burns earlier in the season; the latter would have been more common under a natural lightning-fire regime (Slocum et al. 2003). A controlled burning regime that attempts to simulate a lightning-fire regime would leave significant, but not necessarily extensive, unburned patches, particularly in lower elevation areas, and would maintain a reasonably heterogeneous landscape.

Very fine-grain heterogeneity has been largely overlooked in ecosystems with frequent surface fire regimes, where fuel characteristics and fire behavior vary at scales of around 0.1 to 10 m. Hiers et al. (2009) characterized fine-grain patchiness as "wildland fuel cells" with distinct fuel composition, characteristics, and architecture within what superficially appear to be homogeneous fuel beds. The small size of wildland fuel cells was mirrored by heterogeneity in fire temperatures and residence times at similar scales.

Thaxton and Platt (2006) investigated fine-scale variation in fire intensity and its effects on heterogeneity in a frequently burned longleaf pine savanna in Louisiana. They experimentally manipulated prefire fuel loads to mimic the patchiness that typically occurs as a result of needle fall near mature pines, branch falls, and disturbance from burrowing animals. Experimental fuels supplementation increased fuel consumption and fire temperatures relative to unaltered controls. This supplementation increased damage to shrubs and shrub mortality, while reducing resprouting. The overall effect of fine-grain variation in fire intensity is to increase habitat heterogeneity, which the authors hypothesize contributes to the high species richness of pine savannas. Today, heterogeneity in fire effects is often reduced in comparison with presettlement old-growth conditions, where snags, old downed trees, limbs, cones, and other woody debris created hotspots of high fire intensity, which affect plant recruitment. Conversely, recently toppled trees would have provided fire shadows downwind from moving fires.

Insect Refugia

Insects and other invertebrates that suffer high mortality during fire events provide some of the best evidence for the importance of heterogeneous fires and refugia. Most insects that are sensitive to fire also depend on fire, particularly as mediated by their fire-dependent host plants. Lawrence Earley, in *Looking for Longleaf* (Earley 2004), provides a highly readable account of how insects in pine savannas cope with frequent fire. Flying insects, similarly to birds and many mammals, have the option of fleeing fire as adults, yet their eggs and larvae are immobile and depend on fire microrefugia. Many insects and other invertebrates (e.g., spiders) are flightless through their entire life cycles. Some escape fire by seeking shelter in burrows of many sizes, made by many kinds of animals, in the pine savanna landscape. Some bore into plant roots, pupate underground, or specialize on wet ecotones or microsites that burn less often. Insects and other animals that lack these adaptive options are eliminated from burned sites and must recolonize them from adjacent or nearby unburned patches (Earley 2004).

Orthopterans (grasshoppers and their relatives) are the dominant insect herbivores in longleaf pine savannas. Many orthopteran species are flightless as juveniles or adults and presumably are sensitive to fire. Knight and Holt (2005) recorded insect abundance and herbivory levels on eight plant species common in the understory and interior of three recently burned pine savanna sites. Orthoptera contributed >85 percent of herbivorous insect biomass. As expected, herbivore abundance was much higher at the edge of burned areas than in the interior. In addition to the implications of this study with respect to trophic interactions and control of herbivory, the value of unburned refugia for insects with limited mobility was clearly demonstrated (Knight and Holt 2005).

Insect responses to fire have been relatively better studied in tallgrass prairie remnants than in the Coastal Plain. Some conservationists (e.g., Swengel 2001) have expressed great concern about high mortality of insects from fire, and have recommended alternative management practices (fire surrogates), such as grazing, mowing, and haying as preferable to controlled burning in grasslands. Research in prairie remnants shows the insects most vulnerable to fire inhabit uplands and are remnant-dependent and nonvagile. Nevertheless, although a large proportion (40 percent) of 151 insect species in one prairie remnant study was negatively affected by fire, as opposed to only 26 percent that benefited, 68 percent of populations recovered within one year, and all 163 populations tracked to recovery did so within two years (Panzer 2002). Panzer recommends: "Ecosystem reserves should be burned on a rotational basis, with sufficient unburned refugia maintained each year. Ideally, important microhabitats will be represented in all units." Still, when declining populations of highly imperiled and fire-sensitive insects are the conservation focus, permanent nonfire refugia managed by other means, such as mowing, might be a prudent alternative to burning (Swengel and Swengel 2007).

For fire-sensitive insects and other species the relevance of metapopulation theory to conservation is clear. Multiple sites containing suitable habitat, occupied and unoccupied, and within species-specific dispersal distance, must be protected, so that sites where populations have been extinguished can be recolonized (Hanski 1998).

Oaks and Other Hardwoods in Pine Savannas

The generally positive or neutral effect of lightning-season fire on pines, and the usually negative effects of such fires on oaks and other hardwoods, raises a key question: what is the role of oaks and other hardwoods in the longleaf pine ecosystem? Most fire ecologists and managers in this region consciously or unconsciously follow what has been called the "Tall Timbers model," where the quintessential pine savanna consists of longleaf or other southern pines as the only trees in the overstory and midstory, with oaks and other hardwoods confined to the ground layer. As Glitzenstein et al. (1995) put it, "Our results . . . strongly suggest that tree-sized oaks were a minor component (at best) of the vegetation in presettlement upland savannas. Longleaf pine would have comprised essentially monospecific stands of trees in any of these habitats where spring fires occurred frequently." Others question this interpretation (see below).

Glitzenstein et al. (1995) acknowledge that not all presettlement fires occurred in the early growing season; a significant proportion occurred later in the summer. They cite data from Komarek (1964; Fig. 3.3) as showing that "any given area had a likelihood of burning in July or August about one-third as great as in May and June. Occasional fires may also have occurred in the fall or winter as a result of rare thunderstorms or burning by Native Americans" (Glitzenstein et al. 1995). Still, natural fires outside of the peak lightning-fire season tend to be smaller (Robbins and Myers 1992; Fig. 3.4) and patchier (Slocum et al. 2003). Although the current climate produces a concentration of lightning fires during the transition from dry to wet seasons (Fig. 3.6), or soon thereafter (Duncan et al. 2010), even a small shift in seasonality or frequency of fire could produce conditions under which oak recruitment and survival are enhanced. Topographic fire refugia would also tend to favor oak survival.

Most recent analyses, including reviews of historical accounts, conclude that tree-size pyrophytic oaks are a natural part of the southeastern pine savanna landscape, especially in sandhill and other high pine communities (Greenberg and Simons 1999; Hiers et al. 2014; Stowe 2016; Varner et al. 2016). Romans (1775) observed that "some high pine hills are so covered with two or three species of *quercus* or oak as to make an understory to the lofty pines." Harper (1913), describing longleaf pine communities in Alabama, wrote: "[Longleaf] pine withstands fire better than any other tree we have, but some of the other pines and a few of the oaks and hickories are not much inferior to it in this respect." In central Florida, Harper

(1915) recognized three subtypes of high pine community: open longleaf pine savanna, longleaf pine–turkey oak, and longleaf pine–bluejack oak (*Q. incana*).

Further pointing to the importance of oaks in the longleaf pine landscape, Harper (1927) described "high pine land" on Florida's Lake Wales Ridge: "The dominant tree is the long-leaf pine, and one or more species of medium-sized oaks are nearly everywhere." In the lake region of Highlands County (on the southern end of the Lake Wales Ridge), Harper (1927) characterized the high pine community as "a few small patches on slightly elevated spots in the flatwoods, where they can usually be recognized from the distance by the presence of oaks." Heyward (1939) estimated such "oak ridges" composed about 11 percent of the longleaf pine landscape. As concluded by Greenberg and Simons (1999), "It appears that the commonly described savanna-like high pine ecosystem represents one end of the high pine gradient; oak-dominated sand hills represent the other end." Moreover, in at least some cases oaks contribute to the restoration and persistence of longleaf pine. On xeric sites, midstory oaks (especially turkey oak) can facilitate the establishment of longleaf pine seedlings during their first one to two years after germination by reducing moisture stress during drought (Loudermilk et al. 2016).

How do oaks survive and reproduce in such a high-fire environment? In addition to taking advantage of spatial or temporal fire refugia, many oaks possess thick bark, flammable leaves, and other adaptations to fire that permit them to persist and thrive as trees under a surface fire regime. They need only to get through the vulnerable period of high topkill in their youth, and then grow large enough to develop fire-resistant traits and ultimately reach the midstory or canopy. By so doing, they escape the "fire trap," which confined them to the ground layer (Bond and Midgley 2001; Grady and Hoffmann 2012). Spatial and temporal variability in fire frequency or seasonality would provide those opportunities in at least some landscape positions. In particular, areas within pine savanna landscapes adjacent to riparian areas, ravine slopes, sinkholes, and other "fire shadows" can be expected to have a higher frequency of tree-size oaks and other hardwoods under a natural fire regime (Prince et al. 2016), as would xeric oak ridges within pine flatwoods (Harper 1927). Removal of these hardwoods by managers could have unintended negative consequences.

The presence of specialist animal species that depend on oaks in the pine savanna ecosystem is clear evidence of the ancient status of oaks here. One of the conspicuous species of longleaf and other pine savannas in the southeastern Coastal Plain is the fox squirrel (*S. niger*), four subspecies of which occur within our study region. Fox squirrels depend on oaks as well as pines for food resources, and old hardwood trees provide refugia and nest cavities (Weigl et al. 1989; Perkins et al. 2008). In a study conducted at Fort Bragg, North Carolina, Prince et al. (2016) confirmed previous research showing the dependence of fox squirrels on oak trees. Within the longleaf pine landscape, fox squirrels concentrated their activities near riparian areas, which contain more hardwoods and associated resources. During their study, "fox squirrels selected areas with more oak trees during all seasons of the year, highlighting the importance of oaks for food, cover, and nesting sites" (Prince et al. 2016). Perkins et al. (2008) determined an optimal ratio for Sherman's fox squirrel (*S. niger shermani*) as 88.2 percent mature pine to 11.8 percent hardwood cover.

These studies highlight the importance of seeing pine savannas not as homogeneous stands, but as heterogeneous landscapes with varying fire regimes, species composition, and habitat structure along environmental gradients (Noss 1987a; Hoctor et al. 2006). For any vegetation type, the complex patterns of spatiotemporal heterogeneity in fire regimes at multiple spatial scales can best be achieved in very large reserves managed to maintain or mimic natural disturbance regimes (Noss and Cooperrider 1994; Lindenmayer et al. 2016).

Fire-Adaptive Strategies and Traits

> The set of species growing in fire-prone communities has been filtered in such a way that species without fire-persistent traits have not successfully entered the community.
>
> Pausas and Verdú (2008)

I define "fire-adaptive traits" as traits that permit a species to pass through the environmental filter of fire, regardless of the original selective force for those traits. For taxa such as pines, the evidence for fire as a prominent and ancient selective pressure in macroevolution and microevolution is compelling (Pausas 2015a). Many functional traits, such as thick bark, serotiny, and the "underground tree" (geoxyle) growth form, evolved independently in many unrelated lineages associated with high-fire ecosystems around the world. These traits are splendid examples of convergent evolution in response to a shared selective pressure (Bond 2015). Adaptation to fire is often the most parsimonious explanation (i.e., the simplest explanation, requiring the fewest assumptions) for traits that presently confer fitness in the face of fire. Recent research underscores the evolutionary lability of plant growth forms, such that fire adaptations as striking as the underground tree growth form can evolve rapidly (Pennington and Hughes 2014). Such common fire-adaptive traits as thick bark and the ability to sprout from the root collar may only require changes in gene regulation (Simon and Pennington 2012).

In the text below (summarized in Table 3.1), I present a framework for fire-adaptive traits as ways of avoiding, tolerating, exploiting, or facilitating fire. Some species possess traits that fall into two or more of these categories. Suites of fire-adaptive traits constitute fire-adaptation strategies (Tautenhahn et al. 2016). Distinct strategies are associated with different fire regimes. Fire-adaptive traits should not be viewed in isolation because they are both correlated and interacting for any given fire-adaptation strategy. Suites of correlated fire-adaptive traits are well documented in pines (Schwilk and Ackerly 2001) and oaks (Varner et al. 2016). Fire-adaptation strategies and correlated trait evolution are important considerations for conservation of species in fire-prone ecosystems because "species that exhibit traits that are adaptive under a particular fire regime can be threatened when that regime changes" (Keeley et al. 2011).

One fire-adaptation strategy that is challenging to characterize is adaptation to a mixed or highly variable fire regime. One species that follows this strategy is pond pine (*Pinus serotina*). With relatively few exceptions (e.g., subgenus *Strobus*) pines are associated with fire-prone environments. Keeley and Zedler (1998) propose that fire and its effects "account for much of the diversity in pine life histories." Pond pine is fairly unique in that it possesses traits from both of the two major fire-adaptive complexes seen in pines: one associated with frequent, low-severity surface fires and the other with less frequent, high-severity crown fires (Table 3.2). This combination of traits allows pond pine to exist in crown-fire ecosystems, such as pocosins in the Carolinas with fire-return intervals of up to 30 years or more (Frost 1995), as well as in surface-fire ecosystems, such as wet flatwoods with fire-return intervals on the order of 5–7 years and sometimes less (Landers 1991; FNAI 2010).

Pond pine resprouts from the root collar after fire, unlike other pines in our study region, except shortleaf pine (*P. echinata*), and more like a pyrophytic hardwood. It also sprouts from dormant buds under the bark of the trunk. The latter phenomenon is called epicormic sprouting, and it can allow trees to survive even when the entire crown has been consumed by fire (Weakley and Schafale 1991). Pond pine also retains most branches

Table 3.1. A framework for considering adaptive strategies and traits of plants and animals that allow them to persist or thrive in fire-prone landscapes

Avoiding or escaping fire	Tolerating, resisting, or enduring fire	Exploiting, embracing, or taking advantage of fire	Facilitating or promoting fire
Restriction to habitats that rarely burn	Fire-protective thick bark (often coupled with branch shedding)	Prodigious flowering after fire	Flammable tissues such as live or dead leaves
Relatively nonflammable leaves/litter	Rapid regrowth/ resprouting after fire	Obligate seeding, seed dormancy in soil seed bank, and fire-stimulated germination	
Burrowing behavior or use of subterranean refugia	Underground storage organs and resprouting after topkill	Serotiny and canopy seed storage (often coupled with branch retention)	
Fire-fleeing behavior	Compartmentalization of fire injuries and resin production	Shifting activity into recently burned patches	
	Prolonged seedling development (grass stage)		
	Renesting after nest destruction by fire		

and cones, and as its scientific name suggests, the cones are serotinous—nearly 100 percent so in some populations. All of these traits are typical of an infrequent but intense crown fire regime.

On the other hand, pond pine has moderately thick bark, which protects mature trees from the frequent surface fires characteristic of the wet flatwoods in which it occurs most commonly in the southern portion of its range. In wet flatwoods, pond pine often shares dominance in the overstory with the pine species best adapted to frequent fire, longleaf pine (FNAI 2010). Pitch pine (*Pinus rigida*), the sister species of pond pine, shows variation among populations in several traits, including level of serotiny. This variation has a genetic basis (Ledig et al. 2013). The frequency of trees with serotinous cones also varies geographically in pond pine (Smouse and Saylor 1973). Perhaps the pocosin and flatwoods ecotypes of pond pine are also genetically distinct.

An understanding of how fire-adaptation strategies evolved is useful, but controlled burning should not be postponed until we know the

Table 3.2. Traits of pines commonly interpreted as adaptations to fire

Frequent-fire species (fire tolerators)	Infrequent-fire species (fire embracers)
Long, thin needles (e.g., longleaf pine, slash pine), which are highly flammable and shed frequently (e.g., every 2–3 years in *Pinus palustris*)	Mostly short-needled (e.g., sand pine); needles not very flammable and usually are shed less frequently
Thick bark (protects cambium from fire) Most adults survive fire	Thin bark (no protection from fire) High adult mortality after fire
No serotiny; cones not retained long	Serotiny is common; cones are retained as canopy seed storage and open when exposed to high heat
Germination in bare mineral soil exposed by fire	Germination often in bare mineral soil
Extended seedling stage ("grass stage") and delayed height growth, 5–20 years in some taxa, with protected apical meristems	No extended seedling stage or delayed height growth
Seedlings and adults have low vulnerability to fire, with rapid growth through fire-vulnerable sapling stage	All stages are fire-vulnerable
No basal resprouting in most species	Basal resprouting after fire in some species
Reproduction after stand-maintenance fire varies annually, not stimulated by fire (i.e., "mast years" are years of high seed production) and is patchy, resulting in uneven-aged stands (overlapping generations)	Mass reproduction after stand-replacement fire, creating generally even-aged stands (non-overlapping generations)
Relatively long-lived (>300 years), older age at maturity, tall height	Relatively short-lived (<200 years), younger age at maturity (early reproduction), short height
Generally open-canopied stands with dense herbaceous vegetation (grasslands, savannas, woodlands)	Dense stands with less herbaceous vegetation or other understory except after fire (forests or scrub)
Often self-pruning or thermal pruning of lower dead branches (removes fuel ladders to inhibit spread of fire to crown)	Self-pruning uncommon (i.e., dead branches are retained, enhancing spread of fire to crown)

Two fire-adaptation strategies, illustrated by species associated with frequent surface fire vs. infrequent crown fire, are ends of a continuum but are often distinct. One species (*Pinus serotina*) in the southeastern Coastal Plain shows a mixture of traits from both adaptive complexes. Other pines (e.g., *P. taeda*) lack specific adaptations to fire but readily colonize burned sites. Still other pines are best characterized as fire avoiders because they inhabit sites that rarely burn. Adapted from information in Landers (1991), Keeley and Zedler (1998), Schwilk and Ackerly (2001), Noss (2013), and Pausas (2015a).

intricacies of species-specific adaptations and responses to fire. As an instructive example, Slapcinsky et al. (2010) investigated 14 years of monitoring data for 18 species of rare plants associated with several natural community types in Florida. Even though these species represented 14 plant families and have distinct evolutionary histories, their demographic responses to fire—measured by density, frequency, flowering, and recruitment—clustered into just two groups. Half of the species responded positively, and half showed a neutral or no response to fire. No species was negatively affected or unable to recover from fire. A more recent study in the North Carolina Sandhills shows that fire suppression and burning outside the evolutionary fire season have driven rarity of plants with fire-adaptive traits (Ames et al. 2017).

Regardless of their fire-adaptation strategy or particular fire-adaptive traits, many plants in the southeastern Coastal Plain either tolerate or thrive in a high-fire environment; without fire or the right kind of fire, they decline. Controlled burning, based on current scientific understanding of fire ecology, is unlikely to do harm. Not burning often enough is usually far more risky.

ADAPTATIONS FOR AVOIDING OR ESCAPING FIRE

Although fire avoidance is sometimes presented as a fire-adaptation strategy, it is perhaps more logically viewed as a lack of such strategy. Plants and animals restricted to habitats that rarely burn may be those that failed to evolve fire-adaptive traits. Therefore, by default they are restricted in a region such as the southeastern Coastal Plain to those limited portions of the landscape where fires seldom occur. On the other hand, nonflammable leaves, burrowing behavior, and possibly fire-fleeing behavior could have evolved, at least in part, in response to strong selective pressure from fire.

Restriction to Habitats That Rarely Burn

The "avoider" strategy has been applied to species restricted to habitats that rarely burn (Tautenhahn et al. 2016). Such habitats typically have nutrient-rich and moist to wet soils, which promote the development of dense tree canopies. These canopies create a humid microclimate and protect fuels on the forest floor from drying, thereby reducing the chances of ignition and spread of fire. On the rare occasions when fire does occur in these habitats, typically during prolonged drought, it may be variable in severity but often is stand-replacing in at least some patches. According to Tautenhahn et al. (2016), avoider species rarely survive fire and have no adaptations such as resprouting or aerial seed banks to regenerate after fire. The only way avoider species can recruit into burned patches is through seed dispersal from sources in unburned patches. Varner et al. (2016) suggest that avoiders in the southeastern Coastal Plain include live oak (*Quercus virginiana*) and sand live oak (*Q. geminata*), both of which possess the correlated traits of low leaf flammability, thick bark when mature, rapid growth rate, and high wood density.

In this region, very few terrestrial or wetland sites absolutely never burn. Many sites, however, are topographically protected from fire and burn only rarely and incompletely. In the beech-magnolia (upland hardwood) forest at Woodyard Hammock at Tall Timbers Research Station, fire creeps into the forest from the edges approximately every 50 years (Kevin Robertson, personal communication). Fire rarely burns the entire understory of upland hardwood forest or enters the crown, so species composition is little affected (FNAI 2010). Unlike Tautenhahn et al.'s avoider species, some of the dominant trees in these forests (southern magnolia [*Magnolia grandiflora*], American beech [*Fagus grandifolia*], and some mesic oaks) are capable of resprouting after fire or other damage. Cypress strands (swamps with slowly flowing water) such as Corkscrew Swamp in south Florida, on the other hand, have long hydroperiods and flowing water during the wet season. They therefore burn extremely infrequently, on the order of centuries; nevertheless, bald cypress is somewhat fire-adapted, and cypress strands may be fire-dependent in the long term (Wade et al. 1980; Duever et al. 1984; Ewel 1995). Pond cypress is more fire-adapted and usually exists in frequent-fire communities. Communities that rarely burn are reviewed in chapter 4.

Relatively Nonflammable Leaves/Litter

In a later section I discuss high flammability as the premier adaptation to fire, because flammability promotes fire. Conversely, some plant species may be able to escape damage by fire by possessing nonflammable leaves. Perhaps surprisingly, the peninsula (Ocala) variety of sand pine (*Pinus clausa* var. *clausa*), along with eastern hemlock (*Tsuga canadensis*) and eastern white pine (*P. strobus*), produces the least flammable litter of all trees examined by Varner et al. (2015a). Apparently sand pine evolved to discourage surface fire, instead waiting to self-immolate during the high-intensity fires that quickly move up its retained branches, consume its crown, and stimulate its often-closed cones to open.

Other trees with relatively nonflammable leaves, which in turn produce relatively noncombustible litter/fuels, include several oaks: live oak, sand live oak, sand laurel oak (*Q. hemisphaerica*), and water oak (*Q. nigra*) (Kane et al. 2008). Bluejack oak (*Q. incana*) also has low flammability, which is puzzling for a pyrophytic oak with thick bark even when small. Fire carries very poorly through stands of these nonflammable oaks unless the canopy is open and fuels from more pyrogenic plants, such as long-needled pines and grasses, are abundant. Thus, these trees are often able to avoid fire. Because they also resist ignition and dampen the intensity of fire, they sometimes create alternative stable states resistant to burning. Usually, mesic oaks are present in tree form in pine savannas only on sites with a long history of fire exclusion. Kane et al. (2008) recommended "managers prioritize removal of species that hinder prescribed fire effectiveness for restoration of southeastern USA pine-oak ecosystems."

Burrowing Behavior or Use of Subterranean Refugia

Many animal species in the southeastern Coastal Plain spend much of their life cycle underground, which is likely, at least in part, an adaptation to fire. Some species, such as the gopher tortoise (*Gopherus polyphemus*), southeastern pocket gopher (*Geomys pinetus*), nine-banded armadillo (*Dasypus novemcinctus*), crayfish (family Cambaridae), mole crickets (family Gryllotalpidae), and various rodents excavate their own burrows, whereas other animals use burrows of the excavators or seek shelter in stump holes, tip-up mounds, or other natural subterranean recesses. Flatwoods salamanders are largely associated with crayfish burrows during their nonbreeding season, which encompasses the lightning-fire season. The gopher frog uses gopher tortoise burrows as well as cotton mouse (*Peromyscus polionotus*) burrows and other cavities.

More than 300 species of vertebrates and invertebrates have been recorded in gopher tortoise burrows (Jackson and Milstrey 1989), and I and other field biologists have observed many of these animals taking shelter in burrows during fires. Presumably many of these obligate or facultative commensal species are fire-sensitive and could not persist in the high-fire communities of the southeastern Coastal Plain without burrows as shelters. Bruce Means (2006) states that "subterranean cavities (burrows, rootholes, stumpholes) may be important to more vertebrates than any other physical characteristic of longleaf pine savannas." Although protection from fire is only one use of these cavities, it is an essential one in these landscapes.

Fire-Fleeing Behavior

Quantitative studies of animals fleeing fire in the southeastern Coastal Plain do not exist, to my knowledge. Nevertheless, anyone watching a fire closely has observed this behavior. Nature videos document it. Especially with a fast-moving fire, this option is limited to winged animals and highly vagile terrestrial species such as large snakes, deer, panthers, raccoons, foxes, and coyotes. Fire-fleeing behavior is an adaptive response to fire, but it is unlikely to have evolved solely in response to fire. On a day-to-day basis, predation is a more constant selective pressure.

ADAPTATIONS FOR TOLERATING, RESISTING, OR ENDURING FIRE

The suite of traits considered here are those that improve aboveground or belowground survival during or after fire. The means by which plant or animal species tolerate, resist, or endure fire depend on the nature of the fire regime and the constraints and opportunities of evolution. Several plant life-history strategies are lumped into this category. Thick-barked trees exemplify a "resister" strategy; plants that recruit from resprouting or seed banks represent an "endurer" strategy (Tautenhan et al. 2016). Persistence in soil seed banks is typically coupled with the recruitment strategy of fire-stimulated regeneration, so I include it in a later section on adaptations for exploiting, embracing, or taking advantage of fire. For animals, few species tolerate or resist fire in the immediate sense, but some endure fire by renesting on a site soon after fire.

Fire-Protective Thick Bark (Often Coupled with Branch Shedding)

Thick bark is the earliest known adaptation to fire, with a dated phylogeny of *Pinus* showing emergence of this trait ca. 126 million years ago (He et al. 2012). In the southeastern Coastal Plain thick bark is characteristic of pines and oaks associated with frequent-fire communities. The outer bark of woody plants consists of dead cells and serves to reduce water loss, protect against pathogens and mechanical injury, provide structural support, insulate the stem against cold, and protect against fire (Pausas 2015b).

For trees in communities with stand-maintaining (surface) fire regimes, thick bark is a common trait that protects the meristematic tissues (cambium and buds) from fire-induced injury. Bark also protects the phloem (which transports sap) and xylem (which transports water). Although xylem is composed largely of dead cells, damage to it could cause stem mortality from hydraulic failure (Pausas 2015b). Bark is an extremely effective heat insulator, with a thermal diffusivity of only around 20 percent of wood of the same density (Martin 1963). Species vary in the physical properties of their bark, however, such that tree species with the same bark thickness can differ tremendously in the insulating efficiency and heat resistance of their bark (Hare 1965).

Globally, a large part of the variability among trees in bark thickness is explained by differences in fire regime, with many examples of convergent evolution of thick bark among distantly related taxa. The more frequent fire is in a natural community, typically the thicker the bark of trees in that community. This correlation between fire frequency and bark thickness also holds among populations of single species associated with different fire regimes; thus, it provides a good example of microevolutionary trait divergence and local adaptation (Stephens and Libby 2006; Pausas 2015a). Surface fire regimes typically select for thick bark only at the base of the tree, as this is where tissues are vulnerable to heat from flaming grasses and pine needles. The bark of blackjack oak (*Quercus marilandica*), a pyrophytic oak of sandhill and other upland pine communities, comprises over half of the basal diameter. The bark is thickest at the base of the trees, within the flame zone of approximately 1 m, and then tapers rapidly (Hammond et al. 2015).

Maintaining thick bark above the characteristic flame zone would not be favored by natural selection because thick bark is costly in terms of resource allocation as well as the opportunity costs of reduced diffusion of water, oxygen, CO_2 and light through the stem (Pausas 2015b). No fitness advantage would accrue to trees with thick bark in communities with rare fire or high-severity fire; even the thickest bark would not protect against extreme heat.

Having thick bark is especially important for saplings. Saplings of pines and oaks in communities with surface fire regimes have thick bark, but bark growth subsequently slows relative to growth of the tree stem (i.e., negative allometry), such that bark of adult trees is not necessarily thicker in more fire-prone habitats (Jackson et al. 1999). Species of oaks associated with savannas and woodlands with surface-fire regimes usually allocate more resources to bark at the sapling stage than oaks living in closed forests (Jackson et al. 1999). The adaptive value of thick bark in frequent-fire regimes is vividly demonstrated when fire exclusion permits the entry of thin-barked trees (e.g. *Acer, Prunus, Amelanchier, Liquidambar*) into communities formerly dominated by thick-barked trees (Peterson and Reich 2001; Nowacki and Abrams 2008; Pausas 2015b).

Trees with fire-protective thick bark often have the correlated trait of branch-shedding or self-pruning. This trait is sometimes called "thermal pruning" because heat from surface fires kills lower branches, which the tree then self-prunes. The selective advantage of branch-shedding is that flames from surface fires are not easily carried upward into the canopy by ladder fuels. With branch shedding, a significant fuel gap exists between the herbaceous ground layer and the canopy, which protects the life of the tree. If fire is excluded, branches are not pruned, and the trees usually develop a shorter and broader form more vulnerable to canopy scorch and mortality from subsequent fires (Maynard Hiss, personal communication).

Thick-barked trees in the southeastern Coastal Plain are longleaf pine, South Florida slash pine, slash pine, loblolly pine (*Pinus taeda*), turkey oak, blackjack oak, bluejack oak, post oak (*Q. stellata*), and southern red oak (*Q. falcata*). Trees with moderately thick bark, which also provides considerable fire resistance, include bald cypress (*Taxodium distichum*), pond cypress (*T. ascendens*), pond pine, and shortleaf pine (*P. echinata*) (Hare 1965; Schwilk and Ackerly 2001; Hiers et al. 2014). In general, the insulating efficiency of bark of the same thickness is higher for conifers than for hardwoods. For example, with an equal thickness of bark, longleaf and slash pine have twice the heat resistance of cherry (*Prunus* spp.), holly (*Ilex* spp.), and sweetgum (Hare 1965).

Protection of Buds and Rapid Regrowth after Fire

Plants are able to survive for long periods in frequent-fire environments such as grasslands only if the meristematic tissues that permit growth are insulated from high temperatures and protected from herbivores. Thus, the growth regions of perennial plants in these habitats are situated near or below the soil surface, where temperatures are lower during fires. In grasslands, lethal temperatures are rarely transmitted more than 2 cm below the soil surface (Wright and Clarke 2008). Leaves and other photosynthetic tissues must be rapidly regenerated after fire to permit regrowth. Because these tissues create fine fuels, their rapid regrowth/resprouting facilitates the frequent surface fires characteristic of grasslands.

In a study of the fire ecology of C_3 and C_4 grasses in South Africa, the phylogenetic lineage of grass species was found more important than photosynthetic type in determining responses to fire. These responses also were related to the characteristic fire-return interval (Ripley et al. 2015). The authors concluded "the fire ecology of savanna grasses depends on adaptations acquired in fire-prone environments, while interacting with the background of traits inherited from ancestors." Rapid regrowth of grasses is made possible by high rates of photosynthesis and high specific leaf area rather than by remobilization of carbohydrates stored belowground (see following section). Most graminoids and forbs in the frequently burned savannas and other grasslands of the southeastern Coastal Plain show similar responses.

Underground Storage Organs and Resprouting after Topkill

A common trait in woody and semiwoody plants associated with fire-prone ecosystems around the world is resprouting after topkill or injury by fire, herbivory, or other damage. Resprouting is a generalized response to aboveground tissue loss. The ability to resprout is widespread among plant lineages, dating back to Mesozoic gymnosperms (including *Ephedra* and redwoods) and basal angiosperms. Resprouting is an ancestral state in most lineages (Pausas and Keeley 2014b) and constitutes a "persistence niche" (Bond and Midgley 2001). In the southeastern Coastal Plain fire is the predominant natural disturbance that stimulates resprouting. The woody plants that resprout after fire or other disturbance are too numerous to list here, but include most hardwood trees, shrubs, and subshrubs, as well as pond pine, shortleaf pine, bald cypress, pond cypress, and, as seedlings, longleaf pine.

With the exception of herbaceous plants showing rapid regrowth without underground storage, resprouting after disturbance requires the mobilization of stored nonstructural carbohydrates. Thus, resprouting as considered here is almost invariably paired with another adaptation: underground storage organs. Sometimes these are specialized organs such as lignotubers (woody swellings of the root crown containing dormant buds from which new stems may sprout) or xylopodia (subterranean woody structures derived from roots), which are tightly associated with fire-prone ecosystems globally. In other cases, carbohydrates are stored in less-specialized tap roots, such as those of longleaf pine, or in rhizomes (modified stems, usually found underground but sometimes at or above the ground surface, which send out roots and shoots).

Carbohydrates formed through photosynthesis are shunted to these underground organs and stored as starches, which can then be mobilized after death or injury of aboveground tissues. This is what allows rapid resprouting and growth after fire or other disturbance. Plants with specialized storage organs are usually more effective in storing carbohydrates in high concentrations than are plants with multipurpose organs such as rhizomes (Olano et al. 2006).

Resprouting woody plants in frequent-fire communities are typically caught in a fire trap of repeated topkill followed by resprouting. These plants are maintained in short stature unless or until they experience a fire-free interval long enough for them to reach adult size (Bond and Midgley 2001; Grady and Hoffmann 2012). Carbohydrate storage and resprouting are not restricted to woody species, which have been better studied with respect to this adaptation. Perennial herbaceous plants often possess bulbs, tubers, or corms that store starches and water and permit rapid resprouting after fire or other unfavorable conditions such as drought or cold. Underground carbohydrate stores decline with time-since-fire in Florida scrub for short-statured subshrubs (e.g., *Licania*, *Palafoxia*, *Vaccinium*) because increasing shade reduces photosynthesis, which is necessary to replenish carbohydrate stores. In contrast, dominant woody plants (e.g., *Quercus geminata*) and emergent climbers (e.g., *Smilax auriculata*) show increases in stored carbohydrates after decades without fire because they gain access to the full sunlight of the canopy (Olano et al. 2006).

Underground Forests

One of the most common life forms of fire-swept savannas worldwide is the geoxylic suffrutex, a condition where plants contain most of their biomass in roots, rhizomes, lignotubers, or other organs that are underground or partially so. Such plants have aptly been called "underground trees" (White 1976). In Florida, many of these plants are subshrubs; they appear herbaceous, but they are actually woody.

The underground forests formed by these "geoxyles" constitute old-growth communities just as ancient and venerable as the redwood (*Sequoia sempervirens*) and Douglas-fir (*Pseudotsuga menziesii*) forests of the Pacific Northwest of North America, the mountain ash (*Eucalyptus regnans*) forests of Australia, the alerce (*Fitzroya cupressoides*) of Chile and Argentina, or tropical rainforests. Yet, because geoxyles are old-growth trees that have gone underground (Pennington and Hughes 2014), they are unseen and therefore unappreciated by the vast majority of people (Veldman et al. 2015a).

The geoxylic suffrutex growth form and its evolution have been studied in some detail recently in South American and African savannas (Simon et al. 2009; Simon and Pennington 2012; Maurin et al. 2014), but have been only casually noted in the southeastern Coastal Plain until very recently. Southeastern botanists in the early twentieth century observed that plants in frequently burned habitats often had underground storage organs, which enabled them to quickly resprout after fire by drawing on their stored carbohydrates. But these botanists generally did not investigate this phenomenon much further.

Underground storage organs serve a plant well not only for coping with fire, but with any disturbance that removes or causes death of aboveground organs. Thus, herbivores, drought, flooding, and fire could act independently or synergistically to select for underground storage organs. The geoxylic suffrutex growth form allows plants to minimize annual vegetative growth, leaving more resources to devote to flowers and fruits (Maurin et al. 2014). Effectively, these plants have escaped the fire trap by going underground rather than upward.

Geoxyles with lignotubers, xylopodia, or other underground storage organs are abundant in grasslands and savannas of the southeastern Coastal Plain, especially in Florida. Peninsular Florida is the hotspot of geoxyles in the Southeast (Steve Orzell, personal communication). Examples of geoxyles here include *Gaylussacia dumosa*, *Vaccinium myrsinites*, *Ilex glabra*, *Lyonia lucida*, *Bejaria racemosa*, *Licania michauxii*, *Callicarpa americana*, *Ceanothus americanus*, *Morella carolinensis*, *Photinia pyrifolia*, *Erythrina herbacea*, *Smilax auriculata*, *Zamia pumila*, and the dwarf oaks, *Quercus pumila* and *Q. minima*. *Quercus minima* is one of the dominant plants of Florida dry prairie and also occurs in flatwoods. Milkpeas (*Galactia* spp.) are twining herbaceous plants aboveground but possess a woody lignotuber belowground. The federally Threatened *Clitoria fragrans* and *Euphorbia rosescens*, both endemic to central Florida, possess xylopodia, as apparently does the federally Endangered *Ziziphus celata* (Steve Orzell, personal communication). In addition, many savanna hardwood "trees," such as *Q. laevis*, *Q. marilandica*, *Q. incana*, and *Carya tomentosa*, adopt a geoxyle growth form in frequently burned savannas except when they can grow large enough, with thick enough bark, to tolerate frequent fire.

Maguire and Menges (2011) studied resprouting patterns of 16 scrub species at Archbold Biological Station on Florida's Lake Wales Ridge after a prescribed burn. As commonly observed, the number of resprouted stems postfire exceeded the number of stems before fire. Species differed significantly in resprouting growth rates, with the lowest rates among plants that are both resprouters and seeders. Palmettos (*Serenoa repens* and *Sabal etonia*), however, had the highest resprout growth rates, which was exceptional because they also show substantial seedling recruitment after fire

Figure 3.10. Epicormic branches on pond pine (*Pinus serotina*), which have sprouted from buds under the bark in response to a winter fire. New branches enable the tree to increase photosynthesis after fire to replenish lost carbon. Photo by Reed Noss.

(Maguire and Menges 2011). Palmettos are extreme geoxyles, with a ratio of belowground to aboveground biomass of 7.25, 90 percent of that in carbohydrate-storing rhizomes (Saha et al. 2010). Shrubs had intermediate resprout growth rates, and subshrubs (e.g., the gopher apple, *Licania michauxii*, and *Vaccinium* spp.) the slowest. Maguire and Menges (2011) confirmed that several successful postfire resprouting strategies coexist in Florida scrub, rather than a single optimal strategy. Evolution is versatile.

The rapid postfire resprouting ability of *Serenoa repens* is curious because studies by Warren Abrahamson and colleagues at Archbold Biological Station show that growth of these clonal plants is extremely slow. Although seedling survivorship is high (39 percent in scrubby flatwoods, 57 percent in flatwoods over 19 years), average height increased less than 0.5

cm per year over that period (Abrahamson and Abrahamson 2009). Individuals can take 200 years to mature into adults. Based on growth rates, time to maturity, and DNA analysis, individual clones were determined to live for thousands of years, with some individuals estimated as 10,000 years or older (Takahashi et al. 2011). In contrast to the slow growth rates documented in these studies from relatively dry sites, Huffman and Werner (2000) reported (and many managers have observed) that palmettos can grow rapidly in both cover and height after fire exclusion in wetter flatwoods. The incredible survivorship and longevity of palmettos is related to their superlative capacity for belowground storage and concomitant resistance and resilience to drought, fire, and other disturbances. Recall that *Serenoa* and *Sabal*-type palms are among the earliest plant fossils found in the Coastal Plain, dating from the Middle Eocene ca. 45 million years ago (Graham 1999; see chapter 2).

A special case of sprouting after fire is production of epicormic branches after fire damage. Epicormic buds exist under the bark of the bole and main branches of some tree species—for example, pond pine, as discussed earlier (Fig. 3.10). These buds are stimulated by heat, branch breakage during storms, or other damage to grow new branches. In some species epicormic buds can be dormant for decades until fire or other damage stimulates hormone production, which results in development of new branches (Brommit et al. 2004). As a fire-adaptive trait, epicormic buds and branches allow quick recovery of photosynthetic tissues after fire.

Compartmentalization of Fire Injuries and Resin Production

Thick bark protects the cambium and other tissues of pyrophytic trees. Even the thickest bark, however, will not protect against fires much more intense than usual due to excessive accumulation of litter. Even frequent, low-severity fires, while not usually killing trees, often heat the vascular cambium enough to cause tissue death and leave scars. These fire scars, in turn, are vulnerable to infection by decay fungi and other pathogens. Some trees, including many oaks, are able to contain infection by compartmentalization of the wound following scarring (Shigo 1984; Smith and Sutherland 1999, 2001). Compartmentalization is accomplished by the development of woundwood rings, which close the wound or scar and prevent spread of infection to healthy wood. Smith and Sutherland (1999) concluded that the ability of pyrophytic oaks to rapidly and effectively compartmentalize fire scars helps explain their dominance in hardwood forests subject to low-intensity fire.

In conifers, especially Pinaceae, production of resin is a key component of the tree's defense system. Resin is stored within ducts in the secondary xylem. An interconnected network of axial and radial ducts allows resin to flow to sites of injury or insect attack. In many conifers, attack by bark beetles causes increases in resin production and changes in the chemical composition of resin to become more toxic to the beetles. Resin also helps compartmentalize fungi introduced through beetle attack (Hood et al. 2015). Low-level bark beetle attacks often fail because beetles can be killed by resin flow and because fungal infections introduced by the beetles are sealed off. Recovery from fire damage is also facilitated by resin flow (Ross et al. 1997).

Southern pines vary greatly among species in resin production, with longleaf and slash pine producing the most resin over the longest period (Wahlenberg 1946). The southern pine beetle (*Dendroctonus frontalis*) is the native bark beetle in the southeastern United States. It attacks the inner bark of pine trees. Although this beetle can cause considerable economic damage, it is part of the natural disturbance regime and food web, and thus has ecological benefits when populations are within a natural range of variability. Better studied are western bark beetles of the genera *Dendroctonus* and *Ips*.

In a fascinating study, Hood et al. (2015) showed that frequent, low-severity surface fires induce defense of ponderosa pine (*P. ponderosa*)—the western ecological equivalent, in many respects, of longleaf pine—against bark beetles by stimulating resin duct activity. Stands exposed to frequent fire are more resistant to beetle attack. If fire severity increases—for example, due to fire exclusion and accumulation of fuels—trees injured by fire are more susceptible to beetles (Hood et al. 2015). Southern pines, especially longleaf and slash pine with their high levels of resin production, likely interact similarly with fire and southern pine beetles. Indeed, longleaf pine is more resistant to both fire and bark beetles than other southern pines (Ross et al. 1997). This resistance can be expected to ebb, however, when fire frequency declines due to suppression and fire severity increases.

Prolonged Seedling Development

One of the most specialized fire adaptations in trees is prolonged seedling development. Extended seedlings are often called the "grass stage" because the seedling resembles a perennial bunchgrass. In the southeastern Coastal Plain, this trait is well developed in longleaf pine and persists

Figure 3.11. The terminal bud of a juvenile longleaf pine (*Pinus palustris*), which has just begun to emerge from the grass stage. The bud is protected from fire by a mantle of decumbent needles and fire-resistant scales. Photo by Reed Noss.

from 2 to 15 years or longer (Brockway et al. 2006). South Florida slash pine shows some limited characteristics of a grass stage, but is still often vulnerable to surface fire for up to 10 years (O'Brien et al. 2008; Michael Ross, personal communication). Fires rarely kill pines in the grass stage because the apical meristem is protected by decumbent needles that form a mantle and insulate the bud against the heat (Pausas 2015b); the bud also has fire-resistant scales (Fig. 3.11). If a longleaf pine seedling is top-killed by fire, it can often resprout from the root collar (Brockway et al. 2006).

During the grass stage the juvenile tree rapidly forms a taproot, which stores carbohydrate for subsequent upward growth. A typical one-year-old longleaf pine seedling in the grass stage has a taproot 60–70 cm long and several lateral roots 50–60 cm long in the upper soil layers (Wahlenberg 1946). Ultimately the taproot reaches at least 2 m in length, and the lateral roots extend up to 6 m horizontally from the taproot (Gresham et al. 1991). As one might expect based on depth to water table, the taproot is usually longer in deeper, drier soils. The grass stage terminates with the rapid stem elongation or "bolting" stage, in which the stem can grow more than 1 m in a single growing season. Thus, the juvenile tree can often escape death by fire even in habitats burned at one-to-two-year intervals.

The grass stage evolved independently in several clades of pines globally. In the Sierra Madre Occidental of northwestern Mexico, another high lightning-fire region, *Pinus devoniana* and *P. montezumae* have fire-resistant seedlings similar to those of longleaf pine (Richardson and Rundel 1998). *P. tropicalis* in Cuba also possesses a true grass stage, whereas other Caribbean pine species do not and are vulnerable to fire as seedlings and small saplings (O'Brien et al. 2008). This trait seems to be favored in moderately productive environments (e.g., savannas), perhaps because in more productive environments the higher fire intensities resulting from greater fuel loads would kill even these well-protected seedlings (Keeley and Zedler 1998). For some pines in other regions, the primary selective pressure for evolution of the grass stage appears to have been drought. Given the drought-prone climate of the southeastern Coastal Plain, rainfall limitation could be an additional selective pressure for the grass stage in longleaf pine (Freeman et al. 2017).

Renesting after Nest Destruction by Fire

A common life-history strategy of birds in fire-prone habitats is the ability to renest after a nest is destroyed by fire, another disturbance, or a predator. Virtually every bird species in the southeastern Coastal Plain associated with fire-prone ecosystems will renest up to several times after nest destruction by fire, often very soon after a fire and before much plant regrowth (Cox and Widener 2008). This behavior reduces the conflict between breeding in the spring and summer, when insects to feed the young are most abundant, and the peak of lightning-fire activity during this same season. Fire was likely one of the selective forces favoring renesting behavior.

ADAPTATIONS FOR EXPLOITING, EMBRACING, OR TAKING ADVANTAGE OF FIRE

We can assume that, in a fire-prone environment, most of the plants and animals native to the community either benefit from fire directly or indirectly or possess mechanisms to minimize or recover from tissue damage caused by fire. Species that tolerate, resist, or endure fire, as discussed above, are expected to have a competitive advantage over more fire-sensitive species, yet they are still vulnerable to injury or death from fire. Their functional traits allow them to survive or even prosper in high-fire environments despite the risks and energy expenditure trade-offs for thick bark, resprouting, or other defense or recovery mechanisms.

Farther along the spectrum of adaptation to fire are those species that exploit, embrace, or otherwise take advantage of fire. Such species can be characterized as directly fire-dependent, in contrast to the indirectly fire-dependent species or fire-neutral species discussed above. The adaptive traits of fire-dependent species are reviewed in this and the following main sections. Again, these categories are not absolute or mutually exclusive. A single species may have fire-adaptive traits in two or more of these general categories.

Prodigious Flowering after Fire

Flowering after fire is a trait common to many plants in fire-prone communities. The pioneer of fire ecology in south Florida, William Robertson, reported at the first Tall Timbers fire ecology conference that, in pine rocklands of the Everglades: "In areas unburned for as little as three or four years, most of the low pineland flora is subdued and inconspicuous. Within a few weeks after a late winter or spring burn, however, a tremendous show of bloom occurs in the pinewoods" (Robertson 1962).

Many plants flower most profusely and synchronously when burned during the natural lightning-fire season. Some flower and produce viable seed almost exclusively after fires in this season. Other species are able to take advantage of fires in almost any season, although they may delay flowering until the following year. Examples of species with often spectacular flowering after fire include *Aristida stricta* and *A. beyrichiana*, *Pityopsis spp.*, *Calopogon multiflorus*, *Nemastylis floridana*, *Calydorea coelestina*, and *Harperocallis flava*. *Calopogon multiflorus* will bloom as early as two weeks following fire, including dormant-season fires. More typically, this species flowers three to four weeks after fire, with the peak flowering response from mid-April to early May after March or April fires; however, it can flower in March or April after winter fires (Edwin Bridges, personal communication).

Some plants depart from their usual phenology and opportunistically flower after fire. For example, crow-poison (*Stenanthium densum*) has been observed flowering in late June after fire, long past its typical spring flowering season (Edwin Bridges, personal communication). Other plants may have no fixed flowering phenology, but simply respond to a fire when it occurs. A study of the federally Endangered chaffseed (*Schwalbea americana*) in Georgia found a flexible response to fire. Plants burned in March flowered in May, whereas plants burned in June flowered in late July and August (Kirkman et al. 1998a). Pine lilies (*Lilium catesbaei*) burned in winter or early spring bloom in the summer/fall, while those burned later in the growing season bloom the following summer/fall (Huffman and Werner 2000). Flowering soon after fire has several potential adaptive advantages, including use of fire-released nutrients (e.g., cations such as Ca+, Mg+, and K+, as well as organic nitrogen, ammonium and nitrate; Boring et al. 1990), conspicuous exposure to pollinators, and ability to set seed in an unshaded, fire-exposed seedbed.

Obligate Seeding, Seed Dormancy in Soil Seed Bank, and Fire-Stimulated Germination

A common fire-adaptation strategy is to persist for a period of time in a soil seed bank, then use the heat of fire or other fire cues to break seed dormancy and germinate. This strategy provides not only a way to tolerate or endure fire but also allows plants to exploit fire for recruitment. Whereas resprouting is an ancestral trait in plants, the obligate seeding strategy is a more recently evolved or derived trait (Pausas and Keeley 2014b). Soil seed banks develop when some of the seeds (or fruits containing seeds) that fall from plants become buried in the soil due to wind, rainstorms, the footsteps of animals, or other factors. The fate of these buried seeds is crucial to the persistence of populations.

Seed dormancy, germination, and seedling development are fundamental components of the "regeneration niche" of plants, which is key to the maintenance of species-rich communities (Grubb 1977). Seeds that are buried deep enough are protected from the heat of fire and remain dormant but viable in the soil for long periods of time, in some cases many decades or even for more than a century. Very shallow burial, on the other hand, often results in high mortality of seeds (Bekker et al. 1998). Persistence in a soil seed bank for extended periods of time usually depends on seed dormancy, although the relationship between dormancy and seed persistence is not consistent (Thompson et al. 2003).

Seed dormancy is a condition of delayed or inhibited germination. It has selective value when germination can be timed to coincide with the arrival of conditions suitable for seedling establishment. A fire, by clearing away accumulated live and dead plant biomass, enriching soil nutrients, and exposing bare soil to the sun, often provides such favorable conditions. Delaying germination has been described as a bet-hedging strategy, which spreads the risk of reproductive failure in variable environments (Venable 2007). Ecologically, the importance of a soil seed bank is that it allows a plant population to persist during long periods of unfavorable conditions, such as fire exclusion, and then to germinate in response to fire. In an experiment that simulated the heat of fire vs. hot summer temperatures for six obligate-seeding woody species from the Mediterranean Basin, Moreira and Pausas (2012) showed that heat treatments simulating fire broke physical seed dormancy in all populations of all six species. Simulated summer temperatures had little effect.

Long-lived soil seed banks play a critical role in maintaining genetic diversity of some species, especially obligate seeders or fire ephemerals, in communities with stochastic fire regimes (Bradbury et al. 2016). Moreira and Pausas (2012) note that recurrent fires will shorten the generation time of seeders, which enhances opportunities for natural selection to act on the population. Pausas and Keeley (2014b) likened obligate seeders to annual species with semelparous reproduction (i.e., flowering and fruiting just once before dying), whereas resprouters are analogous to the perennial life history with iteroparous reproduction (flowering and fruiting many times). They summarized the evolutionary advantages of the obligate seeding strategy: "The monopyric life cycle of obligate seeders precludes generational overlapping, which also contributes to increased genetic differentiation among populations and thus enhances evolutionary changes" (Pausas and Keeley 2014b). Obligate seeders are generally more resilient to water stress than resprouters, and they possess a number of anatomical and physiological traits that favor recruitment in dry, open sites (Pausas and Keeley 2014b).

Menges and Kohfeldt (1995) classified 98 species of scrub plants based on their fire-recovery mechanisms and patterns of abundance after fire events. Four recovery mechanisms were most prevalent: obligate seeders or sporers (27 percent), resprouters (25 percent), resprouters and seeders (25 percent), and resprouters and clonal-spreaders (14 percent). A smaller number of species were resprouters/clonal-spreaders/seeders, aerial seeders, or survivors/seeders. The obligate seeders included paper nailwort (*Paronychia chartacea*) and the signature species of rosemary scrub, Florida rosemary. Recovery by seedlings is favored in communities with bare soil patches (gaps), such as rosemary scrub, whereas in more productive habitats, such as scrubby flatwoods, resprouting and clonal spread should be favored because ramets (individual stems of a clone) are competitively superior to seedlings (Menges and Kohfeldt 1995).

Other obligate seeding Florida plants include some of the rare species studied or reviewed by Slapcincky et al. (2010): the previously discussed *Chamaecrista keyensis* and *Hypericum cumulicola*, as well as *Conradina brevifolia*, *Dicerandra frutescens*, *Eryngium cuneifolium*, and *Warea carteri*. Several species, though not obligate seeders, show increased seedling recruitment after fire: *Lechea deckertii*, *L. cernua*, and *Polygala lewtonii*. Species that both resprout and recruit from a seed bank include the sandhill plant *Conradina glabra* and the scrub plant *Eriogonum floridanum*. For apparently all these species, seedling establishment is facilitated by fire's effect of exposing bare mineral soil by removing litter, ground lichens, and potentially competing vegetation.

Besides stimulation of germination by the heat produced by fire (e.g., Moreira and Pausas 2012), other fire cues may prompt seeds to break dormancy. Smoke is well known for this effect and is required by some obligate-seeding plants. Chemicals in smoke, carried beneath the soil surface by rain, are hypothesized to inform seeds in the soil seed bank that environmental conditions are favorable for germination and growth. These chemicals have been identified as karrikins (Flematti et al. 2004) and have been shown to stimulate seed germination in many families of plants. Flematti et al. (2015) suggest that karrikins mimic an unidentified endogenous compound that regulates germination and seedling development. They further speculate that "seed plants could have 'discovered' karrikins during fire-prone times in the Cretaceous when angiosperms (flowering plants) were evolving rapidly" (Flematti et al. 2015). However, microbial activity in disturbed soils can also produce karrikins and stimulate plant germination (Bradshaw et al. 2011).

Somewhat surprisingly, only 3 of 20 plant species from scrub and other communities on the Lake Wales Ridge tested by Lindon and Menges (2008) showed a significant positive germination response to smoke: *Liatris chapmanii*, *Polygala lewtonii*, and *Abrus precatorius* (the last species, rosary pea, is a highly toxic nonnative invasive plant). Rachel King and Eric Menges (unpublished) tested an additional six species of Florida scrub plants for the effects of heat and smoke on seed germination. Smoke treatments stimulated germination for two endemics, *Chrysopsis highlandsensis* and *Eryngium cuneifolium*, but seeds of the other four species (*Hypericum cumulicola*, *Lechea cernua*, *Liatris tenuifolia*, and *Polygonella polygama*) were not stimulated. In contrast to the plants tested in the Mediterranean Basin (Moreira and Pausas 2012), neither a dry heat treatment nor a wet heat treatment stimulated germination in seeds of these Florida scrub plants. Dry heat reduced germination of some species (albeit not at soil temperatures typical for fires in scrub), and wet heat treatments resulted in almost no germination.

Charate, charred wood containing leachable chemicals, stimulates seed germination in some plant species. A study in California showed two species of chaparral herbs are stimulated by charate to germinate. One species, *Emmenanthe penduliflora*, shows a nearly obligate dependence on charate, whereas the other, *Eriophyllum confertiflorum*, is facultatively stimulated (Keeley and Nitzberg 1984). Seeds of the narrow endemic mint of Florida scrub, *Dicerandra christmanii*, also are stimulated to germinate by charate (Eric Menges, personal communication).

Serotiny and Canopy Seed Storage (Often Combined with Branch Retention)

> Canopy seed storage in woody plants is most closely associated with fire-prone habitats. No other disturbance/stress so effectively creates both a colonizing situation in otherwise stable woody vegetation and the cue for propagule release and germination.
>
> Lamont et al. (1991)

Serotiny has been defined as "the retention of seeds in the plant canopy for one to 30 years or more" (Lamont et al. 1991). Serotiny is associated with fire-prone, nutrient-poor, and seasonally dry vegetation subject to intense crown fires, especially in Australia and, to a lesser degree, in South Africa and North America (Lamont et al. 1991). The canopies of serotinous woody species store seeds until the heat of a fire (or in some cases warm days or the heat of the sun) stimulates cones or other seed-bearing structures to open and release their seeds. Heat melts the resins that bind the cone scales together. Serotinous pine cones can retain viable seeds for up to 30 years. Of 95 Northern Hemisphere pine species examined by Lamont et al. (1991), 22 were partially serotinous (polymorphic for serotiny) and another 6 species were obligately serotinous with seeds released only by fire. Being polymorphic for serotiny has the advantage of allowing for recruitment in the absence of disturbance or after disturbances other than fire.

In the southeastern Coastal Plain, only two species, sand pine (Fig. 3.12) and pond pine, are serotinous. A third serotinous species, pitch pine (*Pinus rigida*), occurs in the Appalachians and in the Coastal Plain from Delaware northward, but not in our study area. These pines are short in stature relative to our nonserotinous species. Being short increases the chance of heat from a fire reaching the canopy in cases where ladder fuels (retained branches) are not well developed or the stand is not particularly dense. Typically, however, our serotinous pines occur in denser vegetation

Figure 3.12. Serotinous cones of sand pine (*Pinus clausa*). An intense fire killed the adult tree, but heat from the fire stimulated the cones to open, dropping their seeds into a suitable fire-prepared seedbed of bare mineral soil. Photo by Reed Noss.

than our nonserotinous savanna/woodland pines, and the trees retain dead branches.

The seasonal climate of the southeastern Coastal Plain suits serotinous species well because the peak lightning-fire season is usually followed by heavy rains, which favor seedling establishment. A possible reason why serotiny is not more widespread here is that high lightning frequency has selected primarily for species adapted to frequent stand-maintaining fire rather than the less frequent stand-replacing fire that is most often associated with serotiny. The latter conditions are found, however, in scrub and pocosin communities, and to a limited extent in some wet flatwoods.

What specific conditions favor the evolution of serotiny over the alternative life-history strategies of soil seed storage or no storage of seeds? Lamont et al. (1991) evaluated several hypotheses and concluded that, unlike nonstorage of seeds, serotiny maximizes the number of seeds available for the next postfire generation. Unlike either nonstorage or soil-storage plant species, the synchrony of seed release in serotinous species might satiate seed predators so that some seeds survive (but see below). Seeds stored in the canopy also are better insulated from the heat of a fire than nonstored seeds and probably many soil-stored seeds, although in the latter case this depends on the depth of storage in soil. When seeds are released from serotinous cones, they encounter a postfire seedbed of bare mineral soil suitable for germination and development of seedlings.

Fire-return interval may be particularly influential in the evolution of serotiny as opposed to other seed-storage or nonstorage strategies. When fire intervals exceed lifespan, soil seed storage is favored over canopy storage and serotiny (Lamont et al. 1991). The proportion of closed cones (i.e., the level of serotiny) varies within and among populations in many species of pines, but generally increases with the frequency of crown fires (Pausas 2015a). When seed predators are highly abundant, however, there is selection against storing seeds in the canopy, and the level of serotiny declines (Talluto and Benkman 2013).

The serotiny trait is often correlated with branch retention or non-self-pruning of dead branches (Schwilk and Ackerly 2001). Whereas live branches (or most live fuels) act as a heat sink during fire until they've lost their moisture, dead branches are readily combustible. The ladder fuels of retained dead branches distributed up the bole of a serotinous pine carry fire into the canopy, killing the tree but stimulating the cones to open and release their seeds safely after the fire has passed.

Shifting Activity into Recently Burned Patches

Many species of birds and other vertebrates forage in recently burned areas, where exposed seeds and dead or injured invertebrates and vertebrates, as well as live animals, are available as food. Some species of birds actively shift their territories into recently burned areas. One such example my colleagues and I documented through radiotelemetry is the Florida grasshopper sparrow in the subtropical hyperseasonal grassland (dry prairie) at Kissimmee Prairie Preserve State Park (Fig. 3.13).

Three male Florida grasshopper sparrows with radio transmitters held adjacent territories in the 2006 breeding season in an area of grassland burned the preceding year. An unplanned fire, escaped from a nearby controlled burn, occurred in an adjacent area on 20 February 2007. Shortly after the fire, the three males established new territories in the freshly burned area, which suggests these sparrows prefer very recently burned habitat over habitat burned only two years earlier. No males exclusively occupied the two-year postburn rough where they held territories the previous year. That rough was then control-burned in August 2007. In the subsequent 2008 breeding season, males again occupied territories in this area as well as in the area burned in February 2007 (Noss et al. 2008). Frequent fire maintains conditions of low shrub cover and tree density, which

Figure 3.13. The endemic and highly specialized Florida grasshopper sparrow (*Ammodramus savannarum floridanus*) is potentially the most frequent fire-dependent bird in North America. It is not known to nest successfully in its dry prairie habitat left unburned for more than two years, and has been observed shifting territories into areas shortly after fire. Photo by Reed Noss.

probably reduces predation rates and maintains patches of bare soil that the birds use for foraging and runways. As noted earlier, some ground-nesting bird species nest in burned areas only a few days after fire (Cox and Windener 2008).

ADAPTATIONS FOR FACILITATING OR PROMOTING FIRE

The pinnacle of adaptation to fire is the ability of a plant or animal to facilitate or promote fire to its own advantage at the expense of potential competitors. This is an example of niche construction, where species possess traits that enable them to modify their environment in a way that promotes their fitness (Odling-Smee et al. 2003). Natural selection in the Darwinian sense acts primarily on genes and individuals, although at broader temporal and spatial scales, higher levels of biological organization, including species and even communities, can be favored or disfavored relative to others.

Flammable Tissues Such as Live or Dead Leaves

Ecological success of plants in fire-prone ecosystems is often tied to high flammability of aboveground tissues (Ripley et al. 2015). Flammability is enhanced by such traits as large leaf size, terpenes and other volatile oils in leaves, open fuel beds with low bulk density, low fuel moisture, retention of dead branches, and fine branching patterns (Rundel 1981; Schwilk 2003; Varner et al. 2015b).

Mutch (1970) proposed that "fire-dependent plant communities burn more readily than non-fire-dependent communities because natural selection has favored development of characteristics that make them more flammable." Unfortunately, Mutch's hypothesis was flawed on several counts, which critics jumped on, albeit not as quickly as might be expected. First, the hypothesis does not explain how a mutant genotype that is more flammable could spread through a population of less flammable conspecifics. Second, it does not consider the possibility that increased flammability could have evolved secondarily due to selection for other, more directly beneficial traits, such as defense against herbivores, tolerance of drought, or retention of nutrients. Third, Mutch's hypothesis was presented at the level of the community, whereas evolutionary theory has firmly established that individual selection trumps group or community selection. Finally, critics pointed out that flammability results from many different and probably unrelated traits that are likely affected by different genes (Snyder 1984; Troumbis and Trabaud 1989; Bond and Midgley 1995; Kerr et al. 1999).

Despite mechanistic problems with his hypothesis, Mutch (1970) was on to something, and his basic premise that flammability can increase fitness in fire-prone communities now appears correct. Alternative individual-based hypotheses for the evolution of flammability were offered by Bond and Midgley (1995), Kerr et al. (1999), and Gagnon et al. (2010), among others. Flammability can be favored by natural selection in fire-prone environments because flammable plants can increase their fitness by killing less flammable or less fire-tolerant neighbors and opening space for recruitment of their offspring (the "kill thy neighbor" hypothesis; Bond and Midgley 1995).

Interactions are expected between flammability traits and seedling fitness in the postfire environment, such that in cases where high flammability has evolved, strong selection should exist for regeneration traits that respond to the increased frequency or intensity of fire (Schwilk and

Ackerly 2001). Thus, flammability is most likely to evolve when fire is propagated to neighboring conspecifics in dense populations and when traits that promote flammability confer other fitness benefits (Bond and Midgley 1995). If a species that has evolved flammability contributes strongly to fuel loads in the community, this would then "result in the selective exclusion or admission of other species . . . depending on the compatibility of their pre-existing traits with fire" (Bond and Midgley 1995)—that is, an environmental filter effect.

A species that has become flammable through natural selection has created a new niche for itself, a textbook example of niche construction (Schwilk 2003). This niche, intriguingly, may involve more than killing one's neighbors. Gagnon et al. (2010) raised the titillating possibility that flammable individuals benefit from apparent self-immolation because it protects their belowground organs and nearby propagules. Damage to critical belowground organs and propagules is proportional to soil heating, which in turn depends on the residence times of fires and on fuel location relative to the soil. Given the fine-scale heterogeneity of fires, individual plants can control local soil heating through their fuels. Highly flammable plants gain fitness advantages by burning rapidly and keeping their fuels off the soil surface. This "pyrogenicity as protection" hypothesis (Gagnon et al. 2010) has not yet been explicitly tested, to my knowledge, but it seems highly relevant to rapidly regrowing/resprouting plants of grasslands/savannas or marshes. Gagnon et al. (2010) propose it also applies to seeders and to savanna trees; the latter would benefit from highly flammable litter that does not accumulate long enough to fuel severe fires that could kill the trees. There may be no "self-killing cost" (Schwilk and Kerr 2002) to flammability after all.

C_4 grasses in general are highly flammable, and they are abundant in southeastern grasslands/savannas. The pyrogenicity of wiregrass is legendary, yet no study had quantified its fuel structure and flammability at an individual plant level until recently. Fill et al. (2016) conducted field flammability experiments in a longleaf pine savanna in South Carolina. They determined that flame length increased significantly with increasing biomass of wiregrass. Pine needles and other litter perched on wiregrass tussocks did not affect flame length, but they increased the duration of flaming and smoldering. Surprisingly, wiregrass tussock flammability was not affected by relative humidity, wind speed, air temperature, or even fuel moisture (Fill et al. 2016). The authors suggested that local modification of fire behavior by wiregrass likely scales up to effects on ecosystem dynamics.

Dead resinous needles from longleaf pines dropped into a matrix of wiregrass produce an extraordinarily combustible fuel bed. In an experimental manipulation of litter, shed longleaf pine needles increased fire temperatures and durations of heating more than herbaceous fuels alone (Ellair and Platt 2013). In contrast, leaves of mockernut hickory (*Carya tomentosa*) had no effects on fire characteristics. In the absence of pine needles, most hickories resprouted from aboveground epicormics buds along the stem. When pine needles were present, however, resprouting occurred only from underground root crowns. Hardwoods that persist as resprouting shrubs in pine savannas may get an opportunity to escape the fire trap and grow into the canopy after the death of large pines in their vicinity. The authors reasoned that fine-scale heterogeneity in flammable fuels produced by pyrogenic species controls the size, stand dynamics, and spatial distribution of other trees in pine savanna landscapes (Ellair and Platt 2013).

A follow-up fuel-manipulation experiment showed that longleaf pine needles elevated maximum temperature, duration of heating above 60°C, and consumption of fine fuels considerably beyond the effects of fuels from other plants (Platt et al. 2016). The presence of pine needles reduced the numbers of ground-layer oak stems and grass culms that survived fire and, curiously, depressed postfire flowering of grasses. A structural equation model confirmed and bolstered the results of the field experiment, demonstrating "strong direct and indirect pathways from pine needles to postfire responses of oaks and grasses." Thus, the promotion of frequent fires by patchily distributed resinous pine needles helps maintain humid savannas and their characteristic heterogeneity and high species richness (Platt et al. 2016).

4

Fire in Natural Communities of Florida and the Southeastern Coastal Plain

> In the vegetation of the Southeast, effects of fire are more striking than elsewhere on this continent, and, with few exceptions, more striking than in any other region of the world.
>
> Garren (1943)

Although species respond to their environment in an individualistic manner (Gleason 1926), species with similar sets of tolerances to environmental factors tend to cluster together. Predictable assemblages of species often recur across a regional landscape on sites with similar soil, topography, hydrology, microclimate, and fire regime. For this reason, communities can be classified, described, and mapped with considerable confidence that we are dealing with something more or less real, as opposed to pure abstraction.

The term "natural community" refers to an assemblage of plants and animals and their associated physical habitat and processes—what could also be called an "ecosystem." Natural communities have been classified by state natural heritage programs for most states in the United States beginning in the 1970s, when the natural heritage network was initiated by The Nature Conservancy. The Florida Natural Areas Inventory, which is Florida's heritage program, recognizes natural communities based on a combination of vegetation, landscape position, substrate, and hydrology (FNAI 2010). *The Natural Communities of Georgia* (Edwards et al. 2013) defines a natural community as "an assemblage of native plant and animal species, considered together with the physical environment and associated ecological processes, which usually recurs on the landscape." The Georgia classification of natural communities is based on "ecological systems" (Comer et al. 2003), a typology linked to the National Vegetation

Classification System developed by several cooperating organizations, including federal agencies, The Nature Conservancy, NatureServe, and the Ecological Society of America (http://usnvc.org/).

Natural communities usually are recognizable in the field by competent naturalists, but they often grade into one another. Within grassland/savanna landscapes in the lower southeastern Coastal Plain, environmental gradients are often subtle, to the point that the distinction between "upland" and "wetland" can become virtually meaningless (Noss 2013). A few inches of elevation can make a big difference for species assemblages close to the water table. Plant species composition shows significant turnover (beta diversity) along subtle elevation gradients and associated variation in soil moisture in these landscapes. For example, Orzell and Bridges (2006a) recognize six community types within the Florida dry prairie (subtropical hyperseasonal grassland) landscape: dry-mesic, mesic, wet-mesic spodic, wet-mesic alfic, acidic wet, and calcareous wet prairies. Some of these types would typically be considered upland and others wetland, but they occur together in a complex mosaic on these flat, poorly drained landscapes. To an untrained eye, these various communities look pretty much the same.

In contrast, with a natural fire regime the boundary between dry prairie and forest is typically abrupt, as are the boundaries between high pine (sandhill) and scrub (Myers 1990), pine rockland and rockland hammock (FNAI 2010), and pine flatwoods and various hardwood communities. Such sharp, discontinuous boundaries are typical of savannas worldwide and reflect thresholds in canopy cover vs. grass cover and associated fire-vegetation feedbacks (Staver and Koerner 2015). Distinctly different fire-vegetation feedback systems maintain two alternative stable states closely juxtaposed. Ecotones become less sharp, however, if fire is excluded and forest species invade the grassland/savanna.

When viewed from above, as from an airplane or on Google Earth, communities occur in mosaics. In terms of the ways they appear on a landscape, natural communities can be described as: (1) matrix—the dominant and most extensive community, often pine savanna in the presettlement southeastern Coastal Plain; (2) large patch—typically 125–5,000 acres, such as large floodplain swamps; (3) small patch—typically 1–125 acres, such as depression wetlands; (4) linear—narrow, elongate strips, for example, riparian corridors. Not mutually exclusive to the above categories, communities also can be recognized as: (5) embedded—completely surrounded by another natural community, such as a lake or depression wetland within a pine savanna; (6) interfingered—partially extending into another natural community, for example, headwater woody riparian corridors extending into grassland; (7) transitional to or intergraded with—a broad ecotone with gradual change from one community to another (Comer et al. 2003; Edwards et al. 2013). Natural communities have characteristic sizes, shapes, and patterns of juxtaposition in landscapes, which influence their fire ecology.

Overlaid on the relatively fixed (in a human time frame) pattern of natural communities of various shapes and sizes determined by topography, geology, and other slowly changing abiotic variables is a more rapidly dynamic and spatially shifting mosaic determined by processes of disturbance and recovery. For example, a large fire event creates a more or less heterogeneous patchwork of habitats corresponding to burn severity, from unburned to severely burned. This mosaic changes character as vegetation recovers and regrows after fire. Many human disturbances or land uses, such as intensive agriculture or urban development, do not allow for such recovery.

This chapter summarizes the fire ecology of Florida's terrestrial and wetland natural communities, ranging from those that burn frequently (e.g., sandhill, flatwoods, dry prairie, pine rockland) to those that burn infrequently (e.g., some scrub communities, hardwood forests, floodplain swamp, baygall). I do not attempt a detailed community-by-community account. The historical development of vegetation in the region was reviewed in chapter 2, whereas components of fire regimes and adaptations of species to fire regimes were discussed in chapter 3. More detailed information on natural communities can be found in the *Guide to the Natural Communities of Florida* (FNAI 2010) and comparable texts for other states with recently updated classifications and community descriptions, especially *The Natural Communities of Georgia* (Edwards et al. 2013) and *Guide to the Natural Communities of North Carolina: Fourth Approximation* (Schafale 2012). More comprehensive information on Florida's ecosystems, though somewhat out of date, is in *Ecosystems of Florida* (Myers and Ewel 1990) and in the primary literature (scientific journals).

Types of Fire Regimes

Types of fire regimes are recognized on the basis of combinations of components such as frequency and severity. Fire regimes have been categorized in various ways, with different classifications best suited to different

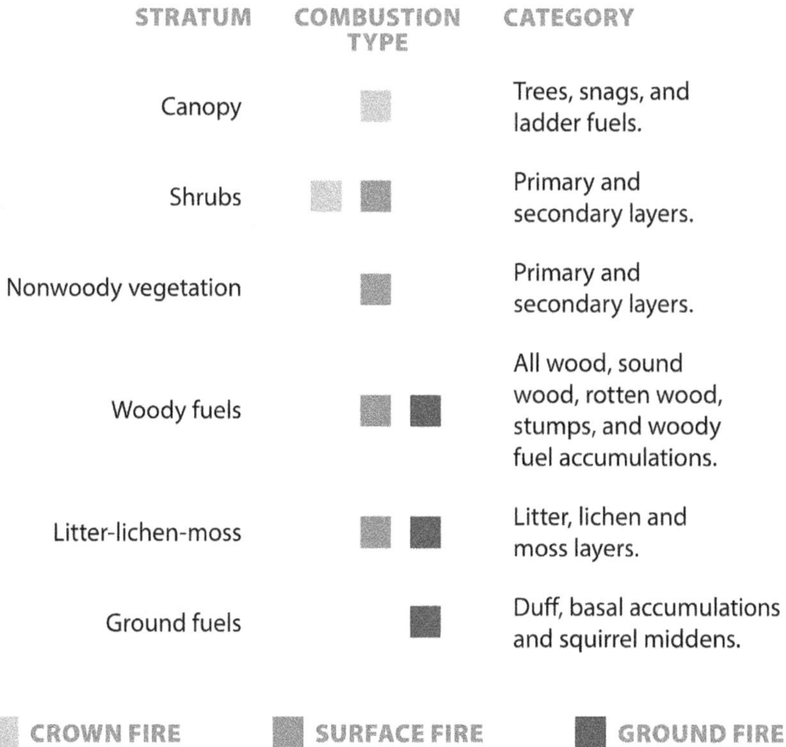

Figure 4.1. Fuel-bed strata and their involvement in different types of combustion. Based on information in Ottmar et al. (2007).

regional vegetation types or specific research objectives. Categorization of phenomena as complex and variable as fire regimes is always somewhat arbitrary, but helps us recognize common patterns and communicate knowledge about them. Davis (1959) and Scott (2000) recognize three categories of global fire regimes based on the types of fuels and the strata the fires burn (Fig. 4.1): *Surface fires* burn accumulated litter as well as smaller-statured living herbaceous plants and shrubs. *Crown fires* burn the crowns of living trees or shrubs. *Ground fires* burn organic-rich soil layers, including peats. (Unfortunately, some authors use "ground fire" as synonymous with "surface fire," although these are very different regimes.) These categories are generally suitable for application to the fire regimes of our region.

Fire ecologists, especially in western North America, often sort fire-prone forest types into three characteristic fire regimes: *high severity*, *low severity*, and *mixed severity* (Noss et al. 2006; DellaSala and Hanson 2015).

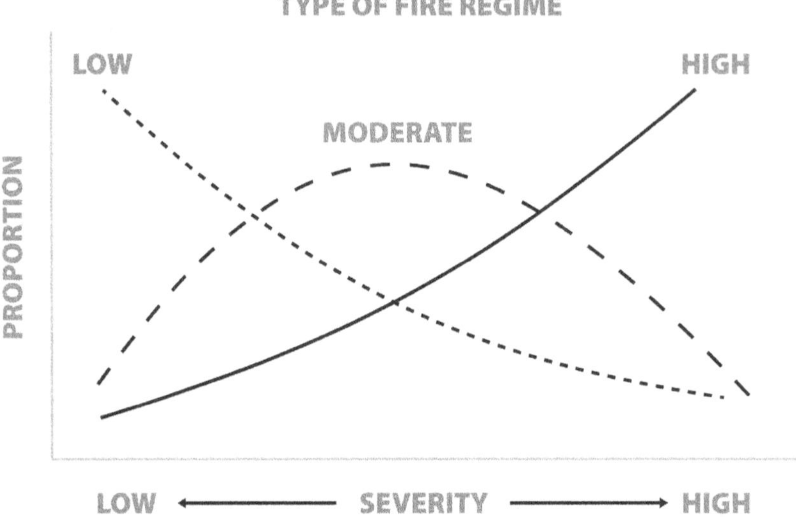

Figure 4.2. A fire regime is not characterized by a single kind of fire. The relative frequency of low-, moderate-, and high-severity fire in an average burn varies along a continuum. Adapted from Agee (1993).

In Florida and the Coastal Plain under natural conditions, high-severity fire regimes are restricted to few community types, such as scrub, while our variable drought conditions and complex fire histories produce a mix of fire severities for each natural community type. A mixture of severities is expected for any general class of fire regime (Fig. 4.2). What we classify as a high-severity fire typically will have patches within it that are unburned or only moderately burned, and a low-severity fire will often have some very hot patches ("hotspots") due to non-uniform distribution of fuels, such as fallen pine needles (see chapter 3). Referring to Figure 4.2, Hutto et al. (2016) noted that "change from one fire regime to the next (movement along the x-axis) is accompanied not by the sudden appearance of a different fire severity, but by continuous changes in the proportions of each fire severity category. Thus, fire regimes blend imperceptibly into one another."

An alternative four-type classification of fire regimes has heuristic value. This synthetic classification corresponds roughly to that of Davis (1959) and Scott (2000), but with names that better reflect the ecological effects of fire on vegetation (Table 4.1). Admittedly, it is tree-centric terminology borrowed from forestry. *Stand-maintaining fires* are usually

Table 4.1. Four ecological fire regimes, named for the effects of fire on dominant vegetation and other ecosystem components

Type of Fire Regime	Characteristics	Comparisons with Other Classifications
Stand-maintaining Fire	Typically frequent and relatively low-intensity (often <400°C) surface fires. Consume live and dead biomass of grasses, forbs, shrubs (topkill), plus duff, small logs (down woody debris), and some roots and bark of trees. Usually cause little mortality of mature thick-barked trees but often kill thin-barked fire-sensitive trees. With longer time-since-fire, accumulation of duff and other biomass leads to higher intensity and severity of fire, often causing substantial mortality of even mature thick-barked trees. Such fires can converge with the stand-replacing fire category.	Essentially equivalent to surface fire and low-severity fire regimes (or occasionally to mixed-severity fire regime or, after unnatural fuel accumulation, to high-severity fire regime).
Stand-replacing Fire	Less frequent fire, but more intense (800–900°C to ca. 1,200°C with abundant dry fuel and wind) and severe, typically with complete or major topkill or total kill of canopy vegetation (trees or, especially in the case of Florida scrub, shrubs). Fires that begin as surface fires may spread up the trunks of trees, burning branches ("ladder fuels") along the way and spreading into the crowns of forest or woodland trees. Burning embers ("firebrands") may fly from the canopy and ignite new fires, a phenomenon known as "spotting."	Essentially equivalent to crown fire and high-severity or mixed-severity fire regimes.
Smoldering Fire	Occur primarily in wetlands when there is a sufficiently dry and thick humic (organic) layer; the limited oxygen supply causes the fire to smolder. Such ground fires may burn thick peat layers when the water table has been lowered, often burning very slowly at a high temperature (ca. 900°C) with nearly complete combustion and conversion of biomass to CO_2. These fires can smolder for many months; hence fire can be "stored" in the landscape for long periods and emerge as a significant fire event long after the original ignition. When the water table returns to its original (higher) level after a peat fire, a lake may form. Smoldering fires may also occur in pine duff or from fire stored in the heartwood and sapwood of the roots of pine stumps.	Essentially equivalent to ground-fire or peat-fire regimes. Flare-ups can converge with high-severity fire regime.
Rare or Infrequent Fire	Sites that are topographically protected from fire such as islands, peninsulas, steep slopes, ravines, sinkholes and grottos, and low inundated floodplains (swamps) may burn infrequently to virtually never, except along boundaries with pyrogenic communities. The low inherent flammability of plants that dominate fire-protected sites contributes to their low probability of fire.	No classified fire regime, but when fires occur they are usually of low or mixed severity. Under extreme drought, some wetlands that normally do not burn could experience peat (ground) fires or high-severity fires.

Information from multiple sources.

low-severity surface fires, which, if they occur frequently enough, maintain a stand in a characteristic herbaceous or open-canopy savanna or woodland condition. *Stand-replacing fires* are fires that burn through the canopy or "crown" of the vegetation, with the crown being either a tree layer or a shrub layer. These fires topkill or complete-kill many or most of the canopy trees and shrubs, which then must be "replaced" by resprouting, germination of seeds from the soil seed bank, reseeding from a canopy seed bank (from serotinous cones), or dispersal from outside the stand.

Stand-replacing fires are mostly high-severity but often (e.g., for scrub) have shorter fire-return intervals than is typical for high-severity forest fires in the western United States, where intervals are often on the order of centuries (Noss et al. 2006). Florida scrub fires are more similar to chaparral fires in California. Some communities in the stand-replacing fire category in our region, however, such as interiors of bald cypress (*Taxodium distichum*) or Atlantic white cedar (*Chamaecyparis thyoides*) swamps, may burn at centuries-long intervals. *Smoldering fires* are ground fires that burn organic soils, often over an extended period. These fires typically occur in wetlands but also in pine duff. A fourth category, *rare or infrequent fire*, comprises natural communities that uncommonly burn under natural conditions, except along their edges or more completely during extreme drought (Table 4.1). This fire regime overlaps with the stand-replacing fire regime.

Probably the single most important component of a fire regime, especially in the southeastern Coastal Plain, is frequency, also described as fire rotation period, fire-return interval (FRI), or mean fire interval (MFI). Time-since-fire is sometimes used to approximate FRI when the longer-term fire history of a site is unknown, but it is an imperfect substitute. The FRI has a profound influence on vertical structure, species composition, and species richness, both within and among communities.

For each natural community type listed in Table 4.2, the FRI is an estimate of the modal range of fire intervals reported in the literature or summarized by FNAI (2010). A given natural community often shows a considerable range of FRI, depending on geographic location, position in the landscape, distance from the coast or other water bodies, topography, details of species composition, fire and land-use history, and chance. Most natural fires are mosaics of fire frequency at one spatial extent or another. Hence, a community with an FRI of one to three years, for example, will usually have some patches that escape fire for longer periods. These less frequently burned patches serve as refugia for species that are sensitive to fire or otherwise benefit from longer FRIs (see discussion of fire heterogeneity in chapter 3). Not all numbers within an FRI range are equally probable; natural fire regimes in our region are often weighted toward the lower end of the FRI range.

Some level of variability in fire frequencies among sites representing a natural community is both natural and desirable, because variability in FRI will tend to enhance among-site (beta) diversity. From a management perspective, aiming toward a characteristic range of variability in all components of a fire regime, but weighted toward the most common portion of that range (e.g., May–July fires for the seasonality component), may be the most defensible strategy biologically (see chapter 5).

Below I review natural communities in Florida, and to a lesser extent across the southeastern Coastal Plain, grouped by their characteristic fire regimes. "Characteristic" refers to "the dominant natural disturbance regime and associated structural characteristics" for a particular natural community or site historically, before significant disruption by humans (Noss et al. 2006). Conversely, "uncharacteristic" refers to disturbances, vegetation structure, or fuel loads outside the historic range of variability. Ideally, the determination of characteristic vs. uncharacteristic will be based on site-specific reconstructions of vegetation and fire history, but in most real cases it must rely on literature surveys and the professional judgment of ecologists. Characteristic should not imply uniformity, because substantial variability occurs within natural community types in fire regimes, species composition, and vegetation structure.

Communities Characterized by Stand-Maintaining Fire

More natural communities in Florida and the southeastern Coastal Plain are characterized by stand-maintaining surface fire than by any other type of fire regime (Table 4.2). The large number of communities in this category reflects the high lightning frequency, high flammability of plants, and generally large fire compartments of the landscapes of this region before anthropogenic fragmentation. Stand-maintaining fires are typically of high frequency and toward the low end of the range of fire intensity and severity.

Table 4.2. The upland, wetland, and estuarine natural communities of Florida (adapted from FNAI 2010), grouped into four categories of fire regime

Stand-maintaining Fire Regime (Surface Fire) (Low-severity or mixed-severity fire)	Stand-replacing Fire Regime (Crown Fire) (High-severity or mixed severity fire)
Upland Mixed Woodland (2–10 FRI)	Scrub (Oak Scrub) (5–20 FRI, 5–12 FRI for yellow sand scrub, 3–20 FRI for Q. *myrtifolia* dominated scrub on Merritt Island)
Upland Pine (1–3 FRI)	
Sandhill (1–3 to 3–5 FRI)	
Wet Flatwoods (1–3, 5–7, to 5–10 FRI)	
Mesic Flatwoods (1–3 to 2–4 FRI)	Rosemary Scrub (peninsula) (10–40, 15–30, or 20–60 FRI)
Scrubby Flatwoods (in part) (5–15 FRI)	Sand Pine Scrub (10–70 FRI in peninsula, often much longer, to 100 or more FRI on Panhandle coast)
Pine Rockland (3–7 FRI)	
Dry Prairie (1–2 FRI)	
Palm savanna (2–3 FRI?)	Scrubby Flatwoods (in part) (5–15 FRI)
Seepage Slope (1–3 FRI)	Shrub Bog/Pocosin (in part) (10–20 FRI)
Cutthroat Seep and Cutthroat Lawn (1–3 FRI)	Atlantic white cedar swamp (50–300 FRI)
Pitcher Plant Prairie/Bog (2–3 FRI)	Strand Swamp (100–300 FRI, occasionally much longer)
Wet Prairie and Calcareous Wet Prairie (2–3 FRI)	
Marl Prairie (2–10 FRI)	
Dwarf Cypress Savanna (ca. 10 FRI)	
Depression Marsh (1–5 FRI?)	
Basin Marsh (5–7 FRI?)	
Floodplain Marsh (ca. 3 FRI?)	
Slough Marsh (1–5 FRI?)	
Glades Marsh (3–10 FRI)	
Canebrake (ca. 1–3, 4–6, 7–12, or 13–25 FRI, depending on site conditions)	
Salt Marsh (in part) (FRI unknown)	
Dome Swamp (Cypress Dome or Gum Pond) (3–5 FRI in outer zones, 100–150 in center)	
Basin Swamp (3–5 FRI in outer zones, 100–150 in center)	
Stringer Swamp (1–3 FRI)	
Basin Swamp (ca. 3–5 FRI in outer zones when in pyrogenic matrix?)	
Strand Swamp (ca. 3–5 FRI in outer zones when in pyrogenic matrix?)	
Hydric Hammock (edges in pyrogenic matrix, FRI unknown)	
Buttonwood Savanna ("Swamp") (FRI unknown)	
Coastal Strand (FRI unknown)	
Upland Glade (FRI unknown, dependent on landscape context)	

Note that some natural community types can be characterized by more than one type of fire regime, depending largely on landscape context, weather conditions, and the fire and management history of the site. In parentheses are estimated natural fire-return intervals (FRI) in years, where known. Recent human activity has generally lengthened FRIs, though some kinds of fires (e.g., peat fires—smoldering fire regime) apparently have become more common due to drainage of wetlands. Information from Myers and Ewel (1990), Frost (1995), FNAI (2010), and other sources.

Smoldering Fire Regime (Ground Fire) (Mixed-severity fire)	Rare or Infrequent Fire Regime
Mesic Hammock (rarely)	Slope Forest
Shrub Bog (in part) (10–20 FRI?)	Upland Hardwood Forest
Depression Marsh (peaty central portion) (FRI unknown)	Mesic Hammock
Basin Marsh (FRI for peat fires unknown)	Rockland Hammock
Slough Marsh (FRI for peat fires unknown)	Xeric Hammock
Glades Marsh (FRI for peat fires unknown)	Rosemary Scrub (Panhandle coast)
Dome Swamp (FRI for peat fires in interior ca. 100–150)	Beach Dune
Basin Swamp (FRI for peat fires in interior decades to centuries, longest when in nonpyrogenic matrix)	Coastal Berm
Strand Swamp (FRI for peat fires in interior unknown, probably centuries)	Coastal Grassland
Baygall (Bayhead) (FRI a few times per century)	Coastal Strand
Salt Marsh (FRI for peat fires unknown)	Maritime Hammock
	Shell Mound
	Sinkhole
	Limestone Outcrop
	Keys Cactus Barren
	Coastal Interdunal Swale
	Slough
	Basin Swamp (in nonpyrogenic matrix)
	Strand Swamp (interior)
	Floodplain Swamp
	Hydric Hammock (centers)
	Bottomland Forest (gaps with canebrakes had ca. 4–25 FRI)
	Alluvial Forest (gaps with canebrakes had ca. 4–25 FRI)
	Salt Flat
	Mangrove Swamp
	Keys Tidal Rock Barren

SAVANNAS, WOODLANDS, AND GRASSLANDS

> The ground vegetation of this type of pine forest is essentially one of true grassland. Such communities would more properly be called the pine grasslands, as the forest itself is not the most pertinent determinant of this community.
>
> Komarek (1965)

Included in this category are several natural communities in the FNAI (2010) classification: sandhill, upland pine, upland mixed woodland, mesic flatwoods, wet flatwoods, pine rockland, and dry prairie (Table 4.2). I also include a few other types recognized by botanists but not in the FNAI classification—for example, palm savanna (Harper 1927; Bridges 2006b), dwarf cypress savanna (Ewel 1990), and canebrake (Platt and Brantley 1997), which are essentially all grasslands (Noss 2013), plus some other communities considered wetlands by FNAI (2010). I treat scrubby flatwoods within the scrub category with a stand-replacing fire regime. Although scrubby flatwoods are intermediate in many respects between mesic flatwoods and scrub, they are more similar to oak scrub in species composition and fire regime, with the important distinction that the pine overstory is normally maintained, not replaced.

Pine Savannas, Palm Savannas, and Pine Rocklands

I use the term "pine savanna" largely in reference to sandhill (inclusive of clayhill and high pine) and flatwoods, as both are dominated most often by longleaf pine (*Pinus palustris*) and usually wiregrass (*Aristida beyrichiana* or *A. stricta*) over most of the southeastern Coastal Plain and have an open-canopy savanna-like physiognomy (Platt 1999). Longleaf pine savannas covered an estimated two-thirds of this region prior to European settlement (Wahlenberg 1946; Hoctor et al. 2006). Savannas are treed grasslands and are most commonly the matrix vegetation of uplands in this region. In some cases, they occur as large patch communities within another vegetation type—for example, within scrub in the Ocala National Forest in north-central Florida and within pocosin in the Green Swamp of southeastern North Carolina. Fire frequency controls species composition and richness within these communities (see chapter 3). For example, the frequent fires in most pine flatwoods of north and central Florida may prevent slash pine (*P. elliottii*) seedlings from invading longleaf-dominated stands (Laessle 1942). Woodlands have higher tree density than savannas and are often ecotonal between savanna (especially sandhill) and forest.

One type of savanna not recognized by FNAI (2010) is the palm savanna described by Harper (1927) for the Indian Prairie region west of the Kissimmee River in southeastern Highlands County and eastern Glades County. Bridges (2006b) notes from Harper's species list and photographs that palm savanna is composed of more wetland species, such as bushy bluestem (*Andropogon glomeratus*), sand cordgrass (*Spartina bakeri*), and sawgrass (*Cladium jamaicense*), than dry prairie, under a scattered canopy of cabbage palm (*Sabal palmetto*), live oak (*Quercus virginiana*), and South Florida slash pine (*Pinus densa*). The FRI for palm savanna is unknown, but is likely slightly to moderately longer than for dry prairie.

Pine rockland also can be considered a savanna. This is a globally critically imperiled (G1) community confined to south Florida on the Miami Rock Ridge, a portion of the Big Cypress region, and the Florida Keys. This is a small-to-large patch community occurring in a mosaic with marl prairie and rockland hammock within the wetland landscape of the Everglades–Big Cypress region, but it is the matrix vegetation on some of the Lower Keys—for example, Big Pine Key. Pine rockland is dominated by South Florida slash pine with a "patchy understory of tropical and temperate shrubs and palms and a rich herbaceous layer of mostly perennial species including numerous species endemic to south Florida" (FNAI 2010). With an FRI of around 3–7 years, it is subject to invasion by hardwood species, softening of the boundary with adjacent rockland hammock (a tropical hardwood forest), and ultimately succession to rockland hammock when fire is excluded for 15–25 years (Wade et al. 1980; FNAI 2010). A pine rockland in good condition can burn with a flame length higher than 1 m, yet except under extreme drought conditions such a fire will extinguish within seconds of reaching a hammock edge (Snyder et al. 1990), thus maintaining the characteristic sharp boundary between the two communities.

Some ecologists (e.g., Harper 1927; Myers 1990) prefer the term "high pine" in reference to sandhill and other well-drained pine savanna or woodland communities. Sandhill and clayhill are two communities of high pine distinguished by their soils (i.e., high content of sand vs. clay) and plant species that respond to this soil difference. High pine and scrub are often treated together in texts because they occur on similar well-drained soils in the same relatively hilly landscapes, whereas flatwoods occur on poorly drained soils on very flat topography (Abrahamson and Hartnett 1990; Myers 1990). Flatwoods, dry prairie, wet prairie, palm savanna, and related communities characteristically (but not entirely) occur on spodosols, which often have a "hardpan" or spodic subsoil layer of organics underlain by clay. This hardpan is up to 1.5 m below the surface and, being relatively impermeable to water, aggravates the already poor drainage of flatwoods (Abrahamson and Hartnett 1990). Shallow inundation during the wet season is common for days to occasionally weeks at a time.

A vegetation classification for pinelands inclusive of treeless dry and wet prairies of the Panhandle and northern and central peninsular Florida recognizes 16 distinct community types, grouped into five ecological series: two xeric sandy upland types, two subxeric sandy upland types, two silty upland types, three flatwoods types, and seven wetland types (Carr et al. 2010). This classification was performed by ordination and cluster analysis of field data, and the types are recognizable in the field. All of these communities are dependent on frequent stand-maintaining fire.

Treeless Grasslands and Dwarf Cypress Savanna

Some grasslands in the southeastern Coastal Plain resemble prairies. Florida dry prairie (subtropical hyperseasonal grassland; Fig. 4.3) is virtually treeless under a natural fire regime with an FRI of one to two years, except for widely scattered cabbage palms (*Sabal palmetto*), which provide nest sites for crested caracara (*Caracara cheriway*). This landscape was described by Harper (1927) as having "views strongly suggestive of the Great Plains." Dry prairie is the matrix vegetation of much of south-central Florida under natural conditions, extending from the Kissimmee River valley north of Lake Okeechobee to areas west of the lake as far as Myakka River State Park in Sarasota County (Bridges 2006a; Orzell and Bridges 2006b). Other virtually treeless grasslands in Florida include seepage slope, cutthroat seep, cutthroat lawn, wet prairie, and marl prairie (Table 4.1). FNAI (2010) classifies all of these communities as wetlands, but this designation is misleading.

In contrast to marsh communities, the wetter portions of seepage slopes may be saturated, but are not inundated, and the dominant graminoids include wiregrass as well as cutthroat grass (*Coleataenia abscissa*). The FRI is one to three years, like mesic flatwoods. Cutthroat grass can form a virtually monotypic groundcover in some wet flatwoods on the Lake Wales Ridge (Fig. 4.4), an exception to the concept of flatwoods as herbaceous biodiversity hotspots.

Wet prairie is found on acidic, nutrient-deficient, saturated soils, and is dominated by wiregrass in its drier portions. Wet prairies occur on gentle slopes between depression marshes or other wetlands on the lower end of

Figure 4.3. Subtropical hyperseasonal grassland, better known as dry prairie, at Kissimmee Prairie Preserve State Park. This vast grassland once dominated much of south-central and parts of southwestern Florida. With a natural fire-return interval of one to two years, it is one of most frequent-fire communities globally. Photo by Reed Noss.

Figure 4.4. A virtually monotypic groundcover of cutthroat grass (*Coleataenia abscissa*) in a South Florida slash pine (*Pinus densa*) wet flatwoods, Archbold Biological Station. This photo was taken in late June, a couple weeks after a lightning-season fire. Such low species richness is extremely rare in southeastern pine savannas. Photo by Reed Noss.

the elevation gradient, and with wet or mesic flatwoods, or dry prairie, on the higher end. In contrast, marl prairie and dwarf cypress savanna (which is marl prairie with scattered, stunted pond cypress [*Taxodium ascendens*]) are inundated for a portion of each year, except very rarely during extreme drought. Soils are shallow marls or sandy marls underlain by limestone, with little or no peat development.

The hydroperiod of marl prairie is two-to-four months in most years, with standing water to a shallow depth averaging about 20 cm; the FRI is 2–10 years (FNAI 2010). The Endangered Cape Sable seaside sparrow (*Ammodramus maritimus mirabilis*), endemic to the Everglades, nests in marl prairie but cannot tolerate FRIs shorter than three years. Local populations of this sparrow have been extinguished by fires (Curnutt et al. 1998; Lockwood et al. 2003). Dwarf cypress savanna presumably has an FRI near the high end of the range for marl prairie.

In contrast to marl prairie and dwarf cypress savanna, true wetlands such as depression marshes or dome and strand swamps have considerably longer hydroperiods, more than half of a typical year in their deeper portions. They show significant development of peat, which during drought can experience smoldering fire. Importantly, both wet and marl prairies have levels of plant species richness an order of magnitude higher than marshes and other true wetlands in the region (FNAI 2010).

Soil Fertility

Fire in savanna and grassland communities in this region is frequent for reasons given earlier but also because the sparse canopies lead to high soil albedo and rapid drying of fine fuels (Keeley and Zedler 1998). Low soil fertility also may partly explain why tree densities are low in savannas. Although a fire-vegetation feedback loop is the predominant factor that allows the persistence of savanna as opposed to closed forest in regions with high precipitation (Staver et al. 2011), low soil productivity also has an influence, often as mediated by fire.

A literature review of tropical savannas shows that nutrients interact with fire to inhibit the invasion of savannas by forest trees (Pellegrini 2016). Fire volatilizes carbon (C) and nitrogen (N) beyond the ability of most plants to compensate quickly for the losses. Fire also pyro-mineralizes phosphorous (P) in plants and soil; some of this P is recycled by plants, but much is lost through leaching or wind transport of ash (Pellegrini et al. 2015). As pointed out in chapter 1, however, most of the nutrients volatilized by a typical fire are recaptured by the ecosystem through rainfall and redeposition of ash (Scott et al. 2014). Still, nutrient limitations persist.

Savanna plants are generally well adapted to nutrient limitation, whereas most forest trees are not. Fast-growing forest species, in particular, have high nutrient demands and are vulnerable to nutrient losses driven by fire. Frequent fire keeps nutrient levels too low for forest trees to grow large enough to become fire-resistant or dense enough to shade out grasses and reduce flammability (Pellegrini 2016). In contrast, savanna plants have experienced frequent fire for many millennia, such that "tropical savanna plant communities can be highly resilient to fire-induced nutrient losses at the ecosystem scale" (Pellegrini et al. 2015).

The low-nutrient status of savannas contributes to their high species richness in the ground layer. A general pattern for grasslands is low plant species richness at sites with high soil fertility, where competitive exclusion by dominant plants is rampant. The pattern of herbaceous species richness increasing as productivity declines is evident for savannas globally (Staver and Koerner 2015), especially on old, strongly weathered soils. The precise mechanisms behind this relationship are still being debated (Laliberté et al. 2013), but it appears that many grasses, at least, evolved to tolerate low nutrient levels and that "species drop out either as a direct result of nutrient availability or as a byproduct of higher biomass and associated light limitation" (Staver and Koerner 2015). Across Earth, accumulation of biomass reduces plant species richness in grasslands (Grace et al. 2016).

Relationships between fire, soil nutrients, and vegetation, though not studied as thoroughly for pine savannas and treeless grasslands of the southeastern Coastal Plain as in tropical savannas, are likely to be similar here. Fires in Florida slash pine plantations increase rates of nutrient turnover, stimulate N fixation, and raise soil pH. The N volatilized by fire was estimated in one study to amount to only 20–40 percent of forest floor N (Gholz and Fisher 1984). Nevertheless, soils in southeastern savanna/grasslands are mostly acidic and nutrient-poor; usually only when limestone is close to the surface or clay, silt, or loam content is high are soils less acidic and somewhat more fertile (Platt 1999). Reductions in fire frequency lead to invasions of pine savannas by nutrient-demanding forest trees, such as mesic oaks and sweetgum (*Liquidambar styraciflua*), which suggests that fire helps keep nutrient levels low, as in tropical savannas (Pellegrini 2016).

An instructive case is our naturally occurring upland mixed woodland, which is ecotonal between upland hardwood forest and sandhill. It is thought to be burned mostly by fires spreading from adjacent sandhill or upland pine communities, which burn through the upland mixed woodland before extinguishing downslope in upland hardwood forest (FNAI 2010). The FRI of upland mixed woodland was estimated as 10 years by Harper (1943), but more recent estimates fall into the 2–10-year range (FNAI 2010).

Sometimes called the "red oak woods" (Harper 1915), upland mixed woodland has a relatively open canopy of southern red oak (*Quercus falcata*), mockernut hickory (*Carya tomentosa*), post oak (*Q. stellata*), blackjack oak (*Q. marilandica*), or black oak (*Q. velutina*), together with shortleaf pine (*Pinus echinata*) or longleaf pine (*P. palustris*). Additional tree species from the adjacent upland hardwood forest also may be present, especially on that lower end of the ecotone. The diversity and density of hardwood trees in upland mixed woodland probably reflect the higher levels of soil nutrients there in comparison with sandhill. These woodlands occur on loamy sands or fine sandy loams, which are richer in phosphorus, potassium, and calcium than typical sandhill soils, and often contain phosphatic pebbles (Thomas et al. 1979; FNAI 2010).

Fire Interactions and Effects of Exclusion

In the case of dry prairie (subtropical hyperseasonal grassland), an interaction of fire and water may largely explain the paucity of trees. This low, flat landscape with a high water table tends to flood relatively early in the wet season, soon after the peak in lightning-season fires. This double-whammy of fire followed closely by flooding is hypothesized to reduce survival and recruitment of pines (Platt et al. 2006). This interaction is strengthened by the large, uninterrupted fire compartments in this landscape, which, prior to alteration by humans, allowed individual fires to spread more widely than in perhaps any landscape type in Florida (Bridges 2006a).

With fire exclusion, live oak (*Q. virginiana*) and swamp laurel oak (*Q. laurifolia*) can invade dry prairie from mesic or hydric hammock (Huffman and Blanchard 1991). At Myakka River State Park, these hammocks naturally occurred as small islands (small patch communities) or relatively narrow borders of the river and lakes (linear or interfingered communities). After decades of fire exclusion or controlled burns only during the winter, an artifactual community of oaks with a dense palmetto (*Serenoa repens*) understory developed on former dry prairie. Only intense fire during spring drought might be capable of killing the large oaks, though unfortunately the unnaturally tall and dense palmettos are resilient even to these fires (Huffman and Blanchard 1991).

Fire frequency patterns in pine savannas and dry prairie, and their relationships to species richness, were discussed in chapter 3, and the estimated FRI for each FNAI natural community type is given in Table 4.2. In general, as evidence from research, especially recent tree-ring/fire-scar research (Huffman 2006; Huffman and Platt 2014), accumulates, the low ends of the estimated FRIs appear to have been most common prior to EuroAmerican settlement. Fires ignited by lightning in pine savannas frequently spread under natural conditions to adjacent fire-dependent communities, such as scrub and seepage slope and the outer zones of hammocks (Myers 1990). In this sense pine savanna functions as a keystone community upon which other communities depend for their source of fire.

In sandhill, reduced fire frequency leads to increased density of shrubs, turkey oak (*Q. laevis*), and other pyrophytic oaks, which is then followed by invasion of off-site trees such as sand live oak (*Q. geminata*), sand laurel oak (*Q. hemisphaerica*), or sand pine (*P. clausa*) if fire is not reintroduced relatively soon. As fire-sensitive plant species overrun the community, the more diminutive fire-dependent species, including grasses, forbs, and subshrubs such as shiny blueberry (*Vaccinium myrsinites*) and gopher apple (*Licania michauxii*), are shaded out and ultimately may be lost from the community if fire is not returned. Regeneration of longleaf or South Florida slash pine may cease with the disappearance of bare mineral soil required for germination. With continued absence of fire, a sandhill community flips to an alternative stable state such as xeric hammock, turkey oak barrens, or an artificial sand pine scrub (Myers 1985).

In cases where sand live oak invades fire-excluded sandhills, often from scrub (where it is a common codominant), the plants commonly form dense clonal oak domes, the interiors of which are usually impenetrable by fire after the domes grow to a height of few meters (Guerin 1993). The leaves of sand live oak are highly nonflammable and impede the spread of fire, allowing the trees to grow large and relatively fire-proof. This transition constitutes a regime shift from sandhill to xeric hammock, a community that naturally occurs on some topographically fire-protected sites but has become much more widespread due to fire exclusion (FNAI 2010).

When a sandhill or other pine savanna shifts to an alternative stable state, restoration through controlled burning alone may have a low

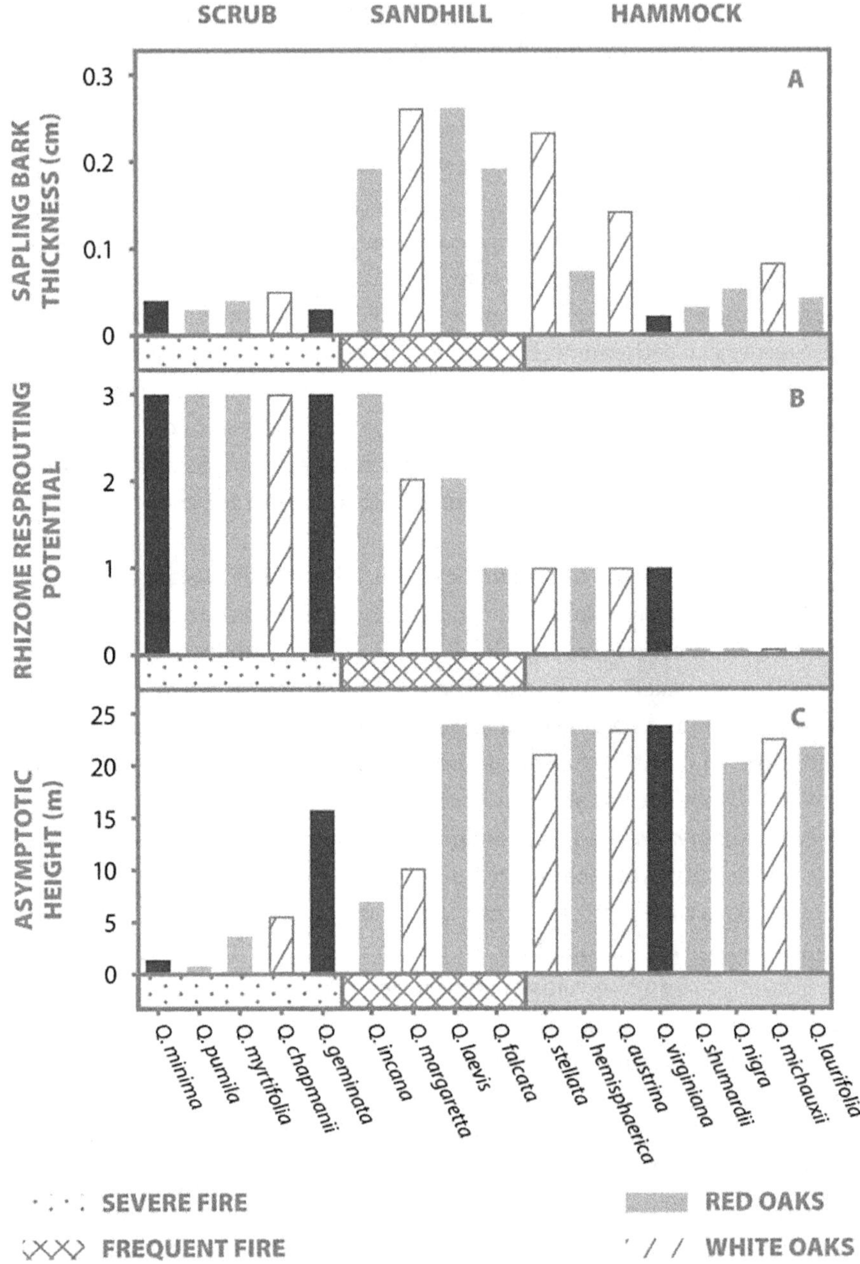

Figure 4.5. Species trait values for (*A*) bark thickness of saplings at 5 cm dbh, (*B*) rhizome sprouting potential, and (*C*) asymptotic height in relation to natural communities (scrub, sandhill, hammock) with dissimilar fire regimes. Phylogenetic lineages of species are indicated by shading. Adapted from Cavender-Bares et al. (2004b).

probability of success. The increased shade has eliminated much of the herbaceous groundcover/fine fuels necessary to carry a surface fire. Even before this threshold is reached, however, reintroducing fire can be tricky, even dangerous. Years of fuel accumulation in the absence of burning can lead to increased heat from fires sufficient to kill adult longleaf pines (Rothermel 1983). After many years of fire exclusion, branch (feeder) roots from the lateral roots of pines grow up into the duff layer. When burned during dry conditions, massive mortality of these roots can kill the adult trees (Varner et al. 2005). As illustrated by the upland mixed-woodland example above, however, somewhat longer FRIs are characteristic of ecotonal pine-hardwood communities. Perhaps the relatively lower grass cover and volume of pine duff in the herbaceous layer of these communities reduce fire intensity sufficiently to protect established pyrophytic pines and hardwoods from death by fire.

The importance of heterogeneity in fire regimes, with respect to the presence of oaks and other hardwoods in pine savannas, was discussed in chapter 3. Invasion of oaks and other hardwoods can be problematic in fire-suppressed sandhills (but see Loudermilk et al. 2016). Nevertheless, pyrophytic oaks and hickories naturally grow to tree size on some sandhill sites due to spatial and temporal heterogeneity in fire regimes and perhaps other factors. Harper (1913) thought "oaks and hickories are not much inferior" to longleaf pine in their tolerance of frequent, low-severity fire. A study of 17 oak species in north-central Florida found oaks sorted along gradients of soil moisture, nutrient availability, and fire regime. Oak species fell into three groups corresponding to major community types: (1) hammock, (2) sandhill, and (3) scrub. Principal components analysis of the functional traits of oak species supported these three groups (Cavender-Bares et al. 2004b; Fig. 4.5).

As portrayed in Figure 4.5, sandhill oaks stand out as having higher sapling bark thickness than scrub or hammock oaks, which reflects the frequent surface-fire regime. Sandhill oaks are intermediate in rhizome sprouting potential, and some are equal to hammock oaks in their adult asymptotic height. Hiers et al. (2014) showed that southeastern oak species vary substantially in their site affinities, fire-adaptive traits, flammability, and ecological roles (Table 4.3). Clearly, pyrophytic oaks are natural components of the southeastern pine savanna landscape, particularly in sandhills and clayhills and in their ecotones with hardwood forest (Greenberg and Simons 1999; Hiers et al. 2014).

Table 4.3. Pyrophytic oaks commonly found in longleaf pine ecosystems (especially sandhill) in comparison with mesic oaks, which may invade from hardwood forest communities

Species	Common name	Site affinity	Fire-adapted traits	Flammability of leaves	Ecological role
PYROPHYTIC OAKS					
Quercus laevis	Turkey oak	Sandhill, poor sandy soils	Thick bark at small size, vigorous resprouting, reproduce at small size	Very high	Exclusively found on sandy, excessively drained sites; facilitates longleaf pine regeneration on xeric sites
Q. incana	Bluejack oak	Sandhill or upland pine, sandy soils	Thick bark at small size, vigorous resprouting, reproduce at small size	Low	Small, isolated tree or shrub of drier longleaf pinelands
Q. margaretta	Sand post oak	Sandhill, poor sites	Vigorous resprouter, forms clonal dome protecting interior stems from topkill by fire	High	May persist as old-growth components of frequently burned uplands. Clones provide habitat heterogeneity but burn less intensely than close relative *Q. stellata*
Q. marilandica	Blackjack oak	Sandhill or upland pine, dry poor soils	Thick bark at small size, vigorous resprouter	High	Similar to turkey oak, but with a wider edaphic range from dry clay hills to sandhills
Q. stellata	Post oak	Dry, poor to rich sites	Thick bark, long-lived; capable of resprouting in the understory for decades	High	Isolated trees found in longleaf and other pine-oak ecosystems

Q. falcata	Southern red oak	Dry woods, upland mixed woodland, abundant across many sites	Thick bark, long-lived; capable of resprouting in the understory for decades	Very high	Isolated tree found in longleaf pine uplands on a variety of sites. Characteristic species of upland mixed woodland ("red oak woods")
Q. geminata	Sand live oak	Sandhill, poor sites	Persistent resprouting if top-killed, fire avoidance through litter deposition	Low	Found exclusively on excessively drained sandy soils; forms clonal domes in xeric sandhill and scrub
MESIC OAKS					
Q. virginiana	Live oak	Moist to dry sites, sandy soils, hammocks	Persistent resprouting if top-killed, fire avoidance through litter deposition	Low	Occurs along lake margins and in riparian borders
Q. hemisphaerica	Sand laurel oak	Hammocks and sandy soils	Persistent resprouting if top-killed, fire avoidance through litter deposition	Low	Upland hardwood forest, mesic and xeric hammock, and slope forest species that invades sandhill and upland pine in the absence of fire
Q. nigra	Water oak	Hammocks, large bottomlands, and wet soils	Persistent resprouting if top-killed, fire avoidance through litter deposition	Low	Wetland ecotonal, hydric hammock, and mesic hammock species that invades uplands in the absence of fire

Modified from Hiers et al. (2014). See original article for further explanation.

FREQUENT-FIRE WETLANDS

Fire is as essential to the ecological integrity of many wetlands as it is to the savannas and grasslands discussed in the preceding section. People are often surprised to learn that some wetlands in the southeastern Coastal Plain burn regularly and even carry fire when inundated. In general, hydrology is second only to fire in its importance to the spatial distribution, species composition, diversity, and resilience of natural communities in this region. The exceptions are mostly edaphic communities that are restricted to unusual substrates (Noss 2013). For wetlands in Florida, fire closely follows hydrology in its effects on vegetation structure and species composition (Duever 1984; Watts et al. 2012).

Figure 4.6 is adapted from an unpublished illustration produced by Florida field biologist Steve Christman. I have modified Christman's figure to reflect current understanding of the arrangement of Florida's natural communities in relation to gradients of fire-return interval vs. hydroperiod (or, alternately, soil moisture). I added several communities to Christman's original diagram. Fire and water are not the only factors that affect the spatial distribution of natural communities, but they are two that are most labile and easily affected by human activities. Both are complex gradients that reflect underlying variation in elevation, topography, soils, and landscape context (e.g., adjacent communities). If either fire frequency or hydroperiod is altered, a community will shift its location in this two-dimensional space, and it will be degraded in quality in comparison with a community occupying its natural position in fire-water space.

Herbaceous Wetlands

Note that herbaceous wetlands (i.e., marsh) in Figure 4.6 occupy the frequent-fire (upper left) portion of the diagram. Frequent-fire wetlands in Florida and the southeastern Coastal Plain include freshwater marshes (depression, basin, floodplain, slough, and glades marshes), the landward zone of salt marshes, and the outer zones or edges of several swamp communities embedded in frequent-fire landscapes, including dome, stringer, basin, and strand swamps. Transitional wetlands such as wet prairie also experience frequent fire.

In chapter 3 we considered the problem of fire exclusion or winter (as opposed to lightning-season) burning in herbaceous wetlands, and the threats these practices pose to animals such as flatwoods salamanders and

Figure 4.6. The approximate arrangement of some of Florida's natural communities in relation to gradients of fire frequency (fire-return interval) and hydroperiod or soil moisture. These gradients control, to a great extent, the spatial distribution of natural communities. Each community type would occupy some two-dimensional space in this diagram (i.e., a polygon); here each community name is placed in the approximate center of that space. Based roughly on Christman (unpublished); see also a similar diagram for south Florida communities in Duever and Roberts (2013).

gopher frogs. Burning during the winter typically leads to shrub invasion and loss of open-water habitat. Usually only fire during the drought of the peak lightning-fire season will maintain herbaceous and open-water zones (Huffman and Blanchard 1991; Kirkman 1995; Means 2006; Martin and Kirkman 2009). In addition, experimental research demonstrates that fire during periods of drought enhances species richness in depression wetlands and adjacent wetland/upland ecotones (Kirkman 1995).

Charcoal in the peat of marshes attests to a history of fire in these communities, with the black color of Everglades peat a persuasive example (Kushlan 1990). Precise estimates of FRI cannot be obtained from charcoal in peat, but rely on observational evidence. Shallow-water marshes, or the shallow outer zones of deeper marshes, burn naturally at one-to-three-year intervals, whereas deep-water marshes have an FRI of approximately three to five years (Wade et al. 1980). Multiple observations suggest that, at least in south Florida, fires regularly pass through marshes that contain standing water (Kushlan 1990); however, usually only during drought can fires clear out invading shrubs and ignite organic soils, known as histosols. A histosol has 40 cm or more of the upper 80 cm as organic material, including muck (sapric soil material), mucky peat (hemic soil material), or peat (fibric soil material). In these soils, organic matter forms at a more rapid rate than it is destroyed, largely due to restricted drainage, which precludes aerobic decomposition. Histosols occur throughout Florida in association with marshes and swamps, but are most extensive in the Everglades. These soils can ignite and result in smoldering fires when moisture content drops below 65 percent (Wade et al. 1980).

Depression marshes are isolated, small-patch communities embedded in frequent-fire "uplands" (Fig. 4.7). If one views an aerial photo of natural or agricultural land in central Florida, an extreme karst landscape, depression marshes are the most common pocks in the pock-marked terrain. They usually form when poor surface drainage results in water moving downward, dissolving limestone and causing overlying sands to slump into the depressions (FNAI 2010), although other origins are possible (Winchester et al. 1985). These marshes are distinguished from the sometimes-adjacent wet prairie by their concentric zones of vegetation, lack of saw palmetto, and lack of wiregrass except in the outer zones (FNAI 2010). Their low FRI can be attributed to the frequent natural fire regime of the flatwoods, sandhill, and grassland communities in which they are embedded. The consequences of reducing fire frequency in depression marshes in central Florida can be readily seen by the encroachment of coastalplain willow (*Salix caroliniana*), common buttonbush (*Cephalanthus occidentalis*), groundsel tree (*Baccharis halimifolia*), wax myrtle (*Morella cerifera*), and other shrubs (Huffman and Blanchard 1991; FNAI 2010).

A study in southwestern Georgia established transects from upland longleaf pine–wiregrass communities into depression marshes to examine changes in plant species composition and richness along this elevation-moisture gradient. Herbaceous species richness is high along the entire gradient, but peaks in the nonhydric ecotonal portion of the gradient (Kirkman et al. 1998b), which is precisely where fire lines are commonly

A

B

Figure 4.7. (*A*) A large (18 ha) depression marsh in Ocala National Forest, Marion County, Florida, November 2015. This marsh, also known as a seasonal pond, is embedded in a mixed matrix of sandhill and scrub. (*B*) A depression marsh in Charles H. Bronson State Forest, Orange County, Florida, December 2010, a drier winter. No standing water is present. The matrix is longleaf pine/pond pine wet flatwoods. Lightning-season fires often burn through such wetlands. Photos by Reed Noss.

plowed. Ostensibly "isolated" depression wetlands do not receive as much protection from federal or state agencies as wetlands associated with flowing or "navigable" waters.

Basin marshes differ from depression marshes in being large-patch communities usually not embedded in a fire-maintained matrix; their FRI is unknown. Other marsh communities in Florida have variable but generally low FRIs (Table 4.2). Fire has been used successfully to control the spread of coastalplain willow in basin marshes of the upper St. Johns River, where artificial drainage facilitated willow invasion (Lee et al. 2005). In the same landscape, a single controlled burn eliminated wax myrtle, red maple (*Acer rubrum*), and groundsel tree, and significantly reduced the cover of buttonbush (Miller et al. 1998).

Fire in slough marshes of central and south Florida is beneficial in promoting growth of sawgrass (*Cladium jamaicense*) and other dominant native plants, including maidencane (*Panicum hemitomon*) and bulltongue arrowhead (*Sagittaria lancifolia*), at least when water levels are just below the surface (Loveless 1959). The probability of peat fires is less for slough marshes because they are underlain by sand (FNAI 2010). Glades marsh, which occurs primarily in the Everglades region, is perhaps to most people the quintessential Everglades. It is usually a low-diversity community, often a dense, tall monoculture of sawgrass. Deeper glades marshes support more species of emergent plants, whereas other herbs benefit from drought (FNAI 2010). Sawgrass is well known for carrying fire over standing water. Although the FRI of glades marsh is estimated as 3–10 years, variability is high (FNAI 2010).

Concerns have been raised about controlled burns during drought in marshes leading to peat fires that kill dominant plant species, especially in situations with artificially lowered water tables (Bacchus 1994; Nyman and Chabreck 1995). By lowering the ground surface and reducing transpiration, however, peat fires typically lead to longer hydroperiods and create valuable open-water zones. Still, peat fires in artificially drained wetlands can cause considerable damage. As always, heterogeneity of burns and provision of refugia for animals are key considerations. Homogeneous burns of marshes can potentially extirpate populations of animals of conservation concern, such as black rails (*Laterallus jamaicensis*), snail kites (*Rostrhamus sociabilis*), and round-tailed muskrats (*Neofiber alleni*), in floodplain marshes of the St. Johns River. Heterogeneous fires here leave small unburned patches of habitat, which are used as refugia by black rails and presumably other animals (Legare et al. 1998; Miller et al. 1998).

With respect to salt marshes, Frost (1995) observed that fire has little impact on floristics in true salt marshes, where plant species richness is very low because few species can tolerate the combination of salinity and standing water. Fire is more important in brackish marshes and is speculated to increase species richness by removing heavy thatch approximately every two to five years (Frost 1995). FNAI (2010) considers data too sparse to estimate the FRI for salt marsh communities, but notes that fires can spread from adjacent uplands or from lightning strikes within the marsh. Salt marshes adjacent to flatwoods and other frequent-fire communities burn regularly. The many barriers to the spread of fire, however, such as tidal creeks and salt flats, would tend to limit the area burned and the frequency of fire in many salt marshes. FRI is unknown for other estuarine communities, but the extensive grassy character of some buttonwood mangrove (*Conocarpus erectus*) "forests" (FNAI 2010; I prefer "buttonwood savannas") along Florida Bay in the Everglades and on some of the larger keys, such as Big Pine Key, makes some kind of regular fire regime plausible.

Swamps and Cypress Domes

Among the swamp communities that qualify for the "stand-maintaining" regime by perpetuating themselves despite—or because of—frequent fire are dome swamps (i.e., cypress domes and gum ponds) and stringer swamps. Dominated by woody vegetation, most swamps burn less frequently than marshes. Often only the outer zones, when in contact with pyrogenic vegetation, will carry fire. Other types, such as cypress domes, usually burn frequently because they are embedded in frequent-fire matrix communities. I place most swamps of our region in the "smoldering" or "rare or infrequent" fire regime categories, although some swamps legitimately belong in the "stand-replacing" category (see below).

Cypress domes (Figure 4.8), like other depression wetlands, probably result, in most cases, from poor surface drainage causing water to move downward and dissolve the underlying limestone. Similar to depression marshes, they are small patch communities embedded in a matrix of pyrogenic vegetation. Cypress domes contain pond cypress, whereas strand swamps and floodplain swamps contain bald cypress (*Taxodium distichum*). Both species of cypress, with their relatively thick fire-protective bark (Hare 1961; see chapter 3) survive frequent surface fire more readily than most hardwoods. A fire passing through a cypress dome in north-central Florida killed most of the hardwoods and pines but only about half

Figure 4.8. A cypress dome (*background*) in a matrix of mesic flatwoods, Three Lakes Wildlife Management Area, Osceola County, Florida, January 2008. Fire (more frequent on the edges) and hydrology (deeper water and longer hydroperiod in the center) probably both explain the domelike shape. These communities are virtually always embedded in pyrogenic vegetation, especially flatwoods. The pond cypress (*Taxodium ascendens*) has moderately thick and fire-resistant bark, and will resprout after topkill by fire (Wade et al. 1980). Photo by Reed Noss.

of the cypress. Three years after fire, the dome was strongly dominated by cypress, but with some hardwood tree and shrub regeneration (Ewel and Mitsch 1978).

Another study examined patterns of fire severity and mortality of cypress domes after a drought-season wildfire in Big Cypress National Preserve. Initial survival of cypress was 99 percent, but after one year postfire 23 percent of the trees had died. Smaller trees, which have thinner bark, and trees in lower areas experienced higher mortality (Watts et al. 2012). Recent estimates of FRI for cypress domes summarized in FNAI (2010) are 3–5 years in the outer zones and 100–150 years in the center of domes. Watts et al. (2012) estimate decadal intervals for surface fires in cypress domes and a 50–100 year FRI for smoldering ground fires.

Cypress domes are true fire-dependent communities. As fire frequency declines, organic matter accumulates, water level drops, and hardwoods such as bays (e.g., loblolly bay, *Gordonia lasianthus*, and swamp tupelo, *Nyssa biflora*), as well as slash pine, gradually take over the dome (Casey and Ewel 2006). Frequent fire is considered partially responsible for their domed shape, with the outer, more frequently burned zone pruned by fire and shorter in stature (FNAI 2010; Watts et al. 2012). With fire exclusion, hardwoods come to dominate domes, and peat accumulates such that when fire occurs it can topkill or complete-kill cypress as well as hardwoods.

Cypress resprouts readily after topkill by fire (Wade et al. 1980). Nevertheless, fire often does complete-kill cypress (Watts et al. 2012). The centers of domes, with their thicker layers of peat, may experience more severe smoldering fire than outer zones, killing the trees even with a natural fire regime (Ewel and Mitsch 1978). Thus, the centers of cypress domes often resemble depression marshes, sometimes with open water and plants and animals associated with ponds. On the other hand, the trees on the edges of cypress domes are smaller and more susceptible to mortality or topkill from surface fires. The characteristic "hat rack" or "bonsai" shape of small pond cypress in pyrogenic landscapes may reflect resprouting from basal or epicormic shoots following topkill (Watts et al. 2012). The trunks of these dwarfed cypress can be hundreds of years old, which suggests that epicormic sprouting after fire-pruning is more typical (Jean Huffman, personal communication).

In Georgia and the Carolinas, depression wetlands including marshes and swamps (e.g., cypress-gum ponds; Edwards et al. 2013) are often associated with Carolina bays. These "bays" are elliptical depressions created by dissolution of limestone beneath deep sands (May and Warne 1999) or through a complex combination of lacustrine and aeolian processes that cause bays to migrate over time (Moore et al. 2016). Dry sandy ridges on bay rims, often occupied by xeric longleaf pine communities, apparently were deposited by winds after the bays were formed (May and Warne 1999). The wetland communities in the interiors of Carolina bays are subject to stand-replacing and smoldering fire regimes and will be discussed in following sections. The bay rims are pyrogenic, especially when the surrounding landscape matrix is pine savanna, but may not carry fire well when fine fuels are sparse, as they often are. Fire enters cypress-gum swamps within Carolina bays during drought, and would probably be frequent in the outer zones of these communities if the pine savanna matrix were intact.

Cypress domes and other depression wetlands in frequent-fire landscapes historically have been managed poorly. The common practice of plowing fire lines around cypress domes in an effort to prevent peat fires has led to profound changes in vegetation structure and function by altering drainage from adjoining uplands, preventing fire from entering the domes, and eliminating the natural ecotone between domes and surrounding communities. Ecotones around depression wetlands of various types are key habitats for some plant species and for turtles, salamanders, anurans, and many invertebrates that require both upland and wetland habitat (e.g., Burke and Gibbons 1995).

Stringer swamps, a variant of dome swamp, are essentially elongate cypress domes. They are linear features along intermittent streams embedded in pyrogenic landscapes; they also have an FRI of one to three years (FNAI 2010). The outer zones of other forested wetland types in the southeastern Coastal Plain, when surrounded by pyrogenic vegetation, can be considered frequent-fire communities. As usual for wetlands, the FRI depends mainly on the presence or absence of adjacent pyrogenic communities. Basin swamps are dominated by pond cypress and swamp tupelo (*Nyssa biflora*). When surrounded by a community such as mesic flatwoods, a basin swamp may burn every three to five years or so in its outer zones, whereas a basin swamp situated within a matrix of hydric hammock may burn no more commonly in its outer zones than in its core, which would place it in the "rare or infrequent fire" category.

Strand swamps, restricted to south Florida, are dominated by bald cypress in the overstory, but with many tropical elements. Although fire is rare in the interiors of these swamps, fires originating in adjacent pine savannas often burn outer zones, such that the cypress trees along these edges are usually smaller and younger than those toward the interior (Duever et al. 1986). Strand edges are often pure cypress, due to the higher vulnerability of other tree species in the community to fire (Wade et al. 1980). Floodplain swamps, dominated by bald cypress and one or more species of tupelo (*Nyssa*), are found along rivers throughout the southeastern Coastal Plain. As with strand swamps, fire is rare except along the edges of floodplain swamps adjacent to a pyrogenic community such as pine savanna. Occasional fires that penetrate into the swamps probably contribute to cypress dominance because the older trees are resistant to fire in comparison with other tree species or younger cypress (FNAI 2010).

Canebrakes

William Bartram (1791) described "an endless wilderness of canes" in parts of the southeastern Coastal Plain. The fire ecology of canebrakes—extensive stands of giant cane (*Arundinaria gigantea*) or switch cane (*A.

tecta) that once occupied alluvial floodplains and some other landscape positions—is not well understood, largely because this ecosystem is almost extinct (Noss et al. 1995; Platt and Brantley 1997). Cane is capable of extremely rapid clonal growth following fire or other disturbance. Such a response is typical of a stand-maintaining grassland fire regime (i.e., cane is a bamboo in the grass family, Poaceae). After several years without disturbance, canebrakes occupying large canopy gaps in bottomland forest become decadent from self-shading or crowding, but they can be rapidly rejuvenated by fire or another new disturbance. A fire will both remove senescing older culms and accelerate the production of new culms (Gagnon and Platt 2008; Gagnon 2009).

Gagnon and Platt (2008) suggest canebrake is an alternative stable state to bottomland forest in floodplains. Based on experimental and observational evidence, they propose that a sequence of disturbance events, especially fire following a canopy-opening disturbance such as windstorm or flood, promotes the development of large canebrakes composed of giant cane. The flammable cane would then enter into a positive feedback relationship with fire, increasing local fire frequency. If this relationship "were sufficiently strong to preclude return to the initial forested state, monotypic giant cane would act as an alternative attractor to forest trees" (Gagnon and Platt 2008).

Frost (1995) estimated the FRI of dense giant canebrakes as 4–6 years, with canebrake and pocosin alternating on sites with FRIs of 7–12 to 13–25 years. Canebrakes composed of switch cane have not been as well studied as giant canebrakes. In the outer Coastal Plain, the FRI of switch canebrakes was estimated as three to five years (Hughes 1966). In the Sandhills of the inner Coastal Plain of the Carolinas and Georgia, switch cane occurs at the heads of drainages or along upper reaches of small streams, where the mean FRI among 13 managed sites was estimated as 1.0–2.3 years (Gray et al. 2016). These authors determined the "optimal fire-return interval for switch canebrakes is 2 yr, with reduced increases in canebrake size associated with fire-return intervals that are either greater or less than 2 yr."

Communities Characterized by Stand-Replacing Fire

Stand-replacing fire, as defined here (Table 3.1), differs from stand-maintaining fire in that it is usually less frequent, but more intense. With stand-maintaining fire most trees survive without topkill, and the dominant grasses, forbs, and shrubs quickly resprout. With stand-replacing fire dominant woody species are topkilled or complete-killed, and must regenerate by resprouting or by germination from a soil or canopy seed bank, thereby replacing the stand. Propagules of some plant species without seed banks must disperse in from offsite.

SCRUB

Scrub, almost entirely restricted to Florida, has a well-studied fire regime. This is largely attributable to researchers at Archbold Biological Station, which sits on one of the largest patches of scrub on the southern tip of the Lake Wales Ridge. This inland ridge and a few other ancient dune systems on the Florida peninsula were coastal and often insular during intermittent high sea-level stands extending back to the Pliocene or earlier and continuing through the Pleistocene (see chapter 2). Younger scrub, with much lower endemicity, occurs on dunes associated with recent Pleistocene shorelines close to the present coast on both the Atlantic and Gulf of Mexico sides of the peninsula as well as in the Panhandle (Myers 1990).

Scrub is a xeric shrubland in many ways similar to California chaparral and other Mediterranean-type shrublands, with the salient difference of receiving at least four times as much annual rainfall as California chaparral and having a reversed dry-wet season. Like sandhill, scrub develops on deep, well-drained sands, which create physiologically xeric conditions for plants and animals. Characteristic plants are several types of scrub oak (*Q. geminata, Q. myrtifolia, Q. chapmanii*, and on the Lake Wales Ridge, *Q. inopina*), palmettos (*Serenoa repens* and *Sabal etonia*), and sand pine (*Pinus clausa*). Florida rosemary (*Ceratiola ericoides*) is distinctive of rosemary scrub but also occurs in sand pine scrub with relatively infrequent fire. The fine-grain endemicity for plants and animals in ancient scrub is one of the highest in North America and globally, especially on the Lake Wales Ridge (Menges 1999; FNAI 2010; Steve Christman, unpublished). Georgia has a similar but comparatively depauperate community (Edwards et al. 2013), and small patches of scrublike vegetation occur in coastal Alabama and Mississippi (Myers 1990).

The fire ecology of scrub was reviewed in chapter 3 with respect to fire frequency and the fire-adaptive strategies and traits of scrub plants and animals. The FRI ranges from as low as 3–20, 5–12, or 5–20 years for types of oak scrub, to 10–40 or 20–60 years for rosemary scrub, to 10–70 years for sand pine scrub, but often considerably longer in the Panhandle (Table 4.2). Scrubby flatwoods are intermediate between mesic flatwoods and oak scrub in several respects, including having a mixed stand-maintaining (i.e., the longleaf or South Florida slash pines mostly persist through fire) and stand-replacing fire regime. Menges (1999) considers oak scrub and scrubby flatwoods essentially the same community. Scrubby flatwoods have an FRI of approximately 5–15 years (Table 4.2).

The fire regime of scrub is primarily a stand-replacing regime because most plants are either topkilled (e.g., oaks, most ericaceous shrubs) or complete-killed (e.g., sand pine, Florida rosemary). Topkilled plants resprout, generally using resources from underground storage organs (Figs. 4.9, 4.10), whereas complete-killed plants are mostly obligate seeders and recruit from either a soil seed bank (herbaceous plants and shrubs such as Florida rosemary) or a canopy seed bank (sand pine). Individual fires are more heterogeneous in more xeric scrub types (e.g., rosemary scrub) due to the inherent patchiness of fuels, with much bare ground (Menges 1999).

The different types of scrub vary not only in their FRIs and other aspects of their fire regimes but also in the life-history traits of their constituent species. As summarized by Menges and Hawkes (1998), rosemary scrub has more herbaceous plants, more endemics, more gap specialists, and more seeders that increase in abundance between fires as compared to other types of scrub. This composition is predictable based on the more open physiognomy of rosemary scrub, with its large and persistent gaps. In scrubby flatwoods and oak scrub, the resprouting strategy prevails over reseeding because open soil space is limited. Gap area in scrubby flatwoods, which is the best predictor of plant species richness, is highest during the first four years postfire. Percent bare sand, which correlates with the richness of obligate seeding species, gradually declines after fire until 26 years postfire, when it drops precipitously (Dee and Menges 2014). Several resprouting species peak in flowering around seven years postfire (Ostertag and Menges 1994). In addition to those species that germinate from a soil seed bank, some species—for example, *Polygonella basiramia*—do not persist in the seed bank and must recolonize burned scrub from nearby unburned patches (Menges 1999).

Like communities that depend on stand-maintaining fire, scrub has suffered from active and passive fire exclusion. Fire exclusion causes changes in the structure and composition of scrub and alters the way the vegetation responds to the next inevitable fire. Because gaps close between fires and the amount of bare sand declines, lengthening the FRI results in dense vegetation cover and reduced species richness (Dee and

Figure 4.9. Fire ecologist Eric Menges on 16 January 2015 in overgrown scrub (Red Hill, Archbold Biological Station) control-burned nine days earlier on 7 January 2015. Although the site appears barren, some saw palmettos (*Serenoa repens*) were already sprouting new green fronds (not shown). Photo by Reed Noss.

Figure 4.10. Same site (but not exactly the same spot) as in Figure 4.9 on 3 October 2015, nine months after the Red Hill fire. Note abundant resprouting of oaks (*Quercus* spp.) and palmettos (*Serenoa repens* and *Sabal etonia*). Resprouting and clonal spread lead to rapid recovery of prefire vegetation cover. Photo by Reed Noss.

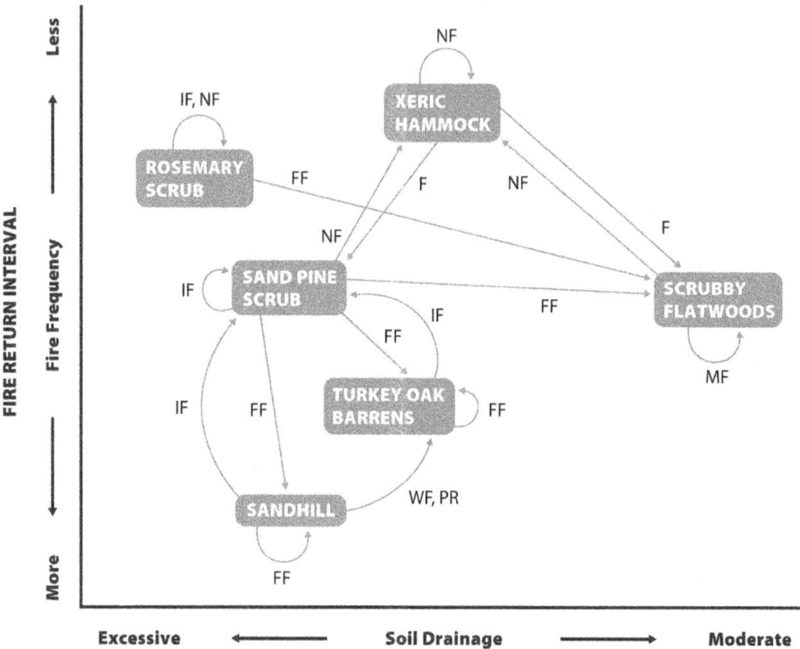

Figure 4.11. A conceptual model for vegetation dynamics of Florida scrub and related upland communities. Community types are grouped by fire-return interval and soil drainage, but may overlap more than portrayed. Arrows indicate hypothesized transitions controlled largely by fire frequency. Although not shown here, sandhill can transition directly to xeric hammock if a seed source for sand live oak (*Quercus geminata*), the dominant tree of xeric hammock, is within dispersal distance (personal observation). FF = frequent fire (1–10 yrs), MF = moderately frequent fire (5–20 yrs), IF = infrequent fire (15–100 yrs), NF = no fire within 100 yrs, WF = winter fire, PR = pine removal, F = fire reintroduction. Adapted from Menges and Hawkes (1998).

Menges 2014). Ultimately the community may undergo a regime shift to xeric hammock. Scrub that is long unburned requires more extreme conditions to burn completely, and the intensity of fire can be unpredictable (Menges and Hawkes 1998).

Figure 4.11 shows hypothesized relationships among three types of scrub and related communities, FRI, and soil drainage. Each community type perpetuates itself with its characteristic range of fire frequency, but can transition to another community if fire increases or decreases in frequency. Well-established xeric hammock is relatively fire-proof except under the most extreme fire-weather conditions.

A fundamental difference between scrub and the frequent-fire pine savannas, grasslands, and herbaceous wetlands of the southeastern Coastal Plain is that scrub has a low cover of grasses and other flammable plants. Therefore, it is ordinarily difficult to ignite. When sand pines are present, they are not of much help, as their leaves are not very flammable (Varner et al. 2015a), although their retained dead branches are. Only in scrubby flatwoods are pines (longleaf or South Florida slash) with highly flammable leaves present, but those needles do not fall into a rich matrix of wiregrass and other bunchgrasses, which enhances combustion and spread of fire through savannas.

The landscape context of scrub is often a pattern of small to large patches of scrub embedded in a matrix of sandhill. Small scrubby ridges often occur within mesic flatwoods or dry prairie; one such ridge at Kissimmee Prairie even contains a small, isolated population of Florida scrub-jays. This juxtaposition is important because scrub is most prone to burn when a fire from the adjacent pine savanna or grassland flows into it during extreme fire-weather conditions. Fires do not always spread from savanna/grassland into scrub, or scrub would generally have an FRI equivalent to sandhill or other highly pyrogenic communities. Under a natural fire regime, the boundary between scrub and sandhill is abrupt, but this boundary breaks down and the two communities homogenize as fire frequency in the sandhill declines (Myers 1990). The Ocala National Forest region of north-central Florida displays a reversed pattern of islands of pine sandhills in a sea of scrub. Ecologists have differed over whether this unusual pattern has led to a fire regime distinct from elsewhere (e.g., Kalisz et al. 1986).

An alternative pattern and set of processes can be seen in the scrub communities of the Florida Panhandle. Most true scrub in the Panhandle is restricted to narrow sand ridges within a mile or two of the coast, often situated on peninsulas or adjacent to wetland barriers, which protect them from landscape-level fires (Edwin Bridges, personal communication). With FRIs sometimes exceeding 100 years, the sand pines in these scrubs have an uneven-age structure. They may also have more nonserotinous cones that open and release seeds without the help of fire, although this difference is not consistent (Drewa et al. 2008). Some old-growth sand pines in the Panhandle exceed 200 years, and these communities have diverse lichen assemblages both on the ground and in the oaks and pines (Jean Huffman, personal communication). It is crucial that fire managers take into consideration such regional differences in fire regimes. Burning

Panhandle scrub on short intervals typical of peninsular scrub destroys a unique community variant found nowhere else.

FORESTED WETLANDS

Some natural community types defy pigeonholing and could be placed into two or more categories of fire regimes. Among the better examples are certain swamp types that rarely burn, but when they do the fires are both severe and stand-replacing. Intense past logging and drainage of cypress swamps make natural fire histories and regimes challenging to reconstruct. Only a few preserves in south Florida, such as Corkscrew Swamp Sanctuary and Fakahatchee Strand Preserve, contain extensive old-growth bald cypress strand swamps. The edges of cypress and other swamps may burn frequently if the adjacent community is pyrogenic. Fire creeps a variable distance into the stand, depending on weather conditions.

Extreme drought events occurring every century or two, but now apparently becoming more frequent with climate change, can burn through the heart of cypress strands, killing the old-growth trees. In such situations, a coastalplain willow commonly often establishes as a thicket (FNAI 2010). One can observe such a thicket from the overlook along the boardwalk at Corkscrew Swamp Sanctuary. The time required to reestablish an old-growth cypress strand is variable, but this stand-replacement fire regime has an FRI on the order of 100–300 or more years for entire strands (Snyder 1991). In Corkscrew Swamp, the old-growth cypress trees are about 700 years old, the last time a major fire burned through the swamp and apparently killed most of the cypress (Michael Duever, personal communication).

Atlantic white cedar swamps are characterized by a stand-replacing fire regime and occasional smoldering fire. These swamps occur on deep peats over sandy soils in narrow riparian bands as well as in more extensive peatlands and Carolina bays. The trees grow in even-age stands that usually do not survive fire. They regenerate from fires severe enough to remove competing woody vegetation but not so severe as to destroy the soil seed bank (Schafale 2012). Seeds stored in the peaty soils germinate in large numbers if the upper peat layers are not consumed by fire. The fire regime of Atlantic white cedar swamps has been described as low-frequency and moderate-severity related to "marginally moist soil conditions" (Christensen 1981). Given the stand-replacing fire regime, most stands are of uniform age (Christensen 2000). Fire exclusion ultimately leads to decline of Atlantic white cedar and an increase in competing hardwoods, including bay species. Frost (1995) estimates the FRI of Atlantic white cedar swamp as approximately 50–300 years.

In the Carolinas, Atlantic white cedar stands are often associated with pocosins, which are essentially equivalent to shrub bogs in the FNAI (2010) classification. Shrub bogs range throughout Florida except extreme south Florida and also occur in the southern Coastal Plain of Alabama and Mississippi. Fires usually ignited in surrounding pinelands burn through shrub bogs during drought every 10–20 years. The shrubs, which include titi (*Cyrilla racemiflora*), black titi (*Cliftonia monophylla*), fetterbush (*Lyonia lucida*), large gallberry (*Ilex coriacea*), gallberry (*I. glabra*), wax myrtle, sweet pepperbush (*Clethra alnifolia*), and others, respond to what is usually stand-replacing fire by resprouting. Bays, including loblolly bay (*Gordonia lasianthus*), sweetbay (*Magnolia virginiana*), and swamp bay (*Persea palustris*) (FNAI 2010), similarly resprout after severe fire. Pond pine is often present and is characteristic of some pocosins in the Carolinas. As noted earlier, pond pine shows a mixed adaptive response to fire, both resprouting from the root collar and bole and reseeding from serotinous cones.

Several examples have been offered in this book of flammable (pyrophilic) and comparatively nonflammable (pyrophobic) community types coexisting in the same landscapes, often directly adjacent to each other. With a natural fire regime, the ecotones between these communities are typically abrupt. Just et al. (2015) showed that positive feedbacks between vegetation and fire explain the coexistence of longleaf pine–wiregrass savanna and shrub bog (streamhead pocosin; Weakley and Schafale 1991) in the Sandhills of North Carolina. Dense C_4 grasses in the savanna facilitate fire spread.

Streamhead pocosins are often too wet to carry fire, but with their long, narrow shape, fires often creep into the edges, and under dry conditions they can burn through. Just et al. (2015) found 43 percent of controlled burns ignited in pine savanna continued through the pocosins. Weakley and Schafale (1991) suggest streamhead pocosins are more fertile than peatland pocosins because nutrients released by fire in adjacent uplands reach the pocosin through runoff, airborne ash, or groundwater. Shrub bog/pocosin often invades adjacent wet prairies or flatwoods if fire frequency is reduced (FNAI 2010).

Communities Characterized by Smoldering Fire

Smoldering fire results from a ground fire, when pine duff or a thick humic layer within the soil is dry enough to ignite. With restricted oxygen supply, the fire then smolders (Scott et al. 2014). Many of the wetland communities characterized by stand-maintaining fire or stand-replacing fire, discussed in previous sections, are subject to smoldering fires that originate during drought. As noted, deep organic soils such as peat or muck (histosols) are most common in marshes and swamps, and can ignite and create smoldering fires when moisture content drops below approximately 65 percent. Cypress domes, which are dry for up to half of a typical year, are subject to smoldering fire (Watts et al. 2012). These fires also can occur in hammocks within their organic soil horizon.

In some regions, subsurface peat fires can burn for decades under circumstances of lowered water tables (Ellery et al. 1989). Duff fires, which often occur in pine savannas that have been fire-excluded and can kill adult pines by killing their branch roots, represent another kind of smoldering combustion (Varner et al. 2005, 2009). Ground fires tend to burn significantly deeper around the base of trees, compounding the tissue damage (Watts and Kobziar 2013 and references therein).

Ground or smoldering fires offer one of the best examples of how fire intensity and fire severity can sometimes be negatively correlated. Smoldering is a flameless form of combustion that usually generates much less heat (intensity) than flaming combustion (500–700°C vs. 1,500–1,800°C, respectively; Rein et al. 2008). Smoldering fires, however, often have residence times measured in days, weeks, months, or potentially years. Their effects on vegetation can be severe because they transfer more heat to soils and plants (Kreye et al. 2011; Watts and Kobziar 2013). Smoldering ground fires are difficult to control because they can occur deep beneath the surface and go undetected, only to flare up at some later date. Organic soils also tend to become hydrophobic when desiccated, which means they repel water, making firefighting difficult (Watts and Kobziar 2013).

Despite the safety hazards of smoldering fires, they have ecological benefits. Fire burning through depression wetlands during the drought of the lightning-fire season kills shrubs that have invaded due to fire exclusion or winter burning. By burning away accumulated peats and mucks, they lower the ground surface, lengthen the hydroperiod, and produce open water areas, enriching overall biodiversity. A severe peat fire during drought in a swamp, such as cypress strand, may kill large trees whose roots are deep in the peat and transform the swamp to a marsh (Watts and Kobziar 2013) or a willow thicket (FNAI 2010). While this state change would be undesirable if it results in loss of old-growth bald cypress, not much of which remains, in a natural landscape it would enrich diversity at species and community levels.

The Okefenokee Swamp, 181,300 ha (700 mi^2) in southern Georgia and a small portion of adjacent Florida, is the largest single area in the southeastern Coastal Plain (except perhaps the Everglades) subject to extensive smoldering fire as well as stand-maintaining and stand-replacing fire. The "swamp" is a mosaic of pinelands, cypress and other forested wetlands, shrub bogs, wet prairies, and other communities, but much of it is peatland. The term "Okefenokee" comes from an indigenous word meaning "land of the trembling earth" (Harper and Presley 1981) because the floating peat mats shake when walked upon. Peat formation here is a slow process, with about 1.25 cm accumulating over a 19–20-year period (Edwards et al. 2013).

Throughout history the Okefenokee Swamp has experienced a series of large and long-lasting fires during periods of drought. Charcoal in peat cores shows a history of relatively frequent peat fires extending back thousands of years, for as long as peat has accumulated. Recent major fires date to 1844, 1911, 1932, 1954, 2007, and 2011. Many of the lakes in Okefenokee are attributable to the 1844 fire (Edwards et al. 2013). The 2007 fire ultimately covered 240,000 ha of Okefenokee and surrounding lands, making it the largest fire in the modern history of both Georgia and Florida. The fire lasted eight months, given the extent of fire storage in the organic soils. In some areas ground fire progressed slowly underneath trees, causing them to topple in large numbers without being burned (Edwards et al. 2013).

Communities Characterized by Rare or Infrequent Fire

Some of the community types described above under the stand-replacing fire category experience moderate- to high-severity fires that kill or topkill most of the trees or shrubs, but stimulate regeneration from the seed bank or resprouting if the fire is not too severe. These are truly fire-dependent ecosystems, even though they burn infrequently. Another category of community characterized by rare or infrequent fire comprises those that show no known dependence on fire and few or no fire-adaptive traits among their fire-avoider species. As community types, they do not require burning to persist on the landscape. Indeed, they fall somewhat closer to the Clementsian model by exhibiting successional recovery from fire, windstorms, and other disturbance events, except they are not regenerating on a blank slate (see chapter 1). These include the hardwood-dominated communities described by Harper (1911) as restricted to sites topographically protected from fire.

Hardwood forests or hammocks are typically characterized by gap-phase replacement, where holes of various sizes in the canopy resulting from treefalls or tree deaths are filled by recruitment of young trees. These gap dynamics promote species coexistence and richness. Across many forest types around the world, treefalls open up about 0.5–2.0 percent of the forest canopy each year (Runkle 1985).

Even seemingly shade-tolerant "climax" species, such as beech (*Fagus grandifolia*), maple (*Acer saccharum* and *A. floridanum*), and southern magnolia (*Magnolia grandiflora*), which can persist in a suppressed state under a closed canopy for many years, may require multiple gap episodes in their vicinity while young to eventually grow into the canopy (Canham 1989). Shade-intolerant trees, such as tulip-tree (*Liriodendron tulipifera*) and white ash (*Fraxinus americana*), seldom regenerate in gaps smaller than 400 m^2 or larger (Runkle 1984; Poulson and Platt 1989). Gap dynamics are part of the reason why the hardwood forests of northern Florida contain the highest number of tree and shrub species per unit area in the continental United States (Platt and Schwartz 1990), with the occasional exception of tropical rockland hammocks in extreme south Florida.

Given the high occurrence of lightning in Florida and the southeastern Coastal Plain, and the frequent-fire landscape matrix vegetation, our hardwood forests are exposed to more fire than their northern counterparts (Platt and Schwartz 1990). Even many of those more northern forest types, however, are more fire-frequented and resilient to fire than many ecologists have assumed (Stambaugh et al. 2015). Fires commonly creep into southern upland hardwood forests, mesic hammocks, bottomland forests, and related communities from their edges with more pyrogenic communities. These are usually low-severity fires that probably have minor effects on species composition. Fires that burn completely through the understory are infrequent, and crown fires are very rare (Batista and Platt 1997; FNAI 2010). High-severity fires do occur in live oak-cabbage palm hammocks (i.e., mesic or hydric hammock) of southern and coastal

Florida (Platt and Schwartz 1990). Many observers have noted the ability of the fire-adapted cabbage palms to "throw" flaming fronds ("firebrands") into the oak canopy, killing or pruning back oaks to the palm's advantage.

As noted earlier, canebrake, a frequent-fire community, may exist as an alternative stable state to bottomland forest, an infrequent-fire community, in alluvial floodplains (Gagnon and Platt 2008). Canopy gaps created by treefalls or other events favor cane and other grasses in the groundcover, increasing flammability and encouraging ignition after a lightning strike.

Xeric hammock was mentioned as developing on scrub or sandhill sites with histories of fire exclusion. Xeric hammocks often develop from oak domes, which are clones of often single genetic individuals with a domelike shape due to fire-pruning of their edges. Oak domes have become common in sandhills in central and north Florida. After a decade or two of fire exclusion, clones of sand live oak and myrtle oak can become fire-resistant.

Guerin (1993) determined that oak domes in the Riverside Island sandhill in Ocala National Forest ranged from 30 m^2 to 1,000 m^2 in area and up to 10 m in height. Domes were usually spatially associated with turkey oaks, perhaps because less pine litter in pine canopy gaps (where turkey oaks mostly occur) reduces fire intensity and favors oak survival (Guerin 1993). Domes greater than 2–3 m in height can maintain their aboveground structures during and after fire, whereas domes less than 2 m tall are topkilled, but then resprout from rhizomes. The sand live oak domes in this study were more fire-resistant than domes of myrtle oak. Winter burning has undoubtedly favored survival of the oak domes. These oaks apparently increase carbohydrate storage in their rhizomes and roots in the fall, making them more capable of resprouting after winter burns than late spring or summer burns, when carbohydrate stores are depleted by new growth (Guerin 1993; see also Robertson and Hmielowski 2014).

Two alternative stable states on the rocklands of the Miami Rock Ridge and in the Florida Keys are pine rockland, a frequent-fire community, and rockland hammock, a rare-fire community. The latter is a tropical hardwood forest, with more than 120 native tree and shrub species in the canopy and shrub layers. Most of the tropical taxa in this community reach the northern extent of their ranges in south or central Florida (FNAI 2010). Snyder et al. (1990) observed fires originating in pine rockland enter the edges of rockland hammocks during drought. Although these fires usually extinguish rapidly in the humid environment of the hammock, fires can propagate through hammocks when they are very dry, killing trees and consuming the organic soil layer. These fires move slowly and can smolder in the leaf litter and interstices of the limestone for weeks or longer (Sah et al. 2006). Rockland hammock can reestablish within 25 years after fire, but its full species richness and structure may require more than 100 fire-free years to recover.

For other natural communities characterized by rare and infrequent fire (Table 4.2) but not discussed here, some combination of topographic protection from fire (e.g., slope forest, upland hardwood forest, mesic hammock, hydric hammock, sinkhole), water-barrier protection (e.g., coastal berm, coastal grassland, coastal strand, coastal interdunal swale, mangrove swamp), or limited flammable fuels (e.g., beach dune, coastal berm, limestone outcrop, Keys cactus barren, salt flat, Keys tidal rock barren) keeps these communities from burning regularly under natural conditions. Anthropogenic fires are known to occur in some of these communities, but given their landscape context and limited flammability, fires are usually not extensive. With global climate change and the anticipated increases in moisture stress (and for coastal communities, sea-level rise), these communities could undergo profound and unpredictable change.

Fire Management

Can We Maintain or Mimic the Evolutionary Environment?

> Only recently has the idea emerged that conservation of entire fire-adapted floras and ecosystems depends on understanding historical fire regimes and on managing human fire regimes so that they mimic historical fire regimes.
>
> Platt et al. (2015)

Modern humans evolved in fire-dependent African savannas. There and elsewhere we used fire for hunting and to manage landscapes for tens of thousands of years. We continue these traditions today (Pyne 1982, 2001). In Florida and the southeastern Coastal Plain humans have applied fire to the land since at least late Archaic time, ca. 4,000 years ago (Milanich 1998; see chapter 2). People burn landscapes for many reasons: pest control, promoting favored plants or wildlife, improving visibility for hunting or predator/enemy detection, surrounding or driving prey, reducing the chance of undesired wildfire, and for aesthetics, ritual, and entertainment.

Given this long and intimate relationship between humans and fire, one might expect our understanding of how to manage fire would be exquisite and fine-tuned. Instead, fire is one of the most controversial topics in the entire arena of land management. In the southeastern Coastal Plain, the good news is that virtually all land managers recognize the need for fire in our pyrogenic ecosystems. Moreover, as discussed later, some states in this region have the most fire-friendly liability laws in the nation. The bad news is that the need to burn in a way that simulates evolutionarily relevant fire regimes is often not appreciated and is sometimes ridiculed by fire managers. Even when evolutionary context is appreciated, various practical, cultural, and political constraints preclude ideal fire management.

Field biological research and natural history observations continually improve our knowledge of the effects of fire on species and ecosystems. Nevertheless, scientists and managers disagree among themselves about such fundamental questions as how frequently to burn, at what time(s) of year, under what weather conditions, and how patchily. Little consensus exists on such fundamental questions as under what conditions—if ever—lightning-ignited or other wildfires should be permitted to burn or what to do after a high-severity wildfire—for example, apply salvage logging or allow natural recovery (Lindenmayer et al. 2004; Lindenmayer and Noss 2006). These latter questions are asked more regularly by fire ecologists and managers in western North America (e.g., Noss et al. 2006; Hessburg et al. 2015), but they are just as applicable to the remaining large natural and seminatural landscapes of the southeastern Coastal Plain.

Lightning fire has very likely been a component of the evolutionary environment of native plants and animals in the Coastal Plain for millions of years, perhaps 10,000 times longer than anthropogenic fire has been present (see chapter 2). The best we might do is try to simulate this evolutionary fire regime with controlled burning. By "simulate" I do not mean replicate precisely some historic fire regime. Fire regimes change naturally over time with changes in climate, abundance of large herbivores, and other factors (see chapter 2). A strict adherence to historical models of fire regimes could constrain understanding of the myriad and dynamic interactions of fire with other environmental variables such as drought and herbivory (Freeman et al. 2017). Rather, what I suggest is we learn all we can about the qualities of the historic fire regime, such as frequency, seasonality, extent, and heterogeneity; the long-term range of variability in these components; and how species and ecosystems in our region respond to this variability. Such knowledge provides the foundation for a modern, managed fire regime that will maintain our native biodiversity and ecosystem functions into the future.

I prefer the term "controlled burning" to "prescribed burning" because the former makes more sense to laypeople, but also because "prescribed burning" is a bureaucratic phrase created by forestry agencies opposed to woodsburning by rural people (Stoddard 1969). It is not unreasonable to suggest that unauthorized burning of rural landscapes by local people can be a good thing ecologically (Putz 2003, 2015), even though today we call it arson. Such traditional burning remains a fairly common and often beneficial practice in the South, although the tradition is in general decline (Coughlan 2016). It would be better still if woodsburners studied fire ecology—not necessarily formally, but through careful observations of fire's effects on species and communities—and sought to simulate natural lightning fires. Still, for fire-dependent ecosystems of this region, virtually any fire is better than no fire.

The focus of this book is on fire as an ecological and evolutionary force that has shaped and continues to shape our native biota. Controlled burning is essential in our modern, altered landscapes because lightning fires are now rarely allowed to burn. Even if they were, such fires would not be frequent enough in a fragmented landscape. With natural and seminatural areas now closely juxtaposed with houses, highways, and other human infrastructure, fires ignited by lightning, accidents, or arson can place human lives and property at risk. Smoke from uncontrolled fires, and sometimes from controlled burns, can and does kill people on highways and often poses health risks. Controlled burns that keep fuel loads relatively low are needed to reduce these very real hazards. Controlled burning designed to mimic the qualities of a lightning-fire regime and carried out by experienced fire managers may be just as beneficial for fire-dependent species and ecosystems as lightning fire.

The efficacy of controlled burns in reducing the threat of wildfire is well established, with both number of wildfires and area burned declining with the extent of controlled burning (Freeman 2004; Mitchell et al. 2009; Addington et al. 2015). Controlled burning also can moderate extreme wildfire behavior (Ryan et al. 2013). Nevertheless, I worry that the bias in favor of strictly controlled burning and against wildfire reflects our domineering attitude toward nature. We want to control nature, rather than respect it and learn from it. A way out of this philosophical dilemma is to use controlled burns to substitute for lightning in human-dominated landscapes where unplanned ignitions are too risky, while allowing wildfires to burn (a practice known as managed wildfire) where and when it is safe to do so. In many landscapes, all wildfire may be deemed too risky. This is increasingly the case as the human population and development explodes in the Southeast (Terando et al. 2014). On the other hand, some large blocks of wild public land still remain. These areas provide an opportunity to observe the behavior and results of truly wild fires and learn how to better mimic the evolutionary fire regime with our management.

In this chapter I summarize how knowledge of the ecological and evolutionary role of fire in Florida and the southeastern Coastal Plain might be translated into guidance for on-the-ground conservation, restoration, and management of fire-prone ecosystems. My premise is that the paramount goal of management in natural and seminatural landscapes is to maintain native biodiversity at all levels of biological organization. I make no apology for this ecocentric worldview. Anthropocentrism is the implicit ideology guiding almost all affairs in our society. For treatment of natural and seminatural lands, however, a more encompassing land ethic is required. As Aldo Leopold (1949) put it, "a land ethic changes the role of *Homo sapiens* from conqueror of the land community to plain member and citizen of it. It implies respect for fellow citizens and respect for the community as such." Respect for a natural community and its citizens demands that we maintain or simulate with management the ecological and evolutionary processes characteristic of that community, including the natural fire regime.

Objectives of Fire Management

In addition to reviewing the literature on fire management for this book, I visited with fire managers and fire ecologists, participated in their meetings and conferences, and accompanied them in the field to hear the opinions, knowledge, and lessons they have to offer. I wanted to identify areas of consensus as well as areas of disagreement or dissent. I also regularly visited managed areas across Florida and other southeastern states to observe the effects of recent and past wildfires and controlled burns, and to watch fires in action. I have been doing this for decades. Because a fundamental attribute of nature is variability, I assumed from the start that no single correct answer exists to any question about fire management. It all depends on the ecological and practical context. Nevertheless, some practices, techniques, and burning regimes have proved more successful in meeting ecological goals than others. My intent is to take an evidence-based approach and summarize the "best practices" for fire management toward ecological goals and objectives in this incomparable region.

The present lack of consensus about best practices for fire management partly reflects differing philosophies and attendant objectives. Some common objectives of controlled burning include: (1) resource management for forestry, range, and wildlife production; (2) reduction of hazardous fuel loads to lower the probability of occurrence or severity of undesired wildfire; (3) creation of suitable habitat for threatened and endangered species; and (4) restoration or maintenance of native ecosystem structure, function, and composition. These objectives are not mutually exclusive, but the emphasis given to each varies widely. An individual fire manager could have all of these objectives generally in mind for a particular site, but must emphasize one or two objectives over others at any given time. This preference may be due to orders from above, rather than to the manager's personal feelings or knowledge about the ecological aspects of controlled burning.

Even where managers agree on general goals, such as burning frequently to promote biodiversity, many burn at inappropriate seasons. This is often due to "burn bans" or other restrictions imposed by forestry agencies. Managers also frequently overapply plowed or disked fire lines, which have a variety of deleterious effects (see below). Hence, managed fire regimes may not fully meet the needs of fire-dependent species, many of which are imperiled in part due to past mismanagement. Many factors can account for these problems. One factor is fire training programs that focus more on equipment, regulations, permits, and techniques than on ecology, evolution, and natural history. Still, the constraints against ecologically informed burning are less onerous in some southeastern states than anywhere in the United States (Wonkka et al. 2015).

Why It Is So Important to Get Fire Management Right in Florida and the Coastal Plain

Suboptimal fire management, such as not burning often enough or burning consistently in the dormant season, might not be especially problematic if carried out in a low-biodiversity region that has little to lose from improper management. That is not the case here. The Coastal Plain is globally significant for its biodiversity, especially its unusually high level of endemism for a mostly nontropical region. Florida is the hottest spot biologically within the 1.13 million km^2 North American Coastal Plain biodiversity hotspot (James 1961; Estill and Cruzan 2001; Sorrie and Weakley 2001; Noss et al. 2015). Some 85 percent of the endemic plants in the Coastal Plain are associated with pinelands and embedded communities such as depression wetlands (Bruce Sorrie and Alan Weakley, unpublished data). All of these communities and many of their rare species depend on very frequent fire and, in some cases, on intense fires during the lightning-fire season (e.g., Ames et al. 2017). In the southeastern Coastal Plain, we have a lot to lose if we fail to mimic the evolutionary fire regime accurately enough.

A nationwide assessment of endangered ecosystems (Noss et al. 1995) determined that although many forest, wetland, and aquatic community types are endangered, generally the most severe conversion and degradation of terrestrial ecosystems have occurred among grasslands, savannas, woodlands, and shrublands. Along with direct conversion to agriculture, a leading cause of habitat loss and degradation among these frequent-fire ecosystems is disruption of the natural fire regime.

As ecosystems become endangered, so do the species associated with them. Although ecosystem conservation and management have many advantages over single-species conservation, especially with respect to cost-efficiency (Noss 1996), an unfortunate reality is that many species now require individual attention to avoid extinction. To save these species we must first identify and protect the sites where they occur. This is not happening nearly quickly enough in Florida and other Coastal Plain states due to political leadership that is hostile to conservation. Funding for land conservation is pathetically low.

In those places where local populations of imperiled taxa are formally or informally protected in managed areas, the crucial next step is managing their habitats to meet their life-history needs and maintain population viability. Such management requires knowledge of the evolutionary environments of those species. Explicit information about evolutionary or historic habitat conditions is lacking for most imperiled species, but their responses to variation in contemporary fire regimes, such as differences in frequency or seasonality of fire (see chapter 3), provide a good sense of the regime with which they evolved. This knowledge provides insights about how to manage fire in their habitats today.

Again, species adapt to the specific disturbance regimes they experience during their evolutionary histories, not to disturbance per se (Keeley et al. 2011). If we alter a fire regime outside the range of variability that species experienced during their evolution, then we can expect many species to decline. In the southeastern Coastal Plain, more species are vulnerable to too little fire than to too much fire. Not enough fire in pyrogenic communities eventually results in a shift to an alternative stable state, such as secondary hardwood forest, which in this region is almost always biologically impoverished and extremely difficult to restore back to a frequent-fire community. If this shift happens in enough places, eventual extinction of fire-dependent species is inescapable.

Given its exceptional biological values, the Coastal Plain deserves and requires the highest quality land management possible, performed by the best-trained and most-experienced personnel. These must be people who place high value on maintaining and restoring native biodiversity.

Otherwise, irreplaceable biological values could be lost forever. I make this point with full realization that fire manager is one of the most challenging professions in the conservation field, and that without these dedicated professionals out there burning the landscape on a regular basis, much more biodiversity would have been lost than already has been.

The Opportunity Exists to Get Fire Management Right in the Southeastern Coastal Plain

Fortunately, we have a better chance of getting fire management right in Florida and the southeastern Coastal Plain than perhaps anywhere in the world. Not only was the science of fire ecology born in Florida and southern Georgia, the practice of fire management appears better established here than anywhere. It is unfortunate that managers are praised when they successfully suppress wildfires, but receive little recognition when they conduct a high-quality prescribed burn or use managed wildfire (Ryan et al. 2013). Still, fire managers collectively are doing a remarkably good job burning natural and seminatural areas, which due to landscape fragmentation and fire suppression would not burn nearly often enough from lightning ignitions or other wildfires.

Landowners and managers in Florida burn more acres each year than in any other state (Scott et al. 2014; Pyne 2016), followed by Georgia. Florida is widely considered a model for "right-to-burn" legislation, and is emulated by other states (Ryan et al. 2013). Prescribed Fire Councils, composed of fire management practitioners, originated in Florida and have subsequently spread to other southeastern states, across much of the United States, and to British Columbia (http://www.prescribedfire.net/).

The Florida Forest Service reports more than 809,371 ha (2,000,000 ac) are prescribe-burned in Florida each year at present (http://www.freshfromflorida.com/Divisions-Offices/Florida-Forest-Service/Wildland-Fire/Prescribed-Fire). The Georgia Forestry Commission reports a current average of 607,028 ha (1,500,000 ac) prescribe-burned annually (http://www.gatrees.org/resources/media/news/2–05–14%20Prescribed%20Fire%20Awareness%20Week.pdf). Curiously, these acreages add up to more than 100 percent of the area prescribe-burned nationally, according to the National Interagency Fire Center (https://www.nifc.gov/fireInfo/fireInfo_stats_prescribed.html). More defensible estimates of the extent

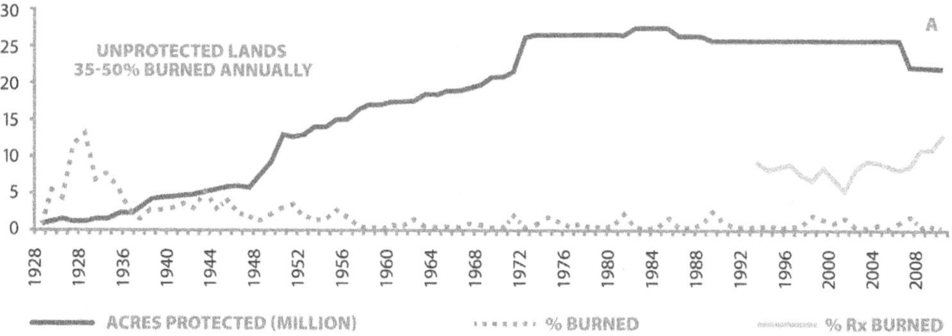

Figure 5.1. Wildfire (percent burned), wildfire suppression (acres protected from fire), and controlled burns (percent Rx burned) in Florida from 1928 through 2010. Wildfire peaked in 1932 but was sharply reduced by enhanced suppression efforts beginning in the mid-1930s. Recording of controlled burning did not begin until 1992; after some fluctuation, this burning has been increasing since the mid-2000s. With more controlled burning, wildfire acres remained low, and not as much fire suppression was needed. Adapted from Scott et al. (2014) based on data from the Florida Forest Service.

of controlled burning are needed and should not be too difficult to obtain through better monitoring.

The southeastern states possess a well-trained army of fire managers equipped to burn the natural communities that need fire. With recent increases in the area control-burned in Florida, the area burned in wildfires has declined and the area protected from fire (fire-suppressed) has begun to decline as well (Fig. 5.1). A further step in the evolution of fire management would be to expand the use of managed wildfire (North et al. 2015) in landscapes where it is safe to do so and under acceptable weather conditions. That is, lightning fires and other wildfires of uncertain origin could be left to burn themselves out, so long as the perimeters of a managed area are well secured and the fire poses minimal risk to human life or property. Although a tradition of managed wildfire exists in western and northern North America, and is growing, it is not well accepted in the East. Although logistically challenging, more managed wildfire would help reduce the "fire deficit" that still plagues the entire United States (Ryan et al. 2013).

Concern about risk of fire escaping a burn unit and the associated liability from any damage incurred is one of the greatest constraints on controlled burning, often discouraging landowners and managers from applying fire (Yoder et al. 2004). Smoke from fires impacting highways is another major source of concern. Different types of liability laws strongly influence the amount of burning that occurs in various states. Current liability laws can be categorized as strict liability, simple negligence, or gross negligence statutes. Arguably the worst places to be a burn boss in the United States are Wisconsin and Minnesota, where strict liability statutes hold a burner liable for any property damage caused by an escaped fire, with no regard to negligence or lack thereof. Many states have simple negligence statutes, which require burners to exercise reasonable care when burning, and which require plaintiffs to show negligence on the part of the burner for the burner to be held liable for damages caused by an escaped fire. About half of U.S. states have no defined prescribed fire liability standard, in which case they follow simple negligence standards from case law (Wonkka et al. 2015).

All states in the southeastern Coastal Plain have simple negligence statutes, except three: Florida, Georgia, and very recently South Carolina (Johnny Stowe, personal communication), which have gross negligence statutes. In these three states alone, if a burner follows standardized procedures when burning, a plaintiff must show "reckless disregard of the duty of care owed others by the burner" (Wonkka et al. 2015). In Florida, the Revised Prescribed Burning Law of 1999 protects the burner from liability if permission or consent of the landowner or the landowner's designee is obtained prior to requesting authorization from the Florida Forest Service to conduct the burn; a written prescription is prepared and authorized by the Florida Forest Service; the written prescription is on-site during the burn; at least one certified prescribed burn manager is present from ignition to completion of the burn; and the individual who is authorized to burn certifies that the area to be burned "has been properly prepared, including adequate firebreaks, and sufficient personnel and firefighting equipment will be on-site to assure control of the fire" (Brenner and Wade 2003).

Gross negligence laws allow—or should allow—for more liberal controlled burning. Comparing adjacent counties across state lines in Florida, Georgia, Alabama, South Carolina, North Carolina, and Tennessee, Wonkka et al. (2015) found that private landowners in counties with gross negligence standards burn significantly more area than those in counties with only simple negligence statutes. They also estimated that a switch from simple negligence to gross negligence laws would result in an additional 7,388 ha (18,256 ac) burned per county in an average year. These authors provide an interesting discussion of the evolution of gross negligence laws in Florida and Georgia. Adoption of gross negligence standards in other southeastern states could do much to stimulate more burning.

Despite the encouraging evolution of right-to-burn legislation in the Southeast, many fire ecologists argue that we are not burning nearly enough. Furthermore, the way we manage fire is often not optimal for meeting the life-history needs of fire-dependent species and for maintaining high-quality natural communities. We can be optimistic that fire management in the southeastern Coastal Plain, given current practices with a few tweaks, has a good chance of benefiting biodiversity, or at least a relatively low chance of doing serious harm. We can also assume that fire regimes have not been constant over evolutionary time; hence, most species should show some flexibility in response to fire. Nevertheless, from a biological perspective and across managed areas, fire management runs a gamut from superb to questionable. There is much room for improvement.

Fire Management Questions

This book is not a "how-to" instruction manual for prescribed burning, nor is it a detailed treatment of the practices of fire management. That topic requires a book of its own, though there are many on-line sources to guide fire management. For example, information on "fire tools" can be found at http://www.freshfromflorida.com/Divisions-Offices/Florida-Forest-Service/Wildland-Fire/Resources/Fire-Tools-and-Downloads. Rather, this book is meant to provide the ecological and evolutionary foundation for designing and implementing fire management plans that simulate, to the extent feasible, the historic fire regimes for the ecosystems of the southeastern Coastal Plain. I organize my discussion of fire management here as questions a conscientious fire manager might ask in attempting to develop and implement a fire management program, consisting of a series of "prescriptions" or burn plans designed to achieve the best results for individual sites.

Implementing an ideal fire regime is challenging. Many fire ecologists and managers accept that controlled burning should attempt to mimic natural lightning-fire regimes. This must be done in a way, however, that fully addresses practical considerations, including human safety and health, as well as liability. Specific management objectives frequently prevent a fire manager from mimicking natural fires precisely. Some fire-excluded natural communities have transitioned to an alternative stable state, perhaps even a novel ecosystem, where restoring a historic fire regime may be prohibitively expensive or even impossible. In cases where nonnative plants have drastically altered fire regimes, attempts to reestablish a historic regime could have negative consequences (Freeman et al. 2017).

A reasonably comprehensive knowledge of the fire ecology of the region where one works should be a requirement for fire managers. Many fire managers possess this knowledge, gained through various sources, including long experience with fire. Others do not. Stepping up the scientific training of fire managers in fire ecology and other aspects of ecology, natural history, and evolutionary biology is critical to improving fire management in Florida and elsewhere. This education must be accompanied by on-the-ground training, observation, tutoring, and experience. As one of Florida's most respected fire managers remarked to me, "It should be much harder to become a fire boss, but much easier to burn once you are one." This comment refers to the often inflexible and overly restrictive requirements for obtaining a burn permit, especially during the lightning-fire season, even in one of the states with the most enlightened right-to-burn legislation.

SHOULD I STRIVE FOR PYRODIVERSITY?

Pyrodiversity is essentially "variation in fire regimes in time and space" and has been recommended as a bet-hedging strategy that promotes coexistence of species with disparate life histories and requirements with respect to fire (Menges 2007). Some authors specifically suggest "pyrodiversity begets biodiversity" (Martin and Sapsis 1992). This assumption must be examined critically and not interpreted casually or naively. Pyrodiversity could be measured in a variety of ways and applied to several distinct components or attributes of a fire regime (Table 5.1). Responses of biodiversity to increases in one measure of pyrodiversity could differ substantially from responses to other measures, and it is unclear how maximum pyrodiversity by any measure would constitute a legitimate management goal.

The history of conservation shows that attempts to achieve maximum diversity of habitats or species are fraught with dangers (Noss 1987b). The highest possible species richness on a local scale is not usually an appropriate goal. As Diamond (1976) warned during the early years of the debate over single large vs. several small reserves, "species must be weighted, not

Table 5.1. Potential applications of the pyrodiversity concept to different attributes of a fire regime

Frequency (Return Interval)	Intensity (Heat) or Severity (Effect on Vegetation)	Seasonality	Patch Size	Patch Interspersion	Internal Patch Structure
Create heterogeneity in fire-return intervals among patches, e.g., an even frequency distribution or one weighted toward shorter intervals	Work with weather and fuel conditions to increase or maximize variability in fire intensity or severity among patches	Burn different patches or burn units in different months or seasons or, alternately, burn each patch in a different season or month at each burn iteration	Vary the fire compartment or size of patches burned, e.g., with an even frequency distribution of very small to very large patches	Intersperse fire treatments, such that patches with different fire-return intervals, fire intensities, seasons, or patch sizes are juxtaposed	Use fire treatments to create variability among patches in habitat structure, e.g., herbaceous vs. shrub or tree cover

Not all of these applications are necessarily appropriate (see text).

just counted." This means that the most extinction-prone species, including narrow endemics and habitat specialists, should be prioritized over more widespread, weedy, or opportunistic species, as the latter usually are not in much danger in a human-dominated landscape.

In fire management, such a precautionary approach would give greatest weight to those specialized endemic species that have demanding requirements with respect to fire, as opposed to species with more generalized needs. Avian examples of highly specialized fire-dependent species include red-cockaded woodpecker (*Leuconotopicus* [*Picoides*] *borealis*), Florida scrub-jay (*Aphelocoma coerulescens*), and Florida grasshopper sparrow (*Ammodramus savannarum floridanus*). Plant examples abound (see chapter 3). A narrow endemic species such as Garrett's mint (*Dicerandra christmanii*), found on only four or five tiny sites (only one of them protected) in a 3 × 6 km area of yellow-sand scrub on the Lake Wales Ridge, must be conserved properly on those sites or it will become globally extinct. The survival of this plant and the fire regime it requires must take precedence over pyrodiversity on those sites.

As a general concept for fire management, pyrodiversity is certainly valid. Several studies show that some measures of pyrodiversity are associated with high richness of native species, including rare ones. Species assemblages in some regions—for example, plants, macrofungi, and several animal taxa in southwestern Australia—appear resilient to considerable variability in fire intervals (Wittkuhn et al. 2011). Hence, this form of pyrodiversity is at least not harmful to species richness there. Similarly, flexible fire regimes are tolerated by savanna termites in Africa (Davies et al. 2012). More positively, in the Brazilian Cerrado biodiversity hotspot, more than 40 percent of overall ant species richness was attributable to differences in species composition among fire treatments, specifically with respect to frequency and seasonality of fire. A diversity of fire regimes on the landscape promotes ant diversity (Maravalhas and Vasconcelos 2014). Research in Yosemite National Park shows that pyrodiversity, as measured by variability in frequency, age, extent, and severity of fire, is positively associated with increased richness of flowering plants, pollinators, and plant-pollinator interactions. Reduction in pyrodiversity, coupled with increased drought, is predicted to reduce this richness (Ponisio et al. 2016).

Pyrodiversity often is implemented as "patch mosaic burning" (Parr and Andersen 2006), where managers create dynamic mosaics of patches representing different fire-return intervals or times-since-fire. In some cases, this approach has succeeded in maintaining a diversity of species, including imperiled taxa with varying habitat requirements and responses to postfire vegetation development (Menges and Quintana-Ascencio 2004). Applying a spatial model, Duncan et al. (2015) showed that patch mosaic burning, expressed as a multiple-age fuel mosaic, reduces fire hazard better than typical fuels-reduction burning in east-central Florida. The rotation of small fires over multiple years produces a shifting mosaic of scrub patches that mimics natural fire regimes and is favorable to Florida scrub-jays and other species requiring heterogeneous fire patterns.

In contrast, Parr and Andersen (2006) found that patch mosaic burning in tropical and subtropical savannas of Africa and Australia is not based on a thorough understanding of natural fire regimes in these ecosystems. They were concerned that controlled burning regimes based on "pyrodiversity rhetoric" lack substance "in terms of operational guidelines and capacity for meaningful evaluation" (Parr and Andersen 2006). Kelly et al. (2015) used extensive field surveys of birds, reptiles, and mammals in Australia; species distribution modeling; composite indices of biodiversity; and decision science to define fire-management objectives based on multiple species. They found that maximizing pyrodiversity in terms of an even allocation of postfire successional stages failed to maximize biodiversity, especially because in this landscape older vegetation is disproportionately important for birds, reptiles, and small mammals. No study of this type has been conducted, to my knowledge, in the southeastern Coastal Plain. I suspect results would be similar here, with the important difference that younger postfire patches would likely be disproportionately valuable, especially to pine savanna and grassland specialists.

The generality of the pyrodiversity-begets-biodiversity hypothesis has been challenged by other studies (e.g., Taylor et al. 2012; Nimmo et al. 2013). Collectively, research demonstrates that increasing landscape heterogeneity by practices such as patch mosaic burning can negatively affect specialized species that require a highly specific postfire habitat structure or patch size. A danger is that managers, in the pursuit of pyrodiversity, might be tempted to try a bit of everything in terms of fire-return interval, season of burn, patchiness of burn, and so on, even if some of the treatments are outside the evolutionary experience of the species in the treated communities. As recognized by Menges and Quintana-Ascencio (2004), "radically long or short fire-return intervals, complete fire suppression, or unusually high fire intensities may be beyond the tolerances of certain species." A more enlightened approach to pyrodiversity might be to view it as "the outcome of the trophic interactions and feedbacks between fire regimes, biodiversity, and ecological processes" (Bowman et al. 2016).

Clearly, a highly regimented or uniform fire regime does not mimic nature, but neither does a randomized approach. Robbins and Myers (1992) suggested a method for scheduling burns to provide pyrodiversity in terms of frequency and seasonality, while still considering (to some extent) the evolutionary environment. They produced several charts to assist fire managers in planning burn schedules that would vary frequency and seasonality of burns, thus hypothetically meeting the needs of multiple species with diverse responses to fire frequency and season. Their intent was "to enable managers to easily develop burning schedules that mimic the variability found in nature, and thus avoid artificially uniform schedules" (Robbins and Myers 1992). The example in Figure 5.2 is a portion of their chart for a mesic longleaf pine–wiregrass community. Such a semirandomized approach to scheduling burns could easily produce a managed fire regime outside the historic range of variability—or at least on the extreme edge of that range—for this or any community type.

3v	8s	8s	2f	6f	7f	3w	10s	9s
4s	2f	7v	4v	7s	4v	2v	4v	1v
9s	9s	4f	10s	2w	10w	10v	3v	2s
4s	1v	3v	1v	9s	2s	4w	4s	5f
3v	3v	10v	2w	2s	3s	2v	5f	2s
2s	9s	10s	2v	5v	8v	4f	9v	3w
4s	5v	1v	3f	8w	1v	7v	9s	2s
7v	6f	2s	4s	1s	8v	2v	7f	4v
6s	1v	10f	1v	7w	7f	6s	4v	2v
8f	6s	6s	1s	3s	9v	9s	2v	4f
1w	2s	1s	10v	10w	1v	7v	2s	10v
10v	6v	10w	3f	4s	4s	1s	3s	1v
1w	3s	10f	10v	3v	5s	5s	7w	10s
10s	8w	7s	6f	6v	2s	3w	6v	1s
4v	3w	8v	6s	7v	3v	2v	4s	3f
3s	2f	1w	5v	5v	4w	3s	3v	9w
1s	3w	9v	2v	2w	3s	1s	4f	4f
10v	4f	3v	3v	3f	6w	4s	2v	6v
2s	1s	4f	2s	10w	4v	1w	9v	8s
4f	7f	5s	4f	1w	10v	1w	3w	9s
4v	2v	2f	6w	4f	4s	5v	9f	10s
8s	7v	9v	7v	8f	5s	7v	7v	5w
9s	4s	5w	3v	1s	2v	4s	9s	8s
3w	3s	2w	4s	1s	4w	2s	10f	8s
3s	4s	9v	10s	9v	2v	8s	4s	8v

HOW DO I DETERMINE HOW FREQUENTLY TO BURN?

Most fire ecologists in the southeastern Coastal Plain agree that frequency is the most critical component of the fire regime for fire-dependent species and ecosystems here. Frequency also is the least difficult aspect of the fire regime to manage within the constraints of society (Kevin Hiers, personal communication). Despite recent concerns about increasing incidence and severity of wildfires, especially in the western United States, the land area burned across the continent in an average year is much smaller today than prior to European settlement (Ryan et al. 2013). The area of the 48 conterminous states burned each year in the late twentieth century has been estimated as 7 to 12 times less than in preindustrial times (Leenhouts 1998). As shown previously (Fig. 5.1), Florida is better off than most states, in that the area control-burned has been generally increasing over recent years. Still, ecological observers are virtually unanimous in the opinion that not nearly enough is being burned each year in Florida or elsewhere in the Coastal Plain, given the ecological objective to burn within the historic range of fire-return intervals. This is a global problem. Fire activity worldwide declined 24 percent from 1998 to 2015, mostly due to less burning in savannas and grasslands because of agricultural expansion and intensification, driven by human population growth (Andela et al. 2017).

Opposite: Figure 5.2. A quest for pyrodiversity, in terms of frequency and seasonality, could lead to perverse outcomes. This example is for a mesic longleaf pine–wiregrass community. Numbers refer to return intervals for controlled burns and letters to season of burn (W = winter; V = spring; S = summer; F = fall). Ten years was assumed to be the maximum fire-return interval, with numbers between 1 and 10 generated randomly but with certain numbers weighted according to moisture class. For a mesic site, lower numbers have twice the probability of being generated. Season of burn was weighted so that spring and summer have twice the probability of appearing as fall or winter. One begins by selecting a number/letter pair randomly—for example, by dropping a pencil (eraser side down) on the chart. Where it hits is your first fire-return interval and season. One can then randomly select the next interval/season or simply proceed down the column. In this example, I randomly chose 1S (circled), so in the first year I would burn the site in the summer. Proceeding down the column (arrow), my next burn is not for 10 years, and in winter; then another 10 years in the fall, and so on. These intervals and seasons are outside the historic range of variability and would lead to degradation of a mesic longleaf pine–wiregrass savanna. Adapted from Robbins and Myers (1992).

Managers of public and private conservation lands in the southeastern Coastal Plain typically attempt to burn pine savannas (flatwoods, sandhill) and other grasslands on fixed intervals of around one to five years. Fire-return intervals of three or fewer years are common for private conservation areas, whereas many public lands have intervals exceeding four to five years for communities characterized by frequent fire (personal observation). This may reflect constraints on burning more than intent. Although natural fire intervals in these communities can range from 1 to around 10 years (Christensen 2000; Frost 2006), in the most fire-exposed landscape positions intervals are strongly skewed, with the most common intervals toward the shorter end of the range (Huffman and Platt 2014). This suggests managers should conduct most burns in savannas and grasslands of the southeastern Coastal Plain toward the more frequent end of the estimated historic range of variability in their subregion, while still allowing for variability in all components of the fire regime.

Within limits, frequent-fire communities in the southeastern Coastal Plain are resilient to variability in fire frequency (Hinman and Brewer 2007). Burning on consistent intervals, although common, is not optimal for maintaining biodiversity. Fire-return intervals substantially longer than the mean or median interval occurred occasionally under a lightning-fire regime, and these windows of time without fire provide temporal refugia to species sensitive to very high fire frequency. Longer intervals enhance seedling and juvenile survival of longleaf and other pines, as well as the pyrophytic oaks and other fire-adapted hardwoods that are legitimate components of many pine savanna and woodland communities in this region (Hiers et al. 2014; Loudermilk et al. 2016; Varner et al. 2016).

A key consideration for determining the optimal range of fire-return intervals is that different landscape positions within the same general area or even the same community type would vary in their natural fire frequency. Extensive flat and undissected fire compartments would normally burn most frequently, whereas fire shadows and higher-moisture sites adjacent to lakes and large wetlands; slopes and some ridges; hilly landscapes; ecotones with hardwood forest, swamp, or water bodies; and other sites topographically protected from fire would burn less frequently. Huffman et al. (2004) and Henderson (2006) documented through fire scars that longleaf and slash pine savannas on islands in the southeastern Coastal Plain have longer fire-return intervals than mainland sites.

Sandhills, because of their lower productivity and sparser fine fuel loads, tend to require more time after fire to become combustible again than do flatwoods or dry prairie (Christensen 1981; Robbins and Myers 1992). Whereas mesic flatwoods and dry-mesic sandhills accumulate enough fine fuels to burn at one-to-two-year intervals, the less productive xeric sandhills may require three to five years to accumulate sufficient fuels to burn, and fires there are often less severe due to the lower density or continuity of fuels (Henderson 2006).

Moderately fire-protected sites can be expected to contain more tree-size pyrophytic oaks and mockernut hickory (*Carya tomentosa*), as well as fox squirrels (*Sciurus niger* ssp.) and other animals associated with pine-hardwood communities. One such community is the "red oak woods" (Harper 1915), or upland mixed woodland, which has a natural fire-return interval in the range of 2–10 years (FNAI 2010; see chapter 4). Because of the tendency of many managers to inadvertently manage for homogeneity by sticking to a fixed fire interval, and with the concept of an open, monotypic pine savanna in mind, such ecotonal communities might be even more reduced in distribution today than are the quintessential vast, open pine savannas.

Debate continues on the question of what range of fire intervals will benefit the largest number of wildlife species. Recall from chapter 3 that Schurbon and Fauth (2003) controversially recommended fire-return intervals of three to seven years in longleaf pine communities to maintain a more diverse amphibian assemblage than what is found with one-to-three-year intervals. Critics claimed that such a low fire frequency would benefit less-threatened forest-associated species or generalists at the expense of longleaf pine specialists (Means 2006).

Darracq et al. (2016) used a comprehensive region-wide database and literature review to examine responses of longleaf pine–associated vertebrates to short (1–3 years), moderate (>3–5 years), and long (>5 years) fire-return intervals. Like Schurbon and Fauth (2003), they found greatest species richness of vertebrates at moderate fire frequency. Longleaf pine specialists were associated with either a short or moderate-frequency fire regime. The highest burn frequencies were optimal for specialists such as flatwoods salamander (*Ambystoma cingulatum*), eastern indigo snake (*Drymarchon couperi*), Bachman's sparrow (*Peucaea aestivalis*), brown-headed nuthatch (*Sitta pusilla*), and red-cockaded woodpecker, which were rare or absent from sites with moderate- or low-frequency fire.

Nevertheless, nearly as many longleaf pine specialists, by their categorization, including eastern diamondback rattlesnake (*Crotalus adamanteus*), Louisiana pine snake (*Pituophis ruthveni*), gopher tortoise (*Gopherus *polyphemus*), and northern bobwhite (*Colinus virginianus*), were either primarily associated with moderately burned sites or used sites across the spectrum of one-to-five-year fire-return intervals. Critics might argue that this second group is not as highly specialized to longleaf pine savannas as the first group. In any case, based on these findings, Darracq et al. (2015) recommended varying the frequency of fire between one and five years in longleaf pine landscapes. Evidence of historic fire frequency from fire scars (e.g., Huffman and Platt 2014) suggests that in at least some subregions of the Coastal Plain, such as central Florida, the more exposed landscape positions have fire-return intervals weighted toward the low end of that range.

Natural fire-return intervals are longer for some other fire-dependent natural communities in our region. Especially for communities with high natural variability in fire frequency, paying attention to the specific type (i.e., plant association) of community and to the needs of the fire specialists found on a given site is critical. For example, the modal fire-return interval for scrub varies from as low as 3–20 years for *Q. myrtifolia*–dominated scrub on Merritt Island, 5–12 years for yellow sand scrub on the peninsula, 10–40 or 20–60 years for rosemary scrub, to 100 years or more for sand pine scrub along the coast of the Panhandle (see chapter 4, Table 4.2). Managers must consider geography, landscape position, species composition, and site history, among other factors, in developing burn schedules.

Nonnative plants sometimes pose a challenge to managers attempting to replicate a historic fire regime. In many ecosystems around the world, exotic grasses have increased fine fuel loads and fire frequency. In pine rocklands and some other communities of south Florida and beyond, the exotic cogon grass (*Imperata cylindrica*) and *Neyraudia neyraudiana* increase the biomass of fine fuels and almost double the litter biomass, which is likely to alter the fire regime strongly (Platt and Gottschalk 2001). In contrast, the invasive shrub Brazilian pepper (*Schinus terebinthifolius*) is sensitive to fire, yet it can cause reductions in fire frequency through a fire-suppression effect when it reaches high density during fire-free intervals in south Florida (Stevens and Beckage 2009). This is yet another example of an alternative stable state mediated by plant-fire feedbacks. These cases suggest that fire managers often must also be exotic species managers to avoid substantial alteration of fire regimes.

HOW DO I DETERMINE THE APPROPRIATE TIME OF YEAR TO BURN?

The optimal season for conducting prescribed burns remains a divisive topic. Fire ecologists and managers argue strenuously about the advantages and disadvantages of burning during the season when most lightning fires occurred historically. This transition or lightning-fire season varies from year to year and geographically, but generally occurs between late April and July, before the peak of the summer wet season (Komarek 1964; Duncan et al. 2010; Platt et al. 2015). Burning in the winter and early spring was the traditional practice of many Indian groups, most early white settlers, some cattleman, and most foresters (see chapter 2), and is still practiced by the majority of modern fire managers, despite evidence that this usually is not the ecologically ideal fire season.

Some fire ecologists and managers advocate high variability in the seasonality of controlled burning, with the assumption that this form of pyrodiversity will maximize biodiversity (e.g., Figure 5.2). Still other managers acknowledge that even though lightning-season fires may be ecologically optimal, practical and sometimes legal constraints often preclude burning during this season. Thus, managers are left with the choice of burning during another season or not burning at all. Most wisely choose the former option to maintain high fire frequency, though the decision to avoid lightning-season fire is perhaps sometimes made too facilely as a matter of convenience or comfort.

Occasional fires outside the lightning-fire season are unlikely to do harm and are virtually always better than no fire. As reviewed in chapter 3, a combination of evolutionary logic and empirical evidence supports burning most often during the lightning-fire season. Nevertheless, just as our pyrophytic communities are resilient to some degree of variability in fire frequency, they are also resilient—probably more so—to some variability in fire season. Although presettlement lightning fires were concentrated in the late spring and early to middle summer, they occurred occasionally in almost every month of the year. In the Gulf Coastal Plain, which has the highest winter thunderstorm activity in the United States, a typical winter has around 10 thunderstorm days (Goodman et al. 2000; Henderson 2006). Seasonal pyrodiversity, if not overdone, could enhance biodiversity within fire-dependent ecosystems of our region, so long as the dominant season of controlled burning roughly matches the dominant season of lightning fire.

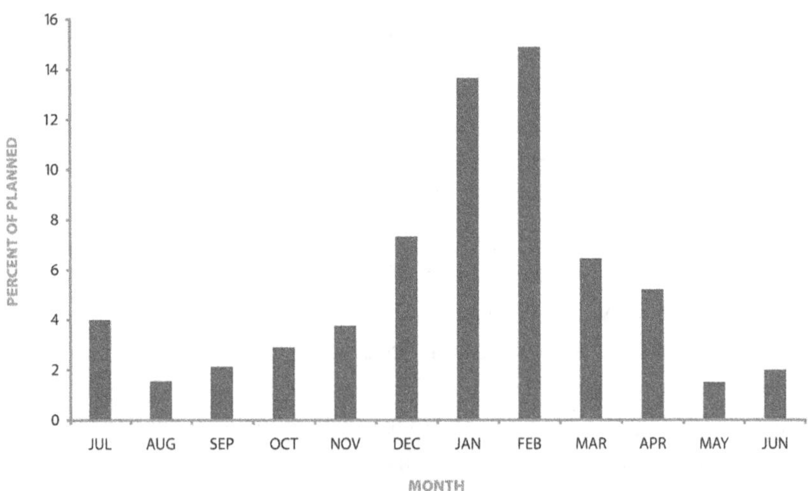

Figure 5.3. The monthly distribution of prescribed burns accomplished in Florida's state forests as a percentage of those annually planned. Percentages shown are averages from fiscal year 2000–2001 through 2007–2008. Adapted from Hardin (2010).

Once land managers in the southeastern Coastal Plain started methodically implementing controlled burns, following the period of active suppression that dominated the early to middle twentieth century, they continued the tradition of earlier woodsburners and burned in the winter and early spring. The eminent state park biologist Jim Stevenson (1993) notes that, for Florida state parks, "during the 1970s, managers traditionally conducted burns during the winter months because they thought it placed less stress on commercially valuable pines." As discussed in chapter 3, the evidence for this assumption is equivocal (e.g., Boyer 1987 vs. Streng et al. 1993; Glitzenstein et al. 1995). Stevenson (1993) reviewed other reasons managers burned in winter: "At this time of year, fires were thought to be easier to control because of predictable winds following the arrival of a cold front and cooler winter air temperatures are more comfortable for fire crews. Furthermore, there is no threat to ground nesting birds during this season."

Beginning with a training burn in 1977, the Florida Park Service started conducting more burns during the growing season because the beneficial effects of such burns, such as profuse flowering of wiregrass, were strikingly apparent. In 1982, the agency officially changed the controlled burning season from winter to the period from 15 April through 31 August. Winter burns were still used to reduce heavy fuel loads, but once fuel loads were close to natural levels, a switch was supposed to made to growing-season fire (Stevenson 1993). Many parks continued to burn largely in winter, however (Jean Huffman, personal communication).

Currently, most fire managers in Florida and elsewhere in the southeastern Coastal Plain are failing to mimic the historic seasonality of fire. In Florida's state forests between the years 2000 and 2009, controlled burns were concentrated in January and February (Hardin 2010), two months when lightning fires virtually never occur (Fig. 5.3). The fewest controlled burns (as percent of annual planned burns) occurred in May, the peak of the lightning-fire season, when the most area would burn under a natural fire régime (Komarek 1964; Robbins and Myers 1992; see Figs. 3.3 and 3.5). Although some in the agency are attempting to conduct more burns during the natural fire season (Brian Camposano, personal communication), the Florida Forest Service still conducts most burns opposite the natural seasonal pattern.

Arguments in Support of Burning during the Evolutionary (Lightning) Fire Season

The crux of the debate over burn season is whether burning during the evolutionary fire season provides conservation and restoration benefits beyond those achieved by burning at the evolutionary fire frequency. In other words, if we burn a site often enough, does it really matter during what season we burn it? Despite imperfect knowledge, powerful logic and a considerable body of empirical evidence support the practice of burning as much as possible during the dry-wet transition season and early wet season, when most lightning fires historically occurred. To summarize the arguments in favor of transition-season burning:

1. Species have evolved adaptations to fire, and if fire incidence varies seasonally, such adaptations should include responses to seasonality of fire.
2. A distinct lightning-fire season (transition season) exists in the southeastern Coastal Plain and presumably has for a long time.
3. Many species characteristic of southeastern ecosystems, especially savannas and other grasslands, respond most positively to fires occurring in the lightning-fire season. These species include the dominant and foundational C_4 grasses.
4. Fires outside the lightning-fire season have less beneficial effects, and sometimes negative effects on natural communities and

species of conservation concern. Negative effects of dormant-season fires include less control of hardwood understory or midstory and declines of fire-dependent and fire-propagating species. In the Sandhills of North Carolina, rare plant species have different functional traits from common species. The most significant difference is the presence in the rare species of traits that promote and intensify fire (Ames et al. 2017). The rare species have declined not only due to fire exclusion, but apparently also because of the dominance of early-season controlled burns, which are less intense than lightning-season fires. Ames et al. (2017) suggest that "the local application of hotter late-season fires may be used as a tool to create conditions favorable to these rare species."

5. Negative effects of late growing-season fires, when water tables are high, include damage to herbaceous vegetation and pines. Some botanists express grave concern about a recent trend among agencies in Florida toward more controlled burns in late summer and early fall, presumably for easier control of fire during a time of high fuel moisture and high water levels. Because most grasses, composites, and many other plants in savanna/grassland habitats flower and bear fruit in late summer and fall, burning then will sharply reduce sexual reproduction for that year. Burns during this time of year also eliminate seeds upon which wintering birds depend.
6. Typically, only lightning-season fires burn across wetlands, reducing woody encroachment and maintaining open water zones.

Platt et al. (2015) pointed out: "It is common for decisions to be made that prescribed fires inside and outside the fire season produce similar effects based on lack of pronounced changes in upland habitat, despite phenological and demographic data from savannas and grasslands suggesting subtle effects." It takes a skilled field naturalist to notice subtle effects on flora and fauna, but they are no less real. Enhanced training of fire managers in natural history might increase their ability to "read the landscape" in a biologically more sophisticated and accurate way.

Arguments in Support of Burning Outside the Lightning Season or for Variable Seasonality of Burning

Because fires in natural landscapes can die back, smolder virtually unnoticed for long periods, then flare back up again, the evolutionary fire season was potentially longer and more variable prior to EuroAmerican settlement than most contemporary fire ecologists assume. Given historically immense fire compartments, sometimes greater than 1,000 km^2 (Ware et al. 1993; Frost 2000) over much of the southeastern Coastal Plain, a fire ignited in one spot could travel for days, weeks, or perhaps even months across a broad landscape (Henderson 2006).

Most fire managers today are not allowed to let fires linger on the landscape. In Florida, daytime burn authorizations permit fires between 9:00 a.m. to one hour after sunset of the same day; nighttime authorizations are uncommon and depend on measures of smoke dispersion potential (Brenner and Wade 2003). Arguably, for ecological reasons the natural fire-storage process should be allowed on large sites if the burn perimeter is well secured and fire behavior is monitored. Fire storage is important in part because the initial fire may burn only the most flammable communities, such as pine savannas, whereas a stored fire may burn wetlands and other communities that are normally less combustible (Maynard Hiss, personal communication).

Season of fire is confounded with several meteorological variables that affect the intensity and severity of fire, including temperature, relative humidity, number of previous rain-free days, and wind speed. Weather conditions that are typical during the demarcated lightning-fire season can occur at other times of year, thus theoretically producing similar fire behavior and effects on vegetation. Variation in fuel loading, fuel moisture, fire weather, and other variables can often "obscure or accentuate seasonal differences" in the effects of fire (Robbins and Myers 1992). Some plant species may respond more strongly to weather and fuel conditions that affect fire intensity than to season, even though the ideal weather conditions most often occur during the transition season. According to a major review, "the latest research suggests that, in many cases, variation in fire intensity exerts a stronger influence on the ecosystem than variation in fire timing" (Knapp et al. 2009).

Some experienced fire managers tell me they can replicate lightning-season fire effects at other times of year by choosing the appropriate weather conditions. Nevertheless, plants and animals whose phenology is tightly tied to season of fire, perhaps because they respond to both photoperiodic and fire cues, may not respond as favorably to fires with identical weather and fuel conditions, but outside of the usual lightning-fire season. In the case of resprouting hardwoods, the physiological status of the plants—that is, lowest carbohydrate reserves—at the time of lightning-season fires leads to reduced resprouting potential and more effective control (Drewa et al. 2002; Robertson and Hmielowski 2014).

To summarize the arguments in support of the alternative hypotheses that burning during the evolutionary fire season does not provide conservation and restoration benefits beyond those achieved by burning at the evolutionary fire frequency, or that a seasonally variable fire regime is optimal:

1. Fires outside of the lightning-fire (transition) season are critical for adding burn days, keeping fire "on the ground." Frequency is the most important component of the fire regime in this region, and burning outside as well as inside the evolutionary fire season is necessary to achieve the required fire frequency on many managed areas.
2. For many species, evidence for strong differences in responses to lightning-season fires vs. fires in other seasons is lacking, at least if the latter are as frequent or as intense as lightning-season fires.
3. Some species (e.g., certain legumes; Hiers et al. 2000) apparently experience greater demographic success when burned outside of the lightning-fire season. Therefore, a strict formulaic lightning-season burn policy may be counterproductive.
4. Some evidence exists of negative effects of lightning-season fire on certain species (including pines, on sites with a history of fire exclusion), air quality, and other resources.
5. Burn bans imposed by state forestry agencies or the U.S. Forest Service prohibit burning during extreme drought, which is common in this region and typically coincides with the lightning-fire season. Fire crews are often occupied with firefighting during this season and are sometimes sent to western states to fight fires, so they are unavailable for prescribed burning.
6. Other legal requirements (e.g., Endangered Species Act) may discourage burning during the lightning-fire season. For example, due to the presence of the federally Endangered Cape Sable seaside sparrow, managers at Everglades National Park are required to suppress or reduce the probability of fire in sparrow habitat from March through August (Rick Anderson, personal communication).
7. Burning during the lightning-fire season can be more dangerous due to extreme fire weather conditions, including drought,

unpredictable winds, and increased probability of spotting (i.e., flaming embers traveling through the air and igniting fires outside the burn unit).

8. Smoke is a public health threat, especially to elderly people, and smoke dispersion issues may restrict burning during the drought of the lightning-fire season. Such fires often produce abundant smoke through smoldering combustion of duff.
9. Other practical considerations (e.g., liability, fire crew safety and comfort, politics, agency culture and tradition) restrict burning during the lightning-fire season.

Most of the above are legitimate arguments for conducting at least some burns outside the lightning-fire season. Especially important are #1 and #5. It is impossible for staff of many managed areas to conduct all their controlled burning during the evolutionary fire season, especially given burn bans and the limited availability of fire crews. Fire frequency is the most critical consideration. The least defensible argument for burning largely outside the lightning-fire season is agency culture and tradition. These can and should change with evolving knowledge of fire ecology.

Time of Year for Restoration Burns vs. Maintenance Burns

In the case of sites long fire-suppressed, the conventional approach is to apply restoration burns during cool or moist conditions with predictable winds. After a few low-intensity restoration burns in winter, which reduce fuel loads, the prescription can then shift to transition-season burns. This conventional approach is not based on much empirical evidence, however. Some managers have found transition-season fires just as easy to apply as first or second fires for restoration purposes. It is not the time of year that matters so much for restoration burns, but rather the specific weather, fuel, and vegetation conditions.

Pine mortality is a serious concern when fire is reintroduced to fire-excluded pine savannas. The restoration goal of reducing duff accumulation and forest floor depth on long-unburned sites (Hiers et al. 2007) may conflict, in the short term, with the goal of retaining mature pines. After fire is excluded for a long period of time, highly flammable pine duff accumulates around mature pines. When fire is reintroduced, excessive mortality of pines can occur, along with abundant smoke from smoldering fires. This can be a severe problem when duff is dry. Burns during drier conditions consume the most duff, but also cause the highest pine mortality from smoldering combustion (Varner et al. 2007). Therefore, restoration fires on fire-excluded sites with pines should occur when duff is very wet, wetter than with typical lightning-season fires. This process must be repeated through a series of low-intensity fires over many years to minimize pine mortality while removing the duff layer (Varner et al. 2005, 2007, 2009).

Often suggested in cases of high duff accumulation is choosing fire weather conditions and firing techniques that allow fire to move quickly through stands rather than smoldering (Outcalt 2006; Rickey et al. 2013). A series of such restoration burns, each one consuming a small amount of duff, will eventually eliminate the duff layer. Mortality of pines tends to be higher after burns in late summer or fall than during other seasons (Robbins and Myers 1992; Menges and Deyrup 2001), which is counterintuitive because this is the wet season, with wet duff. Pine mortality might be reduced in some cases by applying low-intensity backing fires rather than heading fires (Menges and Deyrup 2001), although flanking fires may be most appropriate for restoration burns in most pine savannas (Jean Huffman, personal communication).

In some cases, extremely intense fires are required to confront hysteresis and push a community such as invasive hammock (an alternate stable state) back to sandhill. Steve ("Sticky") Morrison (personal communication), an experienced fire manager formerly with The Nature Conservancy, uses fire during March through May, but particularly during May droughts with a Keetch-Byram drought index (KBDI) of 500 or higher, for restoring xeric hammock to sandhill. Fires at other times of the year are simply not intense enough. Morrison also recommends burning very frequently during the restoration phase, as often as the community will carry fire. Similarly, fire managers at Myakka River State Park found the only way to reverse the invasion of pyrogenic communities by hammock, which resulted from fire exclusion, was to burn during very dry periods in the transition season, outside the conditions under which prescribed fires are usually authorized (Robert Dye and Paula Benshoff, personal communication). Restoration of wet savanna/wetland ecotones is also accomplished most rapidly by fires under dry conditions.

A Precautionary Approach

Burning during the lightning-fire season will require some rethinking of conventional controlled burning practices and rules of thumb. The following are key points relevant to seasonality of controlled burning:

1. Frequency is more important than seasonality, but perhaps not by much over the long term—for example, as competing hardwoods become increasingly dense and fire-resistant and as fire-propagating plants dependent on intense fires decline.
2. "The burn season leading to an amount of fuel consumed and fire intensity closest to or within the historical range of variability will often have the best outcome" (Knapp et al. 2009).
3. Many years of regularly burning outside the lightning-fire season may have cumulative negative effects on species composition and habitat structure that are not evident in short-term (even decades-long) studies.
4. Accordingly, many years of burning within the lightning-fire season are sometimes required to restore communities and landscapes that have been degraded by too little fire or by burning outside the lightning-fire season.
5. Although many native species may respond equally well to fires during almost any season, the foundation species of southeastern grasslands/savannas—that is, wiregrass and most other C_4 grasses—respond most favorably to lightning-season fires.
6. The objectives of a particular prescribed burn (e.g., restoration vs. maintenance) may dictate burning outside the lightning-fire season, as will some salient practical concerns.

Most native species in Florida and the southeastern Coastal Plain have been unstudied with respect to their response to any component of the fire regime. This ignorance is compounded by the growing deficiency of basic natural history knowledge and identification skills among management personnel and even some staff biologists. When fire managers do not know their plant taxonomy, the responses of most plant species to seasonal or other variation in fire regimes will go unnoticed.

Given uncertainty about the importance of season of burn to the viability of species, a precautionary approach to prescribed fire would, whenever feasible, attempt to mimic the seasonality of lightning fire, as documented by such records as fire scars in pine stumps extending back several centuries. This evidence supports the practice of burning most frequently—but not exclusively—during the natural peak lightning-fire season. The burden of proof lies with those who would burn mostly outside this season. It is incumbent on them to show that fires outside this season produce ecological and biodiversity outcomes as good or better than fires occurring within this season.

Probably in very few cases will fire managers be able to restore a purely natural fire regime, given the constraints illustrated above. Burning exclusively during the evolutionary fire season is seldom possible, and would not be desirable if it results in less-frequent fire. Hermann et al. (2015) argue that managers should "focus on obtaining outcomes of a natural fire regime rather than the process itself." Achieving these outcomes will require adaptive management, involving the use of fire and other tools, to maintain high-quality habitat conditions on those sites where they exist and improving habitat conditions on degraded sites.

SHOULD I STRIVE FOR A CLEAN BURN OR A PATCHY BURN?

> For best results in the long run, we know that there should be well-distributed small areas *not* burned over frequently. Many fruit-bearing shrubs such as Huckleberry, Blueberry, Blackberry, Dewberry, Gooseberry, Gallberry, Ground Oak and Chinquapin, and a few others, cannot fruit the year of a burn, but bear heavily for two to four years when pruned back by fires of preceding years.
>
> Stoddard (1964)

All fires, regardless of the ignition source, have some degree of heterogeneity on one or more spatial grains, from fine (<1 m^2) to coarse (many km^2), largely due to patchiness of fuel consumption. Some degree of heterogeneity in postfire vegetation patterns contributes to biodiversity. Patches of unburned or lightly burned vegetation serve as critical refugia or microrefugia to plant and animal species sensitive to fire (see chapter 3). Maximum patchiness, however, is not an appropriate management goal, and many southeastern landscapes are abnormally patchy due to their history of fire exclusion or controlled burning during cool, wet conditions. The challenge is to create a degree of heterogeneity in vegetation that at least roughly simulates the pattern produced by a natural fire regime.

Lightning fires vary greatly in their patchiness because of variability in prefire fuel loads and continuity, topography, and weather conditions (e.g., wind strength and shifts) at the time of the burn. A controlled burning program that seeks to simulate lightning fires would also vary in pattern within and among sites, producing large and relatively homogeneous patches in some cases and a mosaic of variable postfire patches in other cases. Landscape position and morphology have a critical influence on the patchiness of fires, as well as their frequency and intensity.

The general reduction of fire frequency across the southeastern Coastal Plain has led to accumulations of litter, duff, and woody debris that increase fuel continuity and sometimes produce hotter and more homogeneous burns when fire is finally reintroduced. Probably more commonly, especially where hardwood patches develop due to infrequent fire, patchiness becomes unnaturally high and fire does not spread normally. When the first fires are reintroduced to fire-excluded sites, they are usually very incomplete. Each subsequent fire typically burns more and more area. Artificial heterogeneity is reduced as hardwood shrubs and small trees are replaced by herbaceous vegetation, which is more flammable and better carries the next fire. Large and tall hardwood patches (an alternative stable state), however, can become quite fire-proof.

Controlled burns are usually less patchy than lightning fires and leave more homogeneous postburn conditions (Ryan et al. 2013). Lightning ignitions are spot ignitions, sometimes multiple spots, with fire spreading out often in all directions from ignition points, but most rapidly spreading downwind. That is, they are primarily heading fires. Shifting winds, typical during thunderstorms, tend to increase heterogeneity. As described by Ryan et al. (2013), "wildfires typically ignite landscapes in large fingered fronts or via lofted embers (spotting), both of which lead to substantial heterogeneity in burn patterns." Controlled burns, on the other hand, usually are line ignitions, using only backing fires or some combination of backing, heading, and flanking fires, or sometimes a series of linear strips (strip-heading fires) or spots along a uniform grid (Waldrop and Goodrick 2012; Ryan et al. 2013). These controlled burns may result in uniform patterns with few residual unburned patches. This may be a desirable outcome for some restoration burns, but it is less appropriate for maintenance burns on sites with higher-quality vegetation structure.

Historical descriptions of vast, homogeneous pine savannas in the southeastern Coastal Plain, consisting of not much else but widely spaced longleaf pines, luxuriant grasses, and wildflowers, are probably reasonably accurate as far as they go, but not all landscapes in the region were like that. Early travelers would have taken the paths of least resistance through a region, just as highways do today, so the openness of presettlement landscapes was likely exaggerated in their accounts (Greenberg and Simons 1999; Landers et al. 2001). Within the longleaf pine region, significant variability in pine density, age structure, gap area, size and density of hardwoods, and other aspects of habitat structure existed on a regional and even local landscape scale (Hoctor et al. 2006). Slopes and ecotones or fire shadows with water bodies, wetlands, and hammocks would usually have more heterogeneous burn patterns and vegetation under a natural fire regime. Survival of pyrophytic hardwoods would be enhanced in these landscape positions. Even broad, flat landscapes, however, would normally experience moderately patchy fires that retain some residual refugia, even if small. Before native megaherbivores went extinct in the late Pleistocene, fires were almost certainly patchier, with grazing and fire synergistically producing fine-grained mosaics of vegetation (Bowman et al. 2016).

Too much patchiness of controlled burning in landscape positions that would naturally burn relatively uniformly, however, allows oak domes and other patches of fire-resistant vegetation to develop, resulting in patterns potentially outside the historic range of variability. The patchiest burns are often under wet dormant-season conditions. In such cases, grassy areas burn but shrubby areas "skip," which defeats the purpose of the restoration. A very short time-since-fire will also result in patchy burns due to the limited fuel biomass and continuity.

Habitat heterogeneity can be manipulated by variability in burning. Fire managers can take several steps to either increase or decrease the patchiness of their burns, depending on objectives specific to landscape position, fire history, and other factors. Ryan et al. (2013) call for "greater randomness in ignition, including variable, ground-based firing patterns or aerial ignitions" to "better emulate the complexity that historical burning once produced." Managers should not necessarily return with their drip torches to ignite patches of vegetation that did not burn in the first pass of the fire. Where dense shrubby patches have developed due to fire exclusion, this would be the proper thing to do, but usually not in maintenance fires (i.e., after artificial patchiness has been eliminated through restoration treatments).

SHOULD I CONSIDER MECHANICAL TREATMENTS OR OTHER "FIRE SURROGATES" TO CONTROL COMPETING VEGETATION?

The use of roller chopping, mowing, cutting (including logging), and other mechanical treatments, frequently combined with herbicide application, is increasingly common in managed savannas, grasslands, and some other communities of the southeastern Coastal Plain. Managers often see these treatments as necessary because years of fire exclusion have led to a high density, cover, and height of saw palmetto (*Serenoa repens*) and other shrubs, as well as midstory hardwood trees, which may be difficult if not impossible to reduce through burning alone. Mechanical and herbicide treatments avoid problems with smoke or fire escape, and can be conducted at various times of the year or during weather conditions such as drought, when burning is often prohibited by burn bans.

Mechanical treatments bring fuels into the herbaceous layer, where they can then be consumed by fire, promoting recovery of suppressed groundcover (Menges and Gordon 2010). A review by Stephens et al. (2012) concluded that both controlled burns and mechanical treatments meet short-term objectives for fuel reduction in seasonally dry southern, eastern, and western forests of the United States, which are naturally characterized by frequent fire. Treated stands are generally more resistant and resilient to high-severity wildfire. A range of treatments is recommended across sites (Stephens et al. 2012).

Nevertheless, nonfire treatments are expensive and have ecological costs, often including soil disturbance and compaction, increases in nonnative plants (e.g., cogon grass) and animals (e.g., fire ants, *Solenopsis invicta*), and impacts on native ground- or shrub-dwelling animals. Managers should consider the benefit-cost ratio of presumed fire surrogates carefully, as the potential for damage is considerable. Moreover, mechanical and chemical treatments are evolutionarily novel and do not substitute for fire. For example, they fail to provide the necessary environmental cues for seeds that are stimulated to germinate by heat or smoke. They should be considered short-term restoration treatments, not perennial habitat management treatments.

Information on the ecological impacts, both positive and negative, of mechanical and herbicide treatments with or without fire has increased through recent research. The extensive review by Menges and Gordon (2010) found that mechanical and herbicide treatments often result in desired changes in habitat structure in Florida upland communities, but the benefits are greatest when these treatments are combined with fire. Instead of considering mechanical and herbicide treatments as fire surrogates, they are most favorably applied as pretreatments for fire, and "managers should segue to fire-only approaches as soon as possible" after pretreatment (Menges and Gordon 2010). For example, in a longleaf pine sandhill restoration project on the Lake Wales Ridge, felling of a hardwood subcanopy was a useful pretreatment that accelerated restoration to the point where fire alone could be applied to continue restoration. Felling had some undesirable impacts, however, including increased mortality of pines (Rickey et al. 2013).

Saw palmetto is one of the most problematic "shrubs" that increase in cover, height, and density after fire exclusion. An ancient and venerable endemic plant of the southeastern Coastal Plain—indeed, a living fossil—saw palmetto is nevertheless often troublesome to land managers attempting to restore natural structure and groundcover richness on fire-excluded sites. The problem is that years of fire exclusion in flatwoods, dry prairies, some sandhills, and occasionally other communities often results in palmetto so dense and, eventually, tall that the site becomes unsuitable to many herbaceous plants due to shading and other competitive effects. Characteristic birds of grassland habitat, such as northern bobwhite, Bachman's sparrow, and Florida grasshopper sparrow, decline and ultimately disappear, as do fox squirrels and other animals dependent on open grassy groundcover.

As grass cover declines with increasing palmetto dominance, fire does not carry as well, and the site becomes challenging to burn more frequently than on approximately three-year intervals (Paula Benshoff, personal communication). Fires are also less intense due to the paucity of fine, flashy fuels; palmetto is not knocked back by such weak fires. Like the situation with xeric hammock, which develops after fire exclusion, saw palmetto creates an alternative state with stabilizing positive feedbacks that perpetuate its dominance over herbaceous vegetation (Huffman and Werner 2000). On flatwoods sites with dense palmetto cover, pine needle fuels are critical for allowing fire to carry, so great care must be taken to minimize pine mortality during burning and other treatments. Dry prairie, which lacks pines, presents a potentially more challenging situation if flammable grasses are greatly reduced (Watts et al. 2006). In contrast to savannas and grasslands, in the less combustible scrub vegetation saw palmetto may be necessary for carrying fire; hence, impacts on palmetto from roller chopping should be avoided (Menges and Gordon 2010).

Managers often must resort to a combination of mechanical pretreatments and fire when the goal is to reduce palmetto cover and restore a rich herbaceous groundcover in pine flatwoods or dry prairie. After 12–13 years of restoration treatments in dry prairie at Myakka River State Park, in which controlled burning, roller chopping, and a combination of the two were applied in both summer and winter, a combination of chopping and burning was most effective. The objective of chopping was "to sever saw palmetto rhizomes while incurring minimal soil disturbance" (Watts et al. 2006). Effects of season of treatment were mixed. An earlier study at Myakka River State Park determined that roller chopping treatments of heavy palmetto cover produced a greater flowering response of pine lily (*Lilium catesbaei*) than either burn-only or chop-and-burn treatments (Huffman and Werner 2000). Willcox and Giuliano (2010) found no effect of dormant-season burning on density and height of saw palmetto and only temporary reduction in cover. Fire during the growing season also had no effect on palmetto density, although cover was reduced. A combination of fire and roller chopping reduced palmetto height, but only growing-season roller chopping significantly reduced the density of saw palmetto.

After saw palmetto has been substantially reduced through mechanical treatment, frequent fire is often sufficient to maintain the low palmetto cover characteristic of pine savannas, dry prairies, and other grasslands. A review of saw palmetto biology and management showed that late fall or early winter burns can depress cover, while the more common late winter burns stimulate growth. Fire during the growing season is optimal for stimulating flowering more than vegetative growth (Duever 2011).

Many negative impacts of roller chopping have been observed, including soil disturbance and related increases in cover of native and nonnative weedy plants, substantial mortality of reptiles and other ground-dwelling animals, and depression of palmetto flowering and fruit production. Nevertheless, it seems clear that in cases where palmetto is both high and dense, and fine fuel biomass and connectivity are low, some type of mechanical treatment is often a regrettable necessity if the objective is to restore groundcover integrity. In some cases, use of bush hogs or cutters is as effective as roller chopping, with less soil disturbance (Maynard Hiss, personal communication). The key ingredient in any such a restoration prescription is a skilled practitioner.

Despite general agreement that mechanical or herbicide treatments are sometimes necessary to achieve control of excessive shrubs or hardwood midstory, burning remains the preferred option whenever possible. At Eglin Air Force Base, application of the herbicide hexazinone best achieved the objective of reducing the hardwood midstory, but did not meet groundcover restoration goals. Fire was the least effective treatment for midstory reduction, but it was also the cheapest treatment and led to the greatest increases in species richness and density of herbaceous plants (Provencher et al. 2001). In fire-suppressed Florida scrub, fire alone or in combination with pretreatment mowing was more effective than mowing alone in meeting the restoration goals of increasing bare sand gaps, increasing populations of rare endemic plants, and reducing the cover and height of woody vegetation (Weekley et al. 2011). Although nonfire treatments are sometimes necessary to restore vegetation structure, some species of arthropods, reptiles, amphibians, and plants respond positively only to fire (Menges and Gordon 2010).

In an ambitious landscape-scale experiment at Eglin Air Force Base, Steen et al. (2013a, 2013b) tested the effects of fire alone vs. fire following mechanical or herbicide pretreatments on birds and reptiles in longleaf pine sandhill sites that had been invaded by hardwoods due to fire exclusion. Restoration treatments were compared to reference sites that represented target conditions. After more than a decade of burning at two-to-three-year intervals, bird assemblages (including longleaf pine specialists) within all treatment sites were similar to reference sites, regardless of the type of pretreatment. This result demonstrated that fire alone was sufficient for restoration, with no need for the additional expense and effort of mechanical or chemical pretreatments (Steen et al. 2013a). Results were virtually identical for reptiles, with reptile assemblages converging most rapidly toward reference conditions with prescribed burning, but with assemblages at all sites essentially indistinguishable after a decade of treatment. Again, no added benefit of mechanical or chemical pretreatment was apparent in these sandhill sites (Steen et al. 2013b), which might, however, be easier to restore than flatwoods or dry prairie with high palmetto density. On top of their environmental costs, mechanical and chemical treatments are typically 10 to 20 times more expensive than controlled burning (Waldrop and Goodrick 2012).

HOW SHOULD I CONSTRUCT FIRE LINES OR OTHER FUEL BREAKS?

An almost universal feature of modern prescribed burning is a network of fuel breaks constructed to prevent the spread of fire beyond the area intended for burning. Variously called fuel breaks, firebreaks, fire lanes, fire lines, and plow lines, these usually linear features are too often constructed without much regard to their ecological costs. Indeed, ecological impacts of fire lines have not been studied rigorously in the southeastern United States or hardly anywhere. In a study across California, fuel breaks that expose bare soil promote exotic species invasions, with nonnative plant cover 200 percent higher on fuel breaks than in adjacent wildlands (Merriam et al. 2006). Abundant observations, but not yet published studies to my knowledge, suggest a similar effect of fire lines that expose extensive bare soil in the southeastern Coastal Plain. Other types of fire

Figure 5.4. Combined use of a forest road and a wide and deeply plowed fire line to stop potential fire spread in a Florida state forest. The exact location is not revealed. Photo by Reed Noss.

lines, such as mowed but undisked woods roads, are much less damaging (see below).

Fuel breaks of some sort are a necessary accompaniment of most prescribed burns. Indeed, secured fuel breaks are legally required by many states, such as Florida, for burns to be authorized by the state and protected under liability law (Brenner and Wade 2003). The legitimate ecological concern is that plowed or disked fire lines—the most common and most destructive fuel breaks—are often applied much more intensively and extensively than needed to prevent spread of fire outside burn units (Fig. 5.4). The subsequent negative impacts on ecosystems may compromise the benefits of controlled burns. Less harmful alternatives to plowed/disked fire lines are seldom used on public lands, with some important exceptions such as Apalachicola National Forest, although they often are on private conservation lands.

Observed or Potential Negative Effects of Plowed/Disked Fire Lines

Until rigorous studies of the impacts of various types of fire lines are conducted for this region, we are limited to field observations by ecologists and managers and hypotheses about potential impacts we are not seeing. I've discussed the problem of fire lines with many fire ecologists and managers in the Coastal Plain, particularly in Florida. They confirmed my observations and suspicions, while adding a few new ones. Among the observed and potential effects of fire lines are the following:

1. In frequent-fire communities of the southeastern Coastal Plain, most of the biodiversity resides in the herbaceous layer. Plowed or disked fire lines destroy this native groundcover. Once destroyed, native herbaceous vegetation such as wiregrass is very slow to recolonize and in a practical sense is gone forever. Many managers are focused on pines and seem unaware of this problem.
2. Plowed fire lines disrupt hydrology by interrupting or diverting sheet flow of water, pooling water, or converting sheet flow to channel flow and causing gulley erosion.
3. Fire lines are often constructed in ecotones, such as along seepage slopes and pocosin edges or around the perimeters of cypress domes or depression marshes. Fire lines might do their greatest damage in these ecotones. Natural ecotones are local biodiversity hotspots and provide habitat for many imperiled taxa, populations of which may be destroyed when the line is plowed. In addition, keeping fire out of wetlands converts herbaceous wetlands to low-diversity shrub swamps, eliminating breeding habitat for rare amphibians such as flatwoods salamanders (see chapter 3).
4. Plowed fire lines around wetlands can be death traps to larval amphibians. As the wetland dries, the edges of the plow line serve as a moat, severing the connection between the pooled plow line and the wetland. The plow line usually dries out before the natural wetland and before some amphibians can complete metamorphosis. Bishop and Haas (2005) found more than 500 dead leopard frog (*Rana* [*Lithobates*] *sphenocephala*) tadpoles in a dried-out plow line adjacent to a flatwoods salamander breeding pond.
5. Fire lines can serve as movement barriers to small animals with limited mobility or that behaviorally avoid unvegetated strips, potentially fragmenting populations. The road ecology literature is replete with examples of small flightless vertebrates and invertebrates refusing to cross seemingly innocuous linear features such as narrow unpaved roads (e.g., Swihart and Slade 1984; Mader et al. 1990). Presumably these animals evolved in habitats with

virtually continuous herbaceous cover, which makes a linear strip of bare soil a novel feature that presents a predation risk. Fire lines can also be movement traps; I once observed a hatchling gopher tortoise (*Gopherus polyphemus*) stuck in a deeply plowed fire line.

6. Fire lines with plowed, disked, or otherwise disturbed soil facilitate the invasion of exotic species, including nonnative plants (e.g., the strongly invasive cogon grass) and the red imported fire ant, which thrive on disturbed soil. Permanent populations of the exotic natalgrass (*Melinis repens*) often develop in fire lines in south-central Florida (Eric Menges, personal communication).
7. A wall of shrubs or hardwood trees sometimes develops in the fire shadow adjacent to a fire line. These corridors of woody vegetation are unnatural landscape features that disrupt fire, hydrology, and other ecological flows across the landscape and create artificial edge effects.

Rigorous studies, published in peer-reviewed journals, are needed to test and confirm the generality of these anecdotal observations and hypotheses, and quantify the degree of damage done by plowed or disked fire lines and other types of fuel breaks in the southeastern Coastal Plain. It would be appropriate for state forestry agencies, the U.S. Forest Service, and other public agencies that construct plowed fire lines to provide funding for the necessary research.

Alternatives to Aggressively Plowed or Disked Fire Lines

Observations suggest that the number and width of plowed fire lines have been increasing over recent years in Florida. Hints of escalating impacts suggest fire lines should be constructed as conservatively as possible to minimize their linear extent, width, depth of soil disturbance, and density on the landscape. Reducing the use of plowed or disked fire lines, and using less harmful alternatives, might be one of the simplest yet more meaningful things that can be done to improve fire management in the southeastern Coastal Plain.

The following are some useful alternatives to plowed or disked fire lines, which I have observed or were described to me by fire managers:

1. Natural fire-proof features serve quite well as firebreaks. These include lakes, streams, forested wetlands, and other natural com-

munities that are unlikely to burn under the proposed prescription. They also include recently burned areas, before flammable fine fuel loads are reestablished; these are very effective in stopping the spread of a subsequent fire.

2. Existing roads, powerline rights-of-way, trails, and other linear anthropogenic features often serve perfectly well as firebreaks.
3. Mowing can create highly effective firebreaks with little or no soil disturbance. A lawn mower, bush hog, or tree cutter may be used. Raking is often combined with mowing on private conservation areas to create firebreaks. Mowing can be followed by burning to produce fire-proof "black lines."

Finally, better training, supervision, and rewarding of fire personnel are essential, along with more appropriate equipment and larger fire crews. Today, many fire personnel are evaluated solely on the job of putting out fires. These deficiencies can be corrected, but doing so will require increased funding, training, and the goodwill of the agencies.

Fire Ecology and Management into an Uncertain Future

The science of evolutionary fire ecology (Pausas 2015a) must expand its influence on day-to-day fire management. Ideally, this science should guide controlled burning toward the goal of maintaining or restoring the full richness of native species and ecosystems in every fire-prone region on Earth. Otherwise, we will continue to see declines and extinctions of native species, disruption of ecological and evolutionary processes, and loss of many of the attributes of natural landscapes we find most sublime.

In the fire-prone ecosystems of the southeastern Coastal Plain, achieving this goal will require better understanding of all components of the lightning-fire regime. Maintaining or simulating this natural regime is still possible in the larger and more intact landscapes of the region. In probably more areas than we usually assume, we could quite safely allow lightning fires to run their course. Such managed wildfire is increasingly advocated by fire ecologists, especially in western and northern North America, and it is time this philosophy and practice spread eastward and southward in appropriate places. Of course, Florida and the Coastal Plain are more densely populated than much of the rest of the continent, so managed wildfire will not be possible in many landscapes. On the smaller or more degraded sites, which dominate the Coastal Plain today, a more

Figure 5.5. Land manager Johnny Stowe, South Carolina Department of Natural Resources, standing exasperated next to an anti-fire sign in the Sumter National Forest, August 2016. Such messages perpetuate the public's misunderstanding of fire. Photo by Charles Kemp.

achievable goal is using controlled burning to replicate the effects of natural fires on vegetation structure and species composition. For severely degraded sites, mechanical and chemical treatments sometimes may be needed, but fire must always be the main tool of choice.

Restoring frequency of fire to within the estimated presettlement range of variability for each natural community type is of highest priority, because frequency usually has the greatest impact on the structure and composition of fire-dependent communities and the population viability of fire-dependent species. The more challenging objectives of restoring or mimicking the seasonality, intensity, heterogeneity, and other components of lightning-fire regimes should not be neglected, however. Increasing public awareness and understanding of fire is paramount, as the public will not support an expansion of controlled burning without at least a rudimentary understanding of fire's beneficial effects beyond simple reduction of hazardous fuels. To this end, a good start would be for the land-managing agencies to cease their misleading anti-fire propaganda (such as Smokey Bear and related themes; Fig. 5.5) and replace such messages with more positive images, slogans, and substantive information about fire ecology.

We should not underestimate the importance of goal-setting. If we set goals narrowly focused on fuels reduction or firefighting, biodiversity and ecosystem integrity are sure to suffer. If we instead consider fire management a key part of the conservation movement—which the early fire ecologists such as Roland Harper, Herbert Stoddard, and Edward Komarek clearly did, even if they did not say so explicitly—then we have a much better chance of meeting ecological goals.

The future will bring—or force—changes in fire management. A major area of uncertainty is the effect of climate change on fire regimes over the coming century and beyond, and how change in climate may constrain the ability to burn. As noted in chapter 1, some models predict global increases in lightning activity on the order of 50 percent within the current century (Romps et al. 2014). Because of higher temperatures and increased moisture stress, vegetation generally will be more prone to burn. Warmer temperatures generally favor C_4 grasses (Morgan et al. 2011), which should increase the flammability of vegetation. Concurrent increases in atmospheric CO_2, however, benefit woody plants and other C_3 species over C_4 grasses, and might possibly release woody plants from control by fire (Midgley and Bond 2015). The balance between these opposing forces may depend on the level of moisture stress. Beyond some threshold of moisture stress, C_4 grasses may prosper even in a much higher CO_2 world.

Future precipitation levels are uncertain in the Southeast. Most models, however, predict greater variability in precipitation, including increased incidence of severe drought (Mitchell et al. 2014). Predicted increases in drought might produce shifts from some closed forests to grasslands and savannas, as occurred during the late Miocene (see chapter 2). Increasing drought, however, might also compel agencies to impose longer burn bans that result in less controlled burning (Mitchell et al. 2014), especially under ecologically optimal weather conditions. Increased fire suppression could counteract any gains in fire activity related to climate change. Given human population growth and agricultural expansion, reduction of fire activity in savannas and grasslands is likely to continue despite climatic conditions increasingly favorable to fire (Andela et al. 2017).

The effects of wildland fire on carbon balance—that is, burned landscapes as sources or sinks of atmospheric carbon—is an area of active research. Perhaps surprisingly, fire does not necessarily increase levels of atmospheric CO_2. As noted in chapter 2, buried charcoal, which steadily accumulates after a series of fires, could reduce CO_2 levels and potentially result in global cooling (Scott et al. 2014). Regular controlled burning, by reducing density of woody plants and the occurrence of high-severity fire in communities naturally characterized by frequent fire, increases carbon stability in the ecosystem by preventing large pulses of carbon into the atmosphere (Hurteau and Brooks 2011). Frequent low-severity fire releases less net carbon to the atmosphere than high-severity fire because less woody material burns. Longleaf pine communities are particularly stable and both resistant and resilient to fire. Therefore, managing for longleaf pine on long rotations is one of the best forestry strategies for retaining carbon in the ecosystem. Wood products derived from longleaf pine continue to store carbon for long periods of time (Kush et al. 2004).

Carbon released during a typical prescribed fire in the Coastal Plain is equivalent to less than one to three years of carbon sequestered (Clark et al. 2014). Taking into account carbon losses due to fuel consumption by fire, a longleaf pine savanna is a net carbon source in the short term, but the ecosystem recovers its capacity for carbon uptake within one or two months postfire, which suggests that most frequent-fire communities are likely carbon neutral. Drought, however, can lead to net carbon loss (Whelan et al. 2013). Not taken adequately into account in most carbon balance models, however, are the extensive underground storage organs common in plants of ancient frequent-fire communities in the southeastern Coastal Plain (see chapter 3). These organs collectively store large quantities of secure carbon. Globally, grasslands might store as much carbon as forests (White et al. 2000), and their belowground carbon stores are well protected from loss to the atmosphere when disturbed by fire or other factors (Veldman et al. 2015b).

The Coastal Plain, like other global hotspots of endemism, has been relatively stable climatically over millions of years. This relative stability is predicted to continue, albeit many species could still be lost if warming is extreme (Harrison and Noss 2017). If we want to maintain biodiversity within this global hotspot into the future, keeping fire on the landscape is imperative. If seasonal droughts lengthen and become more intense, as predicted, managers must be trained and authorized to burn during these more extreme conditions. Restoration burns can be used to move fire-excluded sites back to the open conditions where frequent fires can be applied safely by skilled practitioners. Thinning of small-diameter trees that have colonized fire-excluded sites may make these sites more resistant to severe fire and accompanying carbon loss (Hurteau et al. 2008).

Resistance and resilience of populations, communities, and ecosystems are fundamental concepts for adaptation to climate change (Noss 2001). Resistance is the ability to persist with little change despite disturbance or other major environmental change, whereas resilience is the ability to bounce back and recover key properties after a disturbance or change (Connell and Sousa 1983; Pimm 1984; Nimmo et al. 2015). The native species of pyrogenic communities in the southeastern Coastal Plain have evolved both resistance and resilience to fire.

The vegetation-fire feedback model of Fill et al. (2015a) is built on the concept of fire regimes maintaining persistent assemblages of species. That pine savanna, grassland, and scrub communities of some type have persisted in this region for millions of years despite past climate changes (see chapter 2) indicates the powerful capability of these communities to resist or be resilient to change. Moreover, that these communities have been able to recover remarkably well from logging, turpentining, and other recent disturbances shows resilience to even novel change. But resilience has its limits. Wiregrass and some other groundcover plants are extremely sensitive to soil disturbance, such as that produced by site preparation for modern silviculture (Clewell 1989; Hardin and White 1989; Noss 1989), and invasive plants that alter combustibility of communities might permanently alter fire regimes (Freeman et al. 2017).

The following recommendations synthesize existing knowledge for maintaining resistance and resilience of fire-prone ecosystems of the southeastern Coastal Plain during the next century and beyond:

1. Minimize novel disturbances, such as plowed/disked fire lines, which are outside the evolutionary experiences of species in the community and encourage invasion by nonnative species. Novel ecosystems will develop with climate change regardless of how we manage the land (Hobbs et al. 2014). Indeed, every ecosystem over any meaningful span of time is novel; species come and go. Nevertheless, in times of rapid change we must do all we can to maintain communities composed primarily of species native to the region (Hanberry et al. 2015). Restoration should seek to maintain the historical trajectory of ecosystem change. As stated by Falk (1990), "Restoration uses the past not as a goal but as a

reference point for the future. If we seek to recreate the (ecosystems) of centuries past, it is not to turn back the evolutionary clock but to set it ticking again."

2. Accept that fire regimes and other disturbances, including hurricanes and droughts, will be altered by changes in climate, as they have been in the past. Virtually all kinds of disturbances are becoming more extreme. Careful monitoring and adaptive management can assist managers in adjusting their goals, objectives, and practices to these changes.
3. As disturbance regimes and landscape spatial patterns change, it will be even more necessary to manage not individual natural communities, but entire landscape mosaics across environmental gradients. With climate change, many species will need to shift their distributions and find microclimatic refugia that meet their needs. These microrefugia are "sites that support locally favorable climates amidst unfavorable regional climates" (Dobrowski 2011). Managing large landscape mosaics—and keeping them relatively unfragmented by potential dispersal barriers such as roads and fire lines—is critical for facilitating adjustments in species distributions (Noss 2001).
4. Burn within the historic fire-return interval, weighted toward the more frequent end in savannas/grasslands. Frequent surface fire will help communities characterized by this regime retain their carbon, rather than burning with high severity due to fire exclusion and releasing large pulses of CO_2 to the atmosphere.
5. Preserve old-growth forests, savannas, and grasslands. Old-growth ecosystems, composed of long-lived organisms, are expected to show considerable inertia in the face of climate change (Franklin et al. 1991; Noss 2001). They may persist through centuries or millennia of unfavorable climate, perhaps with little reproduction during that period, only to rebound when favorable conditions return. Most people are aware of old-growth forests and value their conservation. We should help people appreciate the values of old-growth grasslands as well (Veldman et al. 2015a; Bond 2016).
6. Maintain high native species richness (within limits characteristic of each natural community), diversity of functional groups, and ecologically optimal populations of keystone, foundation, and

other ecologically pivotal species. Doing so will increase the probability that ecosystems will remain resistant and resilient (Tilman et al. 2006; Ives and Carpenter 2007; Isbell et al. 2011).

7. Be humble, open-minded, and ready to accept new information and change management practices accordingly. Stubborn tradition has hamstrung fire management and other resource management for too long. We are often stuck somewhere around the middle of the twentieth century, applying practices that the last few decades of research show are no longer defensible. With an unpredictably changing climate, it is even more critical to be responsive to new information. In the meantime, recognize that good land management during a time of changing climate differs little from good land management under more static conditions. This is management that does the least harm, yet at the same time is willing to take calculated risks.

LITERATURE CITED

Abrahamson, W. G., and C. R. Abrahamson. 2009. Life in the slow lane: Palmetto seedlings exhibit remarkable survival but slow growth in Florida's nutrient-poor uplands. Castanea 74:123–132.

Abrahamson, W. G., and D. C. Hartnett. 1990. Pine flatwoods and dry prairie. In Ecosystems of Florida, edited by R. L. Myers and J. J. Ewel, 103–149. University of Central Florida Press, Orlando.

Abrams, M. D., and G. J. Nowacki. 2015. Exploring the early Anthropocene burning hypothesis and climate-fire anomalies for the eastern U.S. Journal of Sustainable Forestry 34:30–48.

Addington, R. N., S. J. Hudson, J. K. Hiers, M. D. Hurteau, T. F. Hutcherson, G. Matusick, and J. M. Parker. 2015. Relationships among wildfire, prescribed fire, and drought in a fire-prone landscape in the south-eastern United States. International Journal of Wildland Fire 24:778–783.

Adorno, R., and P. C. Pautz, translators and editors. 1999. The Narrative of Cabeza de Vaca. University of Nebraska Press, Lincoln.

Agee, J. K. 1993. Fire Ecology of Pacific Northwest Forests. Island Press, Covelo, CA.

Aleman, J. C., O. Blarquez, I. Bentaleb, P. Bonté, B. Brossier, C. Carcaillet, et al. 2013. Tracking land-cover changes with sedimentary charcoal in the Afrotropics. Holocene 23:1853–1862.

Alexander, M. E. 1982. Calculating and interpreting forest fire intensities. Canadian Journal of Botany 60:349–357.

Alexander, M. E., and M. G. Cruz. 2012. Interdependencies between flame length and fireline intensity in predicting crown fire initiation and crown scorch height. International Journal of Wildland Fire 21:95–113.

Allen, C. D., M. Savage, D. A. Falk, K. F. Suckling, T. W. Swetnam, T. Schulke, P. B. Stacey, P. Morgan, M. Hoffman, and J. T. Klingel. 2002. Ecological restoration of southwestern ponderosa pine ecosystems: A broad perspective. Ecological Applications 12:1418–1433.

Allen, S. 2015. Summer in Florida means lightning—and lots of it. Orlando Sentinel, June 21, 2015.

Ames, G. M., W. A. Wall, M. G. Hohmann, and J. P. Wright. 2017. Trait space of rare plants in a fire-dependent ecosystem. Conservation Biology 31:903–911.

Andela, N., D. C. Morton, L. Giglio, Y. Chen, et al. 2017. A human-driven decline in global burned area. Science 356:1356–1362.

Anderson, M. K. 2007. Native American uses and management of California's grasslands. In California Grasslands—Ecology and Management, edited by M. R. Stromberg, J. D. Corbin, and C. M. D'Antonio, 57–66. University of California Press, Berkeley.

Archibald, S., D. P. Roy, B. W. van Wilgen, and R. J. Scholes. 2009. What limits fire? An examination of drivers of burnt area in Southern Africa. Global Change Biology 15:613–630.

Ashton, K. G., B. M. Engelhardt, and B. S. Branciforte. 2008. Gopher tortoise (*Gopherus polyphemus*) abundance and distribution after prescribed fire reintroduction to Florida scrub and sandhill at Archbold Biological Station. Journal of Herpetology 42:523–529.

Bacchus, S. T. 1995. Groundwater levels are critical to the success of prescribed burns. Proceedings of the Tall Timbers Fire Ecology Conference 19:117–133.

Baker, W. L. 2009. Fire Ecology in Rocky Mountain Landscapes. Island Press, Washington, DC.

Barden, L. S. 1997. Historic prairies in the Piedmont of North and South Carolina, USA. Natural Areas Journal 17:149–152.

Barrett, S. W., T. W. Swetnam, and W. L. Baker. 2005. Indian fire use: Deflating the legend. Fire Management Today 65(3): 31–34.

Bartram, W. 1791. Travels of William Bartram. Edited by M. Van Doren (1955). Dover, New York.

Batista, W. B., and W. J. Platt. 1997. An old-growth definition for southern mixed hardwood forests. USDA Forest Service General Technical Report SRS-9. Asheville, NC.

Beckage, B., and I. J. Stout. 2000. Effects of repeated burning in Florida pine savannas: A test of the intermediate disturbance hypothesis. Journal of Vegetation Science 11:113–122.

Beckage, B., W. J. Platt, and L. J. Gross. 2009. Vegetation, fire, and feedbacks: A disturbance-mediated model of savannas. American Naturalist 174:805–818.

Beckage, B., W. J. Platt, and B. Panko. 2005. A climate-based approach to the restoration of fire-dependent ecosystems. Restoration Ecology 13:429–431.

Beckage, B., W. J. Platt, M. G. Slocum, and B. Panko. 2003. Influences of the El Niño Southern Oscillation on fire regimes in the Florida Everglades. Ecology 84:3124–3130.

Beerling, D. 2007. The Emerald Planet: How Plants Changed Earth's History. Oxford University Press, New York.

Beerling, D., and C. P. Osborne. 2006. The origin of the savanna biome. Global Change Biology 12:2023–2031.

Behling H. 2002. South and southeast Brazilian grasslands during Late Quaternary times: A synthesis. Palaeogeography, Palaeoclimatology, Palaeoecology 177:19–27.

Bekker, R. M., J. P. Bakker, U. Grandin, R. Kalamees, P. Milberg, P. Posschlod, K. Thompson, and J. H. Willems. 1998. Seed size, shape and vertical distribution in the soil: Indicators of seed longevity. Functional Ecology 12:834–842.

Belue, T. F. 1996. The Long Hunt: Death of the Buffalo East of the Mississippi. Stackpole, Mechanicsburg, PA.

Berna, F., P. Goldberg, J. Horwitz, S. Holt, M. Balmford, and M. Chazan. 2012. Microstratigraphic evidence of in situ fire in Archeulean strata of Wonderwerk Cave, Northern Cape Province, South Africa. Proceedings of the National Academy of Sciences 109(20): E1215–1220.

Berry, E. W. 1924. The Middle and Upper Eocene Floras of Southeastern North America. Professional Paper 92. U.S. Geological Survey, Reston, VA.

Bishop, D. C., and C. A. Haas. 2005. Burning trends and potential negative effects of suppressing wetland fires on flatwoods salamanders. Natural Areas Journal 25:290–294.

Bond, W. J. 2015. Fires in the Cenozoic: A late flowering of flammable ecosystems. Frontiers in Plant Science 5(749): 1–11.

———. 2016. Ancient grasslands at risk. Science 351:120–122.

Bond, W. J., and J. E. Keeley. 2005. Fire as a global "herbivore": The ecology and evolution of flammable ecosystems. Trends in Ecology and Evolution 20:387–394.

Bond, W. J., and J. J. Midgley. 1995. Kill thy neighbor: An individualistic argument for the evolution of flammability. Oikos 73:79–85.

———. 2001. Ecology of sprouting in woody plants: The persistence niche. Trends in Ecology and Evolution 16:45–51.

Bond, W. J., and A. C. Scott. 2010. Fire and the spread of flowering plants in the Cretaceous. New Phytologist 188:1137–1150.

Bond, W. J., F. I. Woodward, and G. F. Midgley. 2005. The global distribution of ecosystems in a world without fire. New Phytologist 165:525–538.

Boring, L. R., J. Hendricks, and M. B. Edwards. 1990. Loss, retention, and replacement of nitrogen associated with site preparation burning in southern pine-hardwood forests. In Fire and the Environment: Ecological and Cultural Perspectives, edited by S. D. Nodvin and T. A. Waldrop, 145–153. USDA Forest Service, General Technical Report SE-69.

Bowen, G. J., B. J. Maibauer, M. J. Kraus, U. Röhl, et al. 2015. Two massive, rapid releases of carbon during the onset of the Paleocene-Eocene thermal maximum. Nature Geoscience 8:44–47.

Bowles, M. L., and M. D. Jones. 2013. Repeated burning of eastern tallgrass prairie increases richness and diversity, stabilizing late successional vegetation. Ecological Applications 23:464–478.

Bowman, D. M. J. S., J. K. Balch, P. Artaxo, W. J. Bond, et al. 2009. Fire in the earth system. Science 324:481–484.

Bowman, D. M. J. S., J. K. Balch, P. Artaxo, W. J. Bond, et al. 2011. The human dimension of fire regimes on Earth. Journal of Biogeography 38:2223–2236.

Bowman, D. M. J. S., G. L. W. Perry, S. I. Higgins, C. N. Johnson, S. D. Fuhlendorf, and B. P. Murphy. 2016. Pyrodiversity is the coupling of biodiversity and fire regimes in food webs. Philosophical Transactions of the Royal Society B 371:20150169.

Bowman, D. M. J. S., G. L. W. Perry, and J. B. Marston. 2015. Feedbacks and landscape-level vegetation dynamics. Trends in Ecology and Evolution 30:255–260.

Boyer, W. D. 1987. Volume growth loss: A hidden cost of periodic prescribed burning in longleaf pine. Southern Journal of Applied Forestry 11:154–157.

Bradbury, D., S. L. Tapper, D. Coates, S. McArthur, M. Hankinson, and M. Byrne. 2016. The role of fire and a long-lived soil seed bank in maintaining persistence, genetic diversity and connectivity in a fire-prone landscape. Journal of Biogeography 43:70–84.

Bradshaw, S., K. Dixon, S. Hopper, H. Lambers, and S. Turner. 2011. Little evidence for fire-adapted plant traits in Mediterranean climate regions. Trends in Plant Science 16:69–76.

Braudel, F. 1817. The Structures of Everyday Life. Vol. 1. Harper and Row, New York.

Braun, E. L. 1950. Deciduous Forests of Eastern North America. Blackburn Press, Caldwell, NJ.

Breininger, D. R., and G. M. Carter. 2003. Territory quality transitions and source-sink dynamics in a Florida scrub-jay population. Ecological Applications 13:516–529.

Breininger, D. R., E. D. Stolen, G. M. Carter, D. M. Oddy, and S. A. Legare. 2014. Quantifying how territory quality and sociobiology affect recruitment to inform fire management. Animal Conservation 17:72–79.

Brender, E. V., and R. W. Cooper. 1968. Prescribed burning in Georgia's Piedmont loblolly pine stands. Journal of Forestry 66:31–36.

Brenner, J., and D. Wade. 2003. Florida's revised prescribed fire law: Protection for responsible burners. In Proceedings of Fire Conference 2000: The First National Congress on Fire Ecology, Prevention, and Management, edited by K. E. M. Galley, R. C. Klinger, and N. G. Sugihara, 132–136. Miscellaneous Publication No. 13, Tall Timbers Research Station, Tallahassee, FL.

Brewer, J. S., and W. J. Platt. 1994. Effects of fire season and herbivory on reproductive success in a clonal forb, *Pityopsis graminifolia*. Journal of Ecology 82:665–675.

Bridges, E. L. 2006a. Landscape ecology of Florida dry prairie in the Kissimmee River region. In Land of Fire and Water: The Florida Dry Prairie Ecosystem, edited by R. F. Noss, 14–42. Proceedings of the Florida Dry Prairie Conference, 5–7 October 2004. E. O. Painter, DeLeon Springs, FL.

———. 2006b. Historical accounts of vegetation in the Kissimmee River dry prairie landscape. In Land of Fire and Water: The Florida Dry Prairie Ecosystem, edited by R. F. Noss, 43–63. Proceedings of the Florida Dry Prairie Conference, 5–7 October 2004. E. O. Painter, DeLeon Springs, FL.

Brockway, D. G., K. W. Outcalt, and W. D. Boyer. 2006. Longleaf pine regeneration ecology and methods. In The Longleaf Pine Ecosystem: Ecology, Silviculture, and Restoration, edited by S. Jose, E. Jokela, and D. Miller, 95–133. Springer-Verlag, New York.

Brommit, A. G., N. Charbonneau, T. A. Contreras, and L. Fahrig. 2004. Crown loss and subsequent branch sprouting of forest trees in response to a major ice storm. Journal of the Torrey Botanical Society 131:169–176.

Brown, C. A. 1938. The flora of Pleistocene deposits in the Western Florida Parishes, West Feliciana Parish and East Baton Rouge Parish, Louisiana. Louisiana Department of Conservation, Geological Survey Bulletin 12:59–96, 121–129.

Brudvig, L. A., J. L. Orrock, E. I. Damschen, C. D. Collins, P. G. Hahn, W. B. Mattingly, J. W. Veldman, and J. L. Walker. 2014. Land-use history and contemporary management inform an ecological reference model for longleaf pine woodland understory plant communities. PLoS One 9(1): e86604.

Burke, V. J., and J. W. Gibbons. 1995. Terrestrial buffer zones and wetland conserva-

tion: A case study of freshwater turtles in a Carolina bay. Conservation Biology 1365–1369.
Byram, G. M. 1959. Combustion of forest fuels. In Forest Fire: Control and Use, edited by K. P. Davis, 61–89. McGraw Hill, New York.
Canham, C. D. 1989. Different responses to gaps among shade-tolerant tree species. Ecology 70:548–550.
Carr, S. C., K. M. Robertson, and R. K. Peet. 2010. A vegetation classification of fire-dependent pinelands in Florida. Castanea 75:153–189.
Casey, W. P., and K. C. Ewel. 2006. Patterns of succession in forested depressional wetlands in north Florida, USA. Wetlands 26:147–160.
Cavender-Bares, J., D. D. Ackerly, D. A. Baum, and F. A. Bazzaz. 2004a. Phylogenetic overdispersion in Floridian oak communities. American Naturalist 163:823–843.
Cavender-Bares, J., K. Kitajima, and F. A. Bazzaz. 2004b. Multiple trait associations in relation to habitat differentiation among 17 Floridian oak species. Ecological Monographs 74:635–662.
Cerling, T. E., J. M. Harris, B. J. MacFadden, M. G. Leakey, and J. Quade. 1997. Global vegetation change through the Miocene/Pliocene boundary. Nature 389:153–158.
Cerling, T. E., Y. Wang, and J. Quade. 1993. Expansion of C_4 ecosystems as an indicator of global ecological change in the late Miocene. Nature 361:344–345.
Cerling, T. E., J. G. Wynn, S. A. Andanje, M. I. Bird, et al. 2011. Woody cover and hominin environments in the past 6 million years. Nature 476:51–56.
Chapman, H. H. 1912. Forest fires and forestry in the southern states. American Forests 18:510–517.
———. 1932. Is the longleaf type a climax? Ecology 13:328–334.
———. 1952. The place of fire in the ecology of pines. Bartonia 26:39–44.
Chen, E., and J. F. Gerber. 1990. Climate. In Ecosystems of Florida, edited by R. L. Myers and J. J. Ewel, 11–34. University of Central Florida Press, Orlando.
Christensen, N. L. 1981. Fire regimes in southeastern ecosystems. In Fire Regimes and Ecosystem Properties, edited by H. A. Mooney, T. M. Bonnicksen, N. L. Christensen, J. E. Lotan, and W. A. Reiners, 112–136. USDA Forest Service General Technical Report WO-26.
———. 2000. Vegetation of the southeastern Coastal Plain. In North American Terrestrial Vegetation, 2nd ed., edited by M. G. Barbour and W. D. Billings, 397–448. Cambridge University Press, Cambridge.
———. 2014. An historical perspective on forest succession and its relevance to ecosystem restoration and conservation practice in North America. Forest Ecology and Management 330:312–322.
Clark, K. L., N. Skowronski, H. Renninger, and R. Scheller. 2014. Climate change and fire management in the mid-Atlantic region. Forest Ecology and Management 327:306–315.
Clements, F. E. 1910. The Life History of Lodgepole Pine Forests. Bulletin 79, U.S. Department of Agriculture Forest Service, Washington, DC.
———. 1916. Plant Succession: An Analysis of the Development of Vegetation. The Carnegie Institution of Washington, Washington, DC.

Clewell, A. F. 1989. Natural history of wiregrass (*Aristida stricta* Michx., Gramineae). Natural Areas Journal 9:223–233.

Coetsee, C., W. J. Bond, and E. C. February. 2010. Frequent fire affects soil nitrogen and carbon in an African savanna by changing woody cover. Oecologia 162:1027–1034.

Colburn, D., and L. deHaven-Smith. 2002. Florida's Megatrends: Critical Issues in Florida. University Press of Florida, Gainesville.

Collins, C. S., R. N. Conner, and D. Saenz. 2002. Influence of hardwood midstory and pine species on pine bole arthropods. Forest Ecology and Management 164:211–220.

Colwell, R. K., and T. F. Rangel. 2009. Hutchinson's duality: The once and future niche. Proceedings of the National Academy of Sciences 106:19651–19658.

Comer, P., D. Faber-Langendoen, R. Evans, S. Gawler, C. Josse, G. Kittel, S. Menard, M. Pyne, M. Reid, K. Schulz, K. Snow, and J. Teague. 2003. Ecological Systems of the United States: A Working Classification of U.S. Terrestrial Systems. NatureServe, Arlington, VA.

Connell, J. H., and W. P. Sousa. 1983. On the evidence needed to judge ecological stability or persistence. American Naturalist 121:789–824.

Cooper, A., C. Turney, K. A. Hughen, B. W. Brook, H. G. McDonald, and C. J. A. Bradshaw. 2015. Abrupt warming events drove Late Pleistocene Holarctic megafaunal turnover. Science 349:602–606.

Cooper, W. S. 1913. The climax forest of Isle Royale, Lake Superior, and its development. Botanical Gazette 55:1–44, 115–140, 189–235.

———. 1922. The Broad-Sclerophyll Vegetation of California: An Ecological Study of the Chaparral and its Related Communities. Carnegie Institution of Washington, Publ. No. 319. Technical Press, Washington, DC.

Coughlan, M. R. 2016. Wildland arson as clandestine resource management: A space-time permutation analysis and classification of informal fire management regimes in Georgia, USA. Environmental Management 57:1077–1087.

Covington, W. W., and M. M. Moore. 1994. Southwestern ponderosa pine forest structure: Changes since European settlement. Journal of Forestry 92:39–47.

Cox, J. A., and B. Widener. 2008. Lightning-season burning: Friend or foe of breeding birds? Miscellaneous Publication 17. Tall Timbers Research Station, Tallahassee, FL.

Crawford, R. L., and W. R. Brueckheimer. 2012. The Legacy of a Red Hills Hunting Plantation: Tall Timbers Research Station and Land Conservancy. University Press of Florida, Gainesville.

Cressler, W. L. 2001. Evidence of earliest known wildfires. Papaios 16:171–174.

Curnutt, J. L., A. L. Mayer, T. M. Brooks, et al. 1998. Population dynamics of the endangered Cape Sable Seaside Sparrow. Animal Conservation 1:11–20.

Daniau, A.-L., M. F. Sánchez Goñi, P. Martinez, D. H. Urrego, V. Bout-Roumazeilles, S. Desprat, and J. R. Marlon. 2013. Orbital-scale climate forcing of grassland burning in southern Africa. Proceedings of the National Academy of Sciences 110:5069–5073.

Darracq, A. K., W. W. Boone, and R. A. McCleery. 2016. Burn regime matters: A review of the effects of prescribed fire on vertebrates in the longleaf pine ecosystem. Forest Ecology and Management 378:214–221.

Daubenmire, R. 1990. The *Magnolia grandiflora-Quercus virginiana* forest of Florida. American Midland Naturalist 123:331–347.

Davies, A. B., P. Eggleton, B. J. van Rensburg, and C. L. Parr. 2012. The pyrodiversity-biodiversity hypothesis: A test with savanna termite assemblages. Journal of Applied Ecology 49:422–430.

Davis, J. H. 1967. General Map of Natural Vegetation of Florida. Institute of Food and Agricultural Sciences, University of Florida, Gainesville.

Davis, K. P. 1959. Forest Fire-Control and Use. McGraw-Hill, New York.

Davis, M. B. 1986. Climatic instability, time lags, and community disequilibrium. In Community Ecology, edited by J. Diamond and T. J. Case, 269–284. Harper and Row, New York.

Dee, J. R., and E. S. Menges. 2014. Gap ecology in the Florida scrubby flatwoods: Effects of time-since-fire, gap area, gap aggregation and microhabitat on gap species diversity. Journal of Vegetation Science 25:1235–1246.

Deevey, E. S., Jr. 1949. Biogeography of the Pleistocene. Bulletin of the Geological Society of America 60:1315–1416.

de Laudonniere, R. 1587. A notable historie containing foure voyages made by certain French Captaines into Florida. In The Principal Navigations Voyages Traffiques and Discoveries of the English Nation. Edited and translated by R. Hakluyt (1904). Vols. 8 and 9. McMillan, New York.

Delcourt, P. A. 1980. Goshen Springs: Late Quaternary vegetation record for southern Alabama. Ecology 61:371–386.

Delcourt, P. A., and H. R. Delcourt. 1977. The Tunica Hills, Louisiana-Mississippi: Late glacial locality for spruce and deciduous forest species. Quaternary Research 7:218–237.

———. 1981. Vegetation maps for eastern North America: 40,000 yr B.P. to the present. Geobotany 2:123–165.

———. 1984. Late Quaternary paleoclimates and biotic responses in eastern North America and the western Atlantic Ocean. Palaeogeography, Palaeoclimatology, Palaeoecology 48:263–284.

———. 1996. Quaternary paleoecology of the Lower Mississippi Valley. Engineering Geology 45:219–242.

DellaSala, D. A., and C. T. Hanson. 2015. The Ecological Importance of Mixed-severity Fires: Nature's Phoenix. Elsevier, Amsterdam, The Netherlands.

Diamond, J. M. 1976. Island biogeography and conservation: Strategy and limitations. Science 193:1027–1029.

DiPietro, J. A. 2013. Landscape Evolution in the United States: An Introduction to the Geography, Geology, and Natural History. Elsevier, Waltham, MA.

Dobrowski, S. Z. 2011. A climatic basis for microrefugia: The influence of terrain on climate. Global Change Biology 17:1022–1035.

Donders, T. H., H. J. de Boer, W. Finsinger, E. C. Grimm, S. C. Dekker, G. J. Reichart, and F. Wagner-Cremer. 2011. Impact of the Atlantic warm pool on precipitation and temperature in Florida during North Atlantic cold spells. Climate Dynamics 36:109–118.

Donoghue, J. F. 2011. Sea level history of the northern Gulf of Mexico coast and sea level rise scenarios for the near future. Climatic Change 107:17–33.

Dowsett, H. J., and T. M. Cronin. 1990. High eustatic sea level during the middle Pliocene: Evidence from the southeastern U.S. Atlantic Coastal Plain. Geology 18:435–438.

Drewa, P. B., W. J. Platt, C. Kwit, and T. W. Doyle. 2008. Stand structure and dynamics of sand pine differ between the Florida panhandle and peninsula. Plant Ecology 196:15–25.

Drewa, P. B., W. J. Platt, and E. B. Moser. 2002. Fire effects on resprouting of shrubs in headwaters of southeastern longleaf pine savannas. Ecology 83:755–767.

Duever, L. C. 2011. Ecology and management of saw palmetto. Report to Florida Fish and Wildlife Conservation Commission, Fish and Wildlife Research Unit, Gainesville, FL.

Duever, M. J. 1984. Environmental factors controlling plant communities of the Big Cypress Swamp. In Environments of South Florida: Present and Past II, edited by P. J. Gleason, 127–137. Miami Geological Society, Coral Gables, FL.

Duever, M. J., and R. R. Roberts. 2013. Successional and transitional models of natural South Florida, USA, plant communities. Fire Ecology 9:110–123.

Duever, M. J., J. E. Carlson, J. F. Meeder, L. C. Duever, L. H. Gunderson, L. A. Riopelle, T. R. Alexander, R. L. Myers, and D. P. Spangler. 1986. The Big Cypress National Preserve. National Audubon Society, New York.

Duever, M. J., J. E. Carlson, and L. A. Riopelle. 1984. Corkscrew Swamp: A virgin cypress strand. In Cypress Swamps, edited by K. C. Ewel and H. T. Odum, 334–348. University Presses of Florida, Gainesville.

Duncan, B. W., and P. A. Schmalzer. 2004. Anthropogenic influences on potential fire spread in a pyrogenic ecosystem of Florida, USA. Landscape Ecology 19:153–165.

Duncan, B. W., F. W. Adrian, and E. D. Stolen. 2010. Isolating the lightning ignition regime from a contemporary background fire regime in east-central Florida. Canadian Journal of Forest Research 40:286–297.

Duncan, B. W., P. A. Schmalzer, D. R. Breininger, and E. D. Stolen. 2015. Comparing fuels reduction and patch mosaic fire regimes for reducing fire spread potential: A spatial modeling approach. Ecological Modelling 314:90–99.

Earley, L. S. 2004. Looking for Longleaf: The Fall and Rise of an American Forest. University of North Carolina Press, Chapel Hill.

Edwards, L., J. Ambrose, and L. K. Kirkman. 2013. The Natural Communities of Georgia. University of Georgia Press, Athens.

Eldredge, I. 1911. Fire problems in the Florida national forest. Proceedings of the Society of American Foresters 164–171.

Ellair, D. P., and W. J. Platt. 2013. Fuel composition influences fire characteristics and understorey hardwoods in pine savanna. Journal of Ecology 101:192–201.

Ellery, W. N., K. Ellery, T. S. McCarthy, B. Cairncross, and R. Oelofse. 1989. A peat fire in the Okavango Delta, Botswana, and its importance as an ecosystem process. African Journal of Ecology 27:7–21.

Estill, J. C., and M. B. Cruzan. 2001. Phytogeography of rare plant species endemic to the southeastern United States. Castanea 66:3–23.

Ewel, K. C. 1990. Swamps. In Ecosystems of Florida, edited by R. L. Myers and J. J. Ewel, 281–323. University of Central Florida Press, Orlando.

Ewel, K. C., and W. J. Mitsch. 1978. The effects of fire on species composition in cypress dome ecosystems. Florida Scientist 41:25–31.

Falcon-Lang, H. J. 2000. Fire ecology in the Carboniferous tropical zone. Palaeogeography, Palaeoclimatology, Palaeoecology 164:355–371.

Falcon-Lang, H. J., V. Mages, and M. Collinson. 2016. The oldest *Pinus* and its preservation by fire. Geology 44:303–306.

Falk, D. 1990. Discovering the future, creating the past: Some reflections on restoration. Restoration and Management Notes 8:71–72.

Ferguson, E. R. 1961. Effects of prescribed fires on understory stems in pine-hardwood stands of Texas. Journal of Forestry 59:356–359.

Field, C. B., et al., editors. 2012. Managing the Risks of Extreme Events and Disasters to Advance Climate Change Adaptation. Special Report of the Intergovernmental Panel on Climate Change. Cambridge University Press, Cambridge.

Fill, J. M., B. M. Moule, J. M. Varner, and T. A. Mousseau. 2016. Flammability of the keystone savanna bunchgrass *Aristida stricta*. Plant Ecology. DOI 10.1007/s11258-016-0574-0.

Fill, J. M., W. J. Platt, S. M. Welch, J. L. Waldron, and T. A. Mousseau. 2015a. Updating models for restoration and management of fiery ecosystems. Forest Ecology and Management 356:54–63.

Fill, J. M., J. L. Waldron, S. M. Welch, J. W. Gibbons, S. H. Bennett, and T. A. Mousseau. 2015b. Using multiscale spatial models to assess potential surrogate habitat for an imperiled reptile. PLoS ONE 10(4): e0123307.

Fill, J. M., S. M. Welch, J. L. Waldron, and T. A. Mousseau. 2012. The reproductive response of an endemic bunchgrass indicates historical timing of a keystone process. Ecosphere 3(7): 61.

Fisher, D. C., and D. L. Fox. 2006. Five years in the life of an Aucilla River mastodon. In First Floridians and Last Mastodons: The Page-Ladson Site in the Aucilla River, edited by S. D. Webb, 343–377. Springer, Dordrecht, The Netherlands.

Fitzpatrick, J. W., G. E. Woolfenden, and M. T. Kopeny. 1991. Ecology and development-related habitat requirements of the Florida scrub jay (*Aphelocoma coerulescens coerulescens*). Nongame Wildlife Program Technical Report 8. Florida Game and Fresh Water Fish Commission, Tallahassee.

Flematti, G. R., K. W. Dixon, and S. M. Smith. 2015. What are karrikins and how were they "discovered" by plants? BMC Biology 13:108. DOI 10.1186/s12915-015-0219-0.

Flematti, G. R., E. L. Ghisalberti, K. W. Dixon, and R. D. Trengove. 2004. A compound from smoke that promotes seed germination. Science 305:977.

Florida Natural Areas Inventory (FNAI). 2010. Guide to the Natural Communities of Florida. http://www.fnai.org/naturalcommguide.cfm.

Foster, J. T., II, and A. D. Cohen. 2007. Palynological evidence of the effects of the deerskin trade on forest fires during the eighteenth century in southeastern North America. American Antiquities 72:35–51.

Francis, J. E. 1984. The seasonal environment of the Purbeck (Upper Jurassic) fossil forests. Palaeogeography, Palaeoclimatology, Palaeoecology 48:285–307.

Franklin, J. F., F. J. Swanson, M. E. Harmon, D. A. Perry, T. A. Spies, V. H. Dale, A. McKee, W. K. Ferrell, J. E. Means, S. V. Gregory, J. D. Lattin, T. D. Schowalter, and D. Larsen. 1991. Effects of global climate change on forests in northwestern North America. Northwest Environmental Journal 7:233–254.

Freeman, D. L. 2004. Lightning-ignited wildfire occurrences in a central-Florida landscape managed with prescribed fire. M.S. thesis, University of Florida, Gainesville.

Freeman, J., L. Kobziar, E. W. Rose, and W. Cropper. 2017. A critique of the historical-fire-regime concept in conservation. Conservation Biology 31:976–985.

Frei, B., K. G. Smith, J. H. Withgott, and P. G. Rodewald. 2015. Red-headed woodpecker (*Melanerpes erythrocephalus*). In The Birds of North America Online, edited by A. Poole. Cornell Lab of Ornithology, Ithaca, NY. Retrieved from the Birds of North America Online: http://bna.birds.cornell.edu.bnaproxy.birds.cornell.edu/bna/species/518. doi:10.2173/bna.518.

Froese, D., M. Stiller, P. D. Heintzman, A. V. Reyes, G. D. Zazula, A. E. R. Soares, M. Meyer, E. Hall, B. J. L. Jensen, L. J. Arnold, R. D. E. MacPhee, and B. Shapiro. 2017. Fossil and genomic evidence constrains the timing of bison arrival in North America. Proceedings of the National Academy of Sciences. doi:10.1073/pnas.1620754114.

Frost, C. C. 1995. Presettlement fire regimes in southeastern marshes, peatlands, and swamps. Proceedings of the Tall Timbers Fire Ecology Conference 19:39–60.

———. 2000. Studies in landscape fire ecology and presettlement vegetation of the Southeastern United States. Ph.D. diss., University of North Carolina.

———. 2006. History and future of the longleaf pine ecosystem. In The Longleaf Pine Ecosystem: Ecology, Silviculture, and Restoration, edited by S. Jose, E. Jokela, and D. Miller, 9–48. Springer-Verlag, New York.

Fuhlendorf, S. D., D. M. Engle, J. Kerby, and R. Hamilton. 2009. Pyric herbivory: Rewilding landscapes through the recoupling of fire and grazing. Conservation Biology 23:588–598.

Gagnon, P. R. 2009. Fire in floodplain forests in the southeastern USA: Insights from disturbance ecology of native bamboo. Wetlands 29:520–526.

Gagnon, P. R., and W. J. Platt. 2008. Multiple disturbances accelerate clonal growth in a potentially monodominant bamboo. Ecology 89:612–618.

Gagnon, P. R., H. A. Passmore, W. J. Platt, J. A. Myers, C. E. T. Paine, and K. E. Harms. 2010. Does pyrogenicity protect burning plants? Ecology 9:3481–3486.

Garren, K. H. 1943. Effects of fire on vegetation of the southeastern United States. Botanical Review 9:617–654.

Germain-Aubrey, C. C., P. S. Soltis, K. M. Neubig, T. Thurston, D. E. Soltis, and M. A. Gitzendanner. 2014. Using comparative biogeography to retrace the origins of an ecosystem: The case of four plants endemic to the central Florida scrub. International Journal of Plant Sciences 175:418–431.

Gholz, H. L., and R. F. Fisher. 1984. The limits to productivity: Fertilization and nutrient cycling in coastal plain slash pine forest. Proceedings of the North American Forest Soils Conference 6:105–120.

Gill, J. L., J. L. Blois, B. Benito, S. Dobrowski, M. L. Hunter Jr., and J. L. McGuire. 2015. A 2.5-million-year perspective on coarse-filter strategies for conserving nature's stage. Conservation Biology 29:640–648.

Gill, J. L., J. W. Williams, S. T. Jackson, K. B. Lininger, and G. S. Robinson. 2009. Pleistocene megafaunal collapse, novel plant communities, and enhanced fire regimes in North America. Science 326:100–1103.

Gilliam, F. S., and W. J. Platt. 1999. Effects of long-term fire exclusion on tree species composition and stand structure in an old-growth *Pinus palustris* (longleaf pine) forest. Plant Ecology 140:15–26.

Glasspool, I. J., and A. C. Scott. 2010. Phanerozoic atmospheric oxygen concentrations reconstructed from sedimentary charcoal. Nature Geoscience 3:270–630.

Glasspool, I. J., D. Edwards, and L. Axe. 2004. Charcoal in the Silurian as evidence for the earliest wildfire. Geology 32:381–383.

Gleason, H. A. 1917. The structure and development of the plant association. Bulletin of the Torrey Botanical Club 44:463–481.

———. 1926. The individualistic concept of the plant association. Bulletin of the Torrey Botanical Club 53:7–26.

Gleason, H. A., and A. Cronquist. 1964. The Natural Geography of Plants. Columbia University Press, New York.

Glitzenstein, J. S., W. J. Platt, and D. R. Streng. 1995. Effects of fire regime and habitat on tree dynamics in North Florida longleaf pine savannas. Ecological Monographs 65:441–476.

Glitzenstein, J. S., D. R. Streng, R. E. Masters, K. M. Robertson, and S. M. Hermann. 2012. Fire-frequency effects on vegetation in north Florida pinelands: Another look at the long-term Stoddard Fire Research Plots at Tall Timbers Research Station. Forest Ecology and Management 264:197–209.

Glitzenstein, J. S., D. R. Streng, and D. D. Wade. 2003. Fire frequency effects on longleaf pine (*Pinus palustris* P. Miller) vegetation in South Carolina and northeast Florida, USA. Natural Areas Journal 23:22–37.

Goodman, S. J., B. E. Buechler, K. Knupp, K. Driscoll, and E. W. McCaul Jr. 2000. The 1997–98 El Niño event and related wintertime lightning variations in the southeastern United States. Geophysical Research Letters 27:541–544.

Gorman, T. A., C. A. Haas, and D. C. Bishop. 2009. Factors related to occupancy of breeding wetlands by flatwoods salamander larvae. Wetlands 29:323–329.

Gould, S. J., and E. S. Vrba. 1982. Exaptation—a missing term in the science of forms. Paleobiology 8:4–15.

Grace, J. B., T. M. Anderson, E. W. Seabloom, et al. 2016. Integrative modelling reveals mechanisms linking productivity and plant species richness. Nature 529:390–393.

Grace, S. L., and W. J. Platt. 1995. Effects of adult tree density and fire on demography of pregrass stage juvenile longleaf pine (*Pinus palustris* Mill.). Journal of Ecology 83:75–86

Grady, J. M., and W. A. Hoffmann. 2012. Caught in a fire trap: Recurring fire creates stable size equilibria in woody resprouters. Ecology 93:2052–2060.

Graham, A. 1993. History of the vegetation: Cretaceous (Maastrichtian)—Tertiary. In Flora of North America North of Mexico. Vol. 1, Introduction, edited by Flora of North America Editorial Committee, 57–70. Oxford University Press, New York.

———. 1999. Late Cretaceous and Cenozoic History of North American Vegetation, North of Mexico. 1999. Oxford University Press, New York.

Gray, J. 1960. Temperate pollen genera in the Eocene (Claiborne) Flora, Alabama. Science 23:808–810.

Gray, J. B. 2016. Canebrakes of the Sandhills region of the Carolinas and Georgia: Fire history, canebrake area, and species frequency. Castanea 81:280–291.

Greenberg, C., and R. Simons. 1999. Age, composition, and stand structure of old-growth oak sites in the Florida high pine landscape: Implications for ecosystem management and restoration. Natural Areas Journal 19:30–40.

Greene, S. W. 1931. The forest that fire made. American Forests 37:583–584.

Gresham, C. A., T. M. Williams, and D. J. Lipscomb. 1991. Hurricane Hugo wind damage to southeastern U.S. coastal forest tree species. Biotropica 23:420–426.

Grime, J. P., and S. Pierce. 2012. The Evolutionary Strategies That Shape Ecosystems. Wiley-Blackwell, West Sussex, UK.

Grimm, E. C., G. L. Jacobson Jr., W. A. Watts, B. C. S. Hansen, and K. A. Maasch. 1993. A 50,000-year record of climate oscillations from Florida and its temporal correlation with the Heinrich events. Science 261:198–200.

Grimm, E. C., W. A. Watts, G. L. Jacobson Jr., B. C. S. Hansen, H. R. Almquist, and A. C. Dieffenbacher-Krall. 2006. Evidence for warm wet Heinrich events in Florida. Quaternary Science Reviews 25:2197–2211.

Grubb, P. J. 1977. The maintenance of species-richness in plant communities: The importance of the regeneration niche. Biological Review 52:107–145.

Guerin, D. N. 1993. Oak dome clonal structure and fire ecology in a Florida longleaf pine dominated community. Bulletin of the Torrey Botanical Club 120:107–114.

Guthrie, R. D. 2006. New carbon dates link climatic change with human colonization and Pleistocene extinctions. Nature 441:207–209.

Guyette, R. P., M. C. Stambaugh, D. C. Dey, and R.-M. Muzika. 2012. Predicting fire frequency with chemistry and climate. Ecosystems 15:322–335.

Hairston, N. G., F. E. Smith, and L. B. Slobodkin. 1960. Community structure, population control and competition. American Naturalist 94:421–425.

Hall, S. P., and D. F. Schweitzer. 1992. A survey of the moths, butterflies, and grasshoppers of four Nature Conservancy preserves in southeastern North Carolina. Report to The Nature Conservancy, North Carolina Chapter, Carrboro, NC.

Halligan, J. J., M. R. Waters, A. Perrotti, I. J. Owens, J. M. Feinberg, M. D. Bourne, B. Fenerty, B. Winsborough, D. Carlson, D. C. Fisher, T. W. Safford, and J. S. Dunbar. 2016. Pre-Clovis occupation 14,550 years ago at the Page-Ladson site, Florida, and the peopling of the Americas. Science Advances 2:e1600375.

Hammond, D. H., J. M. Varner, J. S. Kush, and Z. Fan. 2015. Contrasting sapling bark allocation of five southeastern USA hardwood tree species in a fire prone ecosystem. Ecosphere 6(7): 112.

Hanberry, B. B., R. F. Noss, S. K. Allison, D. C. Dey, and H. D. Safford. 2015. Restoration is preparation for the future. Journal of Forestry 113:425–429.

Hann, J. H. 1996. A History of the Timucua Indians and Missions. University Press of Florida, Gainesville.

Hanski, I. 1998. Metapopulation dynamics. Nature 396:41–49.

Hardin, E. D. 2010. Institutional history of prescribed fire in the Florida Division of

Forestry: Lessons from the past, directions for the future. Proceedings of the Tall Timbers Fire Ecology Conference 24:35–42.

Hardin, E. D., and D. L. White. 1989. Rare vascular plant taxa associated with wiregrass (*Aristida stricta*) in the southeastern United States. Natural Areas Journal 9:234–245.

Hare, R. C. 1965. Contribution of bark to fire resistance of southern trees. Journal of Forestry 63:248–251.

Harper, F., and D. E. Presley. 1981. Okefenokee Album. University of Georgia Press, Athens.

Harper, R. M. 1911. The relation of climax vegetation to islands and peninsulas. Bulletin of the Torrey Botanical Club 38:515–525.

———. 1913. The forest resources of Alabama. American Forestry 19:657–670.

———. 1914. Geography and vegetation of northern Florida. Annual Report of the Florida State Geological Survey 6:163–452.

———. 1915. Vegetation types. In E. H. Sellards, R. M. Harper, E. N. Mooney, W. J. Gunter, and E. Gunter, Natural resources survey of an area in central Florida, 135–188. Seventh Annual Report of the Florida Geological Survey, Tallahassee.

———. 1927. Natural resources of southern Florida. Annual Report of the Florida State Geological Survey 18:27–206.

———. 1943. Forests of Alabama. Geological Survey of Alabama. University of Alabama, Tuscaloosa.

Harrison, S., and R. Noss. 2017. Endemism hotspots are linked to stable climatic refugia. Annals of Botany 119:207–214.

Hart, S. J., T. T. Veblen, K. S. Eisenhart, D. Jarvis, and D. Kulakowski. 2014. Drought induces spruce beetle (*Dendroctonus rufipennis*) outbreaks across northwestern Colorado. Ecology 95:930–939.

Haynes, G. 2013. Extinctions in North America's late glacial landscapes. Quaternary International 285:89–98.

He, T., and B. B. Lamont. 2017. Baptism by fire: The pivotal role of ancient conflagrations in evolution of the Earth's biota. National Science Review. doi:10.1093/nsr/nwx041.

He, T., J. G. Pausas, C. M. Belcher, D. W. Schwilk, and B. B. Lamont. 2012. Fire-adapted traits of *Pinus* arose in the fiery Cretaceous. New Phytologist 194:751–759.

Henderson, J. P. 2006. Dendroclimatological analysis and fire history of longleaf pine (*Pinus palustris* Mill.) in the Atlantic and Gulf Coastal Plain. Ph.D. diss., University of Tennessee.

Hermann, S. M., J. S. Kush, J. C. Gilbert, and R. J. Barlow. 2015. Burning for conservation values: Should the goal be to mimic a natural fire regime? In Proceedings of the 17th Biennial Southern Silviculture Research Conference, edited by A. G. Holley, K. F. Connor, and J. D. Haywood, 164–171. General Technical Report SRS-203. USDA Forest Service, Asheville, NC.

Hermann, S. M., T. Van Hook, R. W. Flowers, L. A. Brennan, J. S. Glitzenstein, D. R. Streng, J. L. Walker, and R. L. Myers. 1998. Fire and biodiversity: Studies of vegetation and arthropods. Transactions of the North American Wildlife and Natural Resources Conference 63:384–401.

Hess, C. A., and F. C. James. 1998. Diet of the red-cockaded woodpecker in the Apalachicola National Forest. Journal of Wildlife Management 62:509–517.

Hessburg, P. F., D. J. Churchill, A. J. Larson, et al. 2015. Restoring fire-prone Inland Pacific landscapes: Seven core principles. Landscape Ecology 30:1805–1835.

Hewitt, R. E., and E. S. Menges. 2008. Allelopathic effects of *Ceratiola ericoides* (Empetraceae) on germination and survival of six Florida scrub species. Plant Ecology 198:47–59.

Heyward, F. 1939. The relation of fire to stand composition of longleaf pine forests. Ecology 20:287–304.

Hiers, J. K., J. J. O'Brien, R. J. Mitchell, J. M. Grego, and E. L. Loudermilk. 2009. The wildland fuel cell concept: An approach to characterize fine-scale variation in fuels and fire in frequently burned longleaf pine forests. International Journal of Wildland Fire 18:315–325.

Hiers, J. K., J. J. O'Brien, R. E. Will, and R. J. Mitchell. 2007. Forest floor depth mediates understory vigor in xeric *Pinus palustris* ecosystems. Ecological Applications 17:806–814.

Hiers, J. K., J. R. Walters, R. J. Mitchell, J. M. Varner, L. M. Connor, L. A. Blanc, and J. Stowe. 2014. Ecological value of retaining pyrophytic oaks in longleaf pine ecosystems. Journal of Wildlife Management 78:383–393.

Hiers, J. K., R. Wyatt, and R. J. Mitchell. 2000. The effects of fire regime on legume reproduction in longleaf pine savannas: Is a season selective? Oecologia 125:521–530.

Hill, C. M., P. J. Fitzpatrick, J. H. Corbin, Y. H. Lau, and S. K. Bhate. 2010. Summertime precipitation regimes associated with the sea breeze and land breeze in southern Mississippi and eastern Louisiana. Weather and Forecasting 25:1755–1779.

Hinman, S. E., and J. S. Brewer. 2007. Responses of two frequently burned pine savannas to an extended period without fire. Journal of the Torrey Botanical Society 134:512–526.

Hobbs, R. J., E. Higgs, C. M. Hall, P. Bridgewater, F. S. Chapin III, E. C. Ellis, J. J. Ewel, et al. 2014. Managing the whole landscape: Historical, hybrid, and novel ecosystems. Frontiers in Ecology and the Environment 12:557–564.

Hoctor, T. S., R. F. Noss, L. D. Harris, and K. A. Whitney. 2006. Spatial ecology and restoration of the longleaf pine evosystem. In The Longleaf Pine Ecosystem: Ecology, Silviculture, and Restoration, edited by S. Jose, E. Jokela, and D. Miller, 377–402. Springer-Verlag, New York.

Hodgkins, E. J. 1958. Effects of fire on undergrowth vegetation in upland southern pine forests. Ecology 39:36–46.

Holdridge, L. R. 1947. Determination of world plant formations from simple climatic data. Science 105:367–368.

Hood, S., A. Sala, E. K. Heyerdahl, and M. Boutin. 2015. Low-severity fire increases tree defense against bark beetle attacks. Ecology 96:1846–1855.

Hoppe, K. A., and P. L. Koch. 2006. The biogeochemistry of the Aucilla River fauna. In First Floridians and Last Mastodons: The Page-Ladson Site in the Aucilla River, edited by S. D. Webb, 379–401. Springer, Dordrecht, The Netherlands.

Huang, Y., B. Shuman, Y. Wang, T. Webb III, E. C. Grimm, and G. L. Jacobson Jr. 2006. Climatic and environmental controls on the variation of C_3 and C_4 plant abundances in central Florida for the past 62,000 years. Palaeogeography, Palaeoclimatology, Palaeoecology 237:428–435.

Hudson, C. 1976. The Southeastern Indians. University of Tennessee Press, Knoxville.

Huffman, J. M. 2006. Historical fire regimes in southeastern pine savannas. Ph.D. diss., Louisiana State University.

Huffman, J. M., and S. W. Blanchard. 1991. Changes in woody vegetation in Florida dry prairie and wetlands during a period of fire exclusion and after dry-growing-season fire. In Fire and the Environment: Ecological and cultural perspectives. Forest Service General Technical Report SE-69, edited by S. C. Nodvin and T. A. Waldrop, 75–83. Southeastern Forest Experiment Station, Asheville, NC.

Huffman, J. M., and W. J. Platt. 2014. Fire history of the Avon Park Air Force Range: Evidence from tree-rings. Unpublished report. Department of Biological Sciences, Louisiana State University, Baton Rouge.

Huffman, J. M., and M. T. Rother. 2017. Dendrochronological field methods for fire history in pine ecosystems of the southeastern Coastal Plain. Tree-Ring Research 73:42–46.

Huffman, J. M., and P. A. Werner. 2000. Restoration of Florida pine savanna: Flowering response of *Lilium catesbaei* to fire and roller-chopping. Natural Areas Journal 20:12–23.

Huffman, J. M., W. J. Platt, J. Grissino-Mayer, and C. J. Boyce. 2004. Fire history of a barrier island slash pine (*Pinus elliottii*) savanna. Natural Areas Journal 24:258–268.

Hughes, R. H. 1966. Fire ecology of canebrakes. Proceedings of the Tall Timbers Fire Ecology Conference 5:149–158.

Hulbert, R. C., Jr., editor. 2001. The Fossil Vertebrates of Florida. University Press of Florida, Gainesville.

Hunter, M. E., and E. S. Menges. 2002. Allelopathic effects and root distribution of *Ceratiola ericoides* (Empetraceae) on seven rosemary scrub species. American Journal of Botany 89:1113–1118.

Hurteau, M. D., and M. L. Brooks. 2011. Short- and long-term effects of fire on carbon in US dry temperate forest systems. BioScience 61:139–146.

Hurteau, M. D., G. W. Koch, and B. A. Hungate. 2008. Carbon protection and fire risk reduction: Toward a full accounting of forest carbon offsets. Frontiers in Ecology and the Environment 6:493–498.

Hutchinson, G. E. 1957. Concluding remarks, Cold Spring Harbor Symposium. Quantitative Biology 22:415–427.

———. 1978. An Introduction to Population Biology. Yale University Press, New Haven, CT.

Hutto, R. L. R. E. Keane, R. L. Sherriff, C. T. Rota, L. A. Eby, and V. A. Saab. 2016. Toward a more ecologically informed view of severe forest fires. Ecosphere 7(2): e01255.

IPCC. 2012. Managing the risks of extreme events and disasters to advance climate change adaptation. A Special Report of Working Groups I and II of the Intergovernmental Panel on Climate Change, edited by C. B. Field, V. Barros, T. F. Stocker,

D. Qin, D. J. Dokken, K. L. Ebi, M. D. Mastrandrea, K. J. Mach, G.-K. Plattner, S. K. Allen, M. Tignor, and P. M. Midgley. Cambridge University Press, Cambridge.

Isbell, F., V. Calcagno, A. Hector, et al. 2011. High plant diversity is needed to maintain ecosystem services. Nature 477:199–202.

Ives, A. R., and S. R. Carpenter. 2007. Stability and diversity of ecosystems. Science 317:58–62.

Jackson, D. R., and E. R. Milstrey. 1989. The fauna of gopher tortoise burrows. In Proceedings of the Gopher Tortoise Relocation Symposium, edited by J. E. Diemer, D. R. Jackson, J. L. Landers, J. N. Layne, and D. A. Wood, 86–98. Florida Game and Fresh Water Fish Commission, Tallahassee.

Jackson, J. F., D. C. Adams, and U. B. Jackson. 1999. Allometry of constitutive defense: A model and a comparative test with tree bark and fire regime. American Naturalist 153:614–632.

Jackson, S. T., E. S. Webb, K. H. Anderson, J. T. Overpeck, T. Webb III, J. W. Williams, and B. C. S. Hansen. 2000. Vegetation and environment in eastern North America during the Last Glacial Maximum. Quaternary Science Reviews 19:489–508.

James, C. W. 1961. Endemism in Florida. Brittonia 13:225–244.

James, F. C., C. A. Hess, and D. Kufrin. 1997. Species-centered environmental analysis: Indirect effects of fire history on red-cockaded woodpeckers. Ecological Applications 7:118–129.

Janis, C. M., J. Damuth, and J. M. Theodor. 2002. The origins and evolution of the North American grassland biome: The story from the hoofed mammals. Palaeogeography, Palaeoclimatology, Palaeoecology 177:183–198.

Jansson, R. 2003. Global patterns in endemism explained by past climate change. Proceedings of the Royal Society of London B 270:583–590.

Jarzen, D. M., and D. L. Dilcher. 2006. Middle Eocene terrestrial palynomorphs from the Dolime Minerals and Gulf Hammock quarries, Florida, U.S.A. Palynology 30:89–110.

Johnson, A. F. 1982. Some demographic characteristics of the Florida rosemary, *Ceratiola ericoides* Michx. American Midland Naturalist 108:170–174.

Johnson, E. I., J. K. DiMiceli, and P. C. Stouffer. 2009. Timing of migration and patterns of winter settlement by Henslow's sparrows. Condor 111:730–739.

Juras, P. 1997. The presettlement Piedmont savanna. Master's thesis, University of Georgia.

Just, M. G., M. G. Hohmann, and W. A. Hoffmann. 2015. Where fire stops: Vegetation structure and microclimate influence fire spread along an ecotonal gradient. Plant Ecology. DOI 10.1007/s11258-015-0545-x.

Kalisz, P. J., A. W. Dorian, and E. L. Stone. 1986. Prehistoric land-use and the distribution of longleaf pine on the Ocala National Forest, Florida: An interdisciplinary analysis. Florida Anthropologist 39:183–193.

Kane, J. M., J. M. Varner, and J. K. Hiers. 2008. The burning characteristics of southeastern oaks: Discriminating fire facilitators from fire impeders. Forest Ecology and Management 256:2039–2045.

Kay, C. E. 2007. Are lightning fires unnatural? A comparison of aboriginal and lightning ignition rates in the United States. Proceedings of the Tall Timbers Fire Ecology Conference 23:16–28.

Keeley, J. E. 2009. Fire intensity, fire severity, and burn severity: A brief review and suggested usage. International Journal of Wildland Fire 18:116–126.

Keeley, J. E., and M. E. Nitzberg. 1984. Role of charred wood in the germination of the chaparral herbs *Emmenanthe penduliflora* (Hydrophyllaceae) & *Eriophyllum confertiflorum*. Madrono 31(4): 208–218.

Keeley, J. E., and P. H. Rundel. 2005. Fire and the Miocene expansion of C_4 grasslands. Ecology Letters 8:683–690.

Keeley, J. E., and P. H. Zedler. 1998. Evolution of life histories in *Pinus*. In Ecology and Biogeography of *Pinus*, edited by D. Richardson, 219–250. Cambridge University Press, Cambridge.

Keeley, J. E., G. H. Aplet, N. L. Christensen, S. G. Conard, E. A. Johnson, P. N. Omi, D. L. Peterson, and T. W. Swetnam. 2009. Ecological foundations for fire management in North American forest and shrubland ecosystems. USDA Forest Service General Technical Report PNW-GTR-779. Pacific Northwest Research Station, Portland, OR.

Keeley, J. E., W. J. Bond, R. A. Bradstock, J. G. Pausas, and P. W. Rundel. 2012. Fire in Mediterranean Climate Ecosystems: Ecology, Evolution, and Management. Cambridge University Press, Cambridge.

Keeley, J. E., J. G. Pausas, P. W. Rundel, W. J. Bond, and R. A. Bradstock. 2011. Fire as an evolutionary pressure shaping plant traits. Trends in Plant Science 16:406–411.

Kelly, L. T., A. F. Bennett, M. F. Clarke, and M. A. McCarthy. 2015. Optimal fire histories for biodiversity conservation. Conservation Biology 29:473–481.

Kerr, B., D. W. Schwilk, A. Bergman, and M. W. Feldman. 1999. Rekindling an old flame: A haploid model for the evolution and impact of flammability in resprouting plants. Evolutionary Ecology Research 1:807–833.

Kerstyn, A., and P. Stiling. 1999. The effects of burn frequency on the density of some grasshoppers and leaf miners in a Florida sandhill community. Florida Entomologist 82:499–507.

Kirkman, K. P., S. L. Collins, M. D. Smith, A. K. Knapp, D. E. Burkepile, C. E. Burns, R. W. S. Fynn, N. Hagenah, S. E. Koerner, K. J. Matchett, D. I. Thompson, K. R. Wilcox, and P. D. Wragg. 2014. Responses to fire differ between South African and North American grassland communities. Journal of Vegetation Science 25:793–804.

Kirkman, L. K. 1995. Impacts of fire and hydrological regimes on vegetation in depression wetlands of southeastern USA. Proceedings of the Tall Timbers Fire Ecology Conference 19:10–20.

Kirkman, L. K., M. B. Drew, and D. Edwards. 1998a. Effects of experimental fire regimes on the population dynamics of *Schwalbea americana* L. Plant Ecology 1115–137.

Kirkman, L. K., M. B. Drew, L. T. West, and E. R. Blood. 1998b. Ecotone characterization between upland longleaf pine/wiregrass stands and seasonally-ponded isolated wetlands. Wetlands 18:346–364.

Kitzberger, T., T. W. Swetnam, and T. T. Veblen. 2001. Inter-hemispheric synchrony of forest fires and the El Niño–Southern Oscillation. Global Ecology and Biogeography 10:315–326.

Klaus, J. M., and R. F. Noss. 2016. Specialist and generalist amphibians respond to wetland restoration treatments. Journal of Wildlife Management 6:1106–1119.

Knapp, E. E., B. L. Estes, and C. N. Skinner. 2009. Ecological effects of prescribed fire season: A literature review and synthesis for managers. USDA Forest Service Gen. Tech. Rep. PSW-GTR-224.

Knight, T. M., and R. D. Holt. 2005. Fire generates spatial gradients in herbivory: An example from a Florida sandhill ecosystem. Ecology 86:587–593.

Koch, P. L., and A. D. Barnosky. 2006. Late Quaternary extinctions: State of the debate. Annual Review of Ecology, Evolution, and Systematics 37:215–50.

Koch, P. L., K. A. Hoppe, and S. D. Webb. 1998. The isotopic ecology of late Pleistocene mammals in North America. Part I. Florida. Chemical Geology 152:119–138.

Koehler, J. T. 1992–1993. Prescribed burning: A wildfire prevention tool? Fire Management Notes 53–54:9–13.

Komarek, E. V., Sr. 1962. Fire ecology. Proceedings of the Tall Timbers Fire Ecology Conference 1:95–107.

———. 1964. The natural history of lightning. Proceedings of the Tall Timbers Fire Ecology Conference 3:139–183.

———. 1965. Fire ecology—grasslands and man. Proceedings of the Tall Timbers Fire Ecology Conference 4:169–220.

———. 1967. The nature of lightning fires. Proceedings of the Tall Timbers Fire Ecology Conference 7:5–43.

———. 1974. Effects of fire on temperate forests and related ecosystems: Southeastern United States. In Fire and Ecosystems, edited by T. T. Kozlowski and C. E. Ahlgren, 251–277. Academic Press, New York.

Korosy, M. G., J. S. Reece, and R. F. Noss. 2013. Winter habitat associations of four grassland sparrows in Florida dry prairie. Wilson Journal of Ornithology 125:502–512.

Kraft, N. J. B., R. Valencia, and D. D. Ackerly. 2008. Functional traits and niche-based tree community assembly in an Amazonian forest. Science 322:580–582.

Kreye, J. K., J. M. Varner, and E. E. Knapp. 2011. Effects of particle fracturing and moisture content on fire behavior in masticated fuelbeds burning in a laboratory. International Journal of Wildland Fire 20:308–317.

Küchler, A. W. 1964. Potential Natural Vegetation of the Conterminous United States. American Geographic Society Special Publication No. 36. Reprinted by U.S. Geological Survey, Reston, VA (1985).

Kush, J. S., C. K. McMahon, and W. D. Boyer. 2004. Longleaf pine: A sustainable approach for increasing terrestrial carbon in the southern United States. Environmental Management 33:S139–S147.

Kushlan, J. A. 1990. Freshwater marshes. In Ecosystems of Florida, edited by R. L. Myers and J. J. Ewel, 324–363. University of Central Florida Press, Orlando.

Laessle, A. M. 1942. The plant communities of the Welaka area with special reference to correlation between soils and vegetational succession. Biological Sciences Series 4. University of Florida Publications, Gainesville.

Laliberté, E., J. B. Grace, M. A. Huston, H. Lambers, F. P. Teste, B. L. Turner, and D. A.

Wardle. 2013. How does pedogenosis drive plant diversity? Trends in Ecology and Evolution 28:331–340.

Lamont, B. B., D. C. Le Maitre, R. M. Cowling, and N. J. Enright. 1991. Canopy seed storage in woody plants. Botanical Review 57:277–317.

Landers, J. L. 1991. Disturbance influences on pine traits in the southeastern United States. Proceedings of the Tall Timbers Fire Ecology Conference 17:39–60.

Landers, J. L., N. A. Byrd, and R. Komarek. 2001. A holistic approach to managing longleaf pine communities. In Proceedings of the Symposium on the Management of Longleaf Pine, edited by R. J. Farrar, 135–169. Long Beach, MS.

Larson, L. H. 1980. Aboriginal Subsistence Technology on the Southeastern Coastal Plain during the Late Prehistoric Period. Ripley P. Bullen Monographs in Anthropology and History. Florida State Museum, Gainesville.

Ledig, F. T., J. L. Hom, and P. E. Smouse. 2013. The evolution of the New Jersey pine plains. American Journal of Botany 100:778–791.

Lee, M. A. B., K. L. Snyder, P. Valentine-Darby, S. J. Miller, and K. J. Ponzio. 2005. Dormant season prescribed fire as a management tool for the control of *Salix caroliniana* Michx. in a floodplain marsh. Wetlands Ecology and Management 13:479–487.

Leenhouts, B. 1998. Assessment of biomass burning in the conterminous United States. Conservation Ecology 2:1.

Legare, M., H. Hill, R. Farinetti, and F. T. Cole. 1998. Marsh bird response during two prescribed fires at the St. Johns National Wildlife Refuge, Brevard County, Florida. Proceedings of the Tall Timbers Fire Ecology Conference 20:114.

Legleu, C. 2012. Modeling gopher tortoise (*Gopherus polyphemus*) habitat in a fire-dependent ecosystem in north Florida. Master's thesis, Louisiana State University.

Leopold, A. 1924. Grass, brush, timber, and fire in southern Arizona. Journal of Forestry 22(6): 1–10.

———. 1949. A Sand County Almanac. Oxford University Press, New York.

Lewis, C. M. 1978. The Calusa. In Tacachale: Essays on the Indians of Florida and Southeastern Georgia during the Historic Period, edited by J. Milanich and S. Proctor, 19–49. University Presses of Florida, Gainesville.

Lewis, H. T. 1982. Fire technology and resource management in aboriginal North America and Australia. In Resource Managers: North American and Australian Hunter-Gatherers, edited by N. M. Williams and E. S. Hunn, 45–67. AAAS Selected Symposium 67. Westview Press, Boulder, CO.

Lindenmayer, D. B., and R. F. Noss. 2006. Salvage logging, ecosystem processes, and biodiversity conservation. Conservation Biology 20:949–958.

Lindenmayer, D. B., W. Blanchard, C. MacGregor, P. Barton, S. C. Banks, M. Crane, D. Michael, S. Okada, L. Berry, D. Florance, and M. Gill. 2016. Temporal trends in mammal responses to fire reveals the complex effects of fire regime attributes. Ecological Applications 26:557–573.

Lindenmayer, D. B., D. R. Foster, J. F. Franklin, M. L. Hunter, R. F. Noss, F. A. Schmiegelow, and D. Perry. 2004. Saving forests or saving fiber? Salvage harvesting policies after natural disturbance impairs ecosystem and species recovery. Science 303:1303.

Lindon, H. L., and E. Menges. 2008. Effects of smoke on seed germination of twenty species of fire-prone habitats in Florida. Castanea 73:106–110.

Liu, H., and E. S. Menges. 2005. Winter fires promote greater vital rates in the Florida Keys than summer fires. Ecology 86:1483–1495

Liu, H., E. S. Menges, and P. F. Quintana-Ascencio. 2005. Population viability analyses of *Chamaecrista keyensis*: Effects of fire season and frequency. Ecological Applications 15:210–221.

Liu, Z., and M. C. Wimberly. 2015. Climatic and landscape influences on fire regimes from 1984 to 2010 in the western United States. PLoS ONE 10(10): e0140839. doi:10.1371.

Lockwood, J. L., M. S. Ross, and J. P. Sah. 2003. Smoke on the water: The interplay of fire and water flow on Everglades restoration. Frontiers in Ecology and the Environment 1:462–468.

Lomolino, M. V., B. R. Riddle, and J. H. Brown. 2006. Biogeography. 3rd ed. Sinauer, Sunderland, MA.

Long, E. C. 1888. Notes on some of the forest features of Florida with items of tree growth in that state. Proceedings of the Seventh Annual Meeting of the American Forestry Congress, 38–41.

Loope, W. L., and J. B. Anderton. 1998. Human vs. lightning ignition of presettlement surface fires in coastal pine forests of the upper Great Lakes. American Midland Naturalist 140:206–218.

Loudermilk, E. L., J. K. Hiers, S. Pokswinski, J. J. O'Brien, A. Barnett, and R. J. Mitchell. 2016. The path back: Oaks (*Quercus* spp.) facilitate longleaf pine (*Pinus palustris*) seedling establishment in xeric sites. Ecosphere 7(6): e01361.

Loveless, C. M. 1959. A study of the vegetation in the Florida Everglades. Ecology 40:1–9.

Lyell, C. 1849. Second Visit to the United States of North America. Vol. 2. Harper and Brothers, New York.

MacFadden, B. J. 1997. Fossil mammals of Florida. In The Geology of Florida, edited by A. F. Randazzo and D. S. Jones, 119–137. University Press of Florida, Gainesville.

MacFadden, B. J., and T. E. Cerling. 1996. Mammalian herbivore communities, ancient feeding ecology, and carbon isotopes: A 10-million-year sequence from the Neogene of Florida. Journal of Vertebrate Paleontology 16:103–115.

MacFadden, B. J., and R. C. Hulbert Jr. 1990. Body size estimates and size distribution of ungulate mammals from the Late Miocene Love Bone Bed of Florida. In Body Size in Mammalian Paleobiology: Estimation and Biological Implications, edited by J. Damuth and B. J. MacFadden, 337–363. Cambridge University Press, Cambridge.

Mader, H. J., C. Schell, and P. Kornacker. 1990. Linear barriers to movement in the landscape. Biological Conservation 54:209–222.

Maguire, A. J., and E. S. Menges. 2011. Post-fire growth strategies of resprouting Florida scrub vegetation. Fire Ecology 7(3): 12–25.

Main, M. B., and M. J. Barry. 2002. Influence of season of fire on flowering of wet prairie grasses in south Florida, USA. Wetlands 22:430–434.

Mann, C. 2005. 1491: New Revelations of the Americas Before Columbus. Knopf, New York.

Maravalhas, J., and H. L. Vasconcelos. 2014. Revisiting the pyrodiversity-biodiversity hypothesis: Long-term fire regimes and the structure of ant communities in a Neotropical savanna hotspot. Journal of Applied Ecology 51:1661–1668.

Marlon, J. R., P. J. Bartlein, M. K. Walsh, S. P. Harrison, et al. 2009. Wildfire responses to abrupt climate change in North America. Proceedings of the National Academy of Sciences 106:2519–2524.

Marquardt, W. H., and K. J. Walker, editors. 2013. The Archaeology of Pineland: A Coastal Southwest Florida Site Complex, A.D. 50–1710. Institute of Archaeology and Paleoenvironmental Studies, Monograph 4. University of Florida, Gainesville.

Marshall, C. H., R. A. Pielke, L. T. Steyaert, and D. A. Willard. 2004. The impact of anthropogenic land-cover change on the Florida peninsula sea breezes and warm season sensible weather. Monthly Weather Review 132:28–52.

Martin, K. L., and L. K. Kirkman. 2009. Management of ecological thresholds to reestablish disturbance-maintained herbaceous wetlands of the south-eastern USA. Journal of Applied Ecology 46:906–914.

Martin, R. E. 1963. Thermal properties of bark. Forest Products Journal 13:419–426.

Martin, R. E., and D. B. Sapsis. 1992. Fires as agents of biodiversity: Pyrodiversity promotes biodiversity. In Proceedings of the Symposium on Biodiversity in Northwestern California, 1991, edited by H. M. Kerner, 150–157. Wildland Resources Center, University of California, Berkeley.

Maurin, O., T. J. Davies, J. E. Burrows, B. H. Daru, K. Yessoufou, A. M. Muasya, M. van der Bank, and W. J. Bond. 2014. Savanna fire and the origins of the "underground forests" of Africa. New Phytologist 204:201–214.

May, J. H., and A. G. Warne. 1999. Hydrogeologic and geochemical factors required for the development of Carolina Bays along the Atlantic and Gulf of Mexico, coastal plain, USA. Environmental and Engineering Geoscience 5:261–270.

McIntosh, R. P. 1985. The Background of Ecology. Cambridge University Press, Cambridge.

Means, D. B. 1972. Notes on the autumn breeding biology of *Ambystoma cingulatum* (Cope) (Amphibia: Urodela: Ambystomatidae). Association of Southeastern Biologists Bulletin 19:84.

———. 2006. Vertebrate faunal diversity of longleaf pine ecosystems. In The Longleaf Pine Ecosystem: Ecology, Silviculture, and Restoration, edited by S. Jose, E. Jokela, and D. Miller, 157–213. Springer-Verlag, New York.

Means, D. B., J. G. Palis, and M. Baggett. 1996. Effects of slash pine silviculture on a Florida population of flatwoods salamander. Conservation Biology 10:426–437.

Menges, E. S. 1999. Ecology and conservation of Florida scrub. In Savannas, Barrens, and Rock Outcrop Plant Communities of North America, edited by R. C. Anderson, J. R. Fralish, and J. M. Baskin, 7–22. Cambridge University Press, Cambridge.

———. 2007. Integrating demography and fire management: An example from Florida scrub. Australian Journal of Botany 55:261–272.

Menges, E. S., and M. A. Deyrup. 2001. Postfire survival in South Florida slash pine: Interacting effects of fire intensity, fire season, vegetation, burn size, and bark beetles. International Journal of Wildland Fire 10:53–63.

Menges, E. S., and D. R. Gordon. 2010. Should mechanical treatments and herbicides be used as fire surrogates to manage Florida's uplands? A review. Florida Scientist 73:147–174.

Menges, E. S., and C. V. Hawkes. 1998. Interactive effects of fire and microhabitat on plants of Florida scrub. Ecological Applications 8:935–946.

Menges, E. S., and J. Kimmich. 1996. Microhabitat and time-since-fire: Effects on the demography of *Eryngium cuenifolium* (Apiaceae), a Florida scrub endemic plant. American Journal of Botany 83:185–191.

Menges, E. S., and N. Kohfeldt. 1995. Life history strategies of Florida scrub plants in relation to fire. Bulletin of the Torrey Botanical Club 122:282–297.

Menges, E. S., and P. F. Quintana-Ascencio. 2004. Population viability with fire in *Eryngium cuneifolium*: Deciphering a decade of demographic data. Ecological Monographs 74:79–99.

Merriam, C. H. 1894. Laws of temperature control of the geographic distribution of terrestrial animals and plants. National Geographic 6:229–238.

Merriam, K. E., J. E. Keeley, and J. L. Beyers. 2006. Fuel breaks affect nonnative species abundance in California plant communities. Ecological Applications 16:515–527.

Meylan, P. A. 1982. The squamate reptiles of the Inglis IA fauna (Irvingtonian: Citrus County, Florida). Bulletin of the Florida State Museum 27:1–85.

Midgley, G. F., and W. J. Bond. 2015. Future of African terrestrial biodiversity and ecosystems under anthropogenic climate change. Nature Climate Change 5:823–829.

Milanich, J. T. 1998. Florida's Indians from Ancient Times to the Present. University Press of Florida, Gainesville.

Milanich, J. T., and C. H. Fairbanks. 1980. Florida Archaeology. Academic Press, New York.

Millar, C. I. 1998. Early evolution of pines. In Ecology and Biogeography of *Pinus*, edited by D. M. Richardson, 69–91. Cambridge University Press, Cambridge.

Miller, J. J. 1998. An Environmental History of Northeast Florida. University Press of Florida, Gainesville.

Miller, S. J., K. J. Ponzio, M. A. Lee, L. W. Keenan, and S. R. Miller. 1998. The use of fire in wetland preservation and restoration: Are there risks? Proceedings of the Tall Timbers Fire Ecology Conference 20:127–139.

Mitchell, R. J., J. K. Hiers, J. J. O'Brien, and G. Starr. 2009. Ecological forestry in the Southeast: Understanding the ecology of fuels. Journal of Forestry 107:391–397.

Mitchell, R. J., Y. Liu, J. J. O'Brien, K. J. Elliott, G. Starr, C. F. Miniat, and J. K. Hiers. 2014. Future climate and fire interactions in the southeastern region of the United States. Forest Ecology and Management 327:316–326.

Mitchener, L. J., and A. J. Parker. 2005. Climate, lightning, and wildfire in the national forests of the southeastern United States: 1989–1998. Physical Geography 26:147–162.

Mittermeier, R. A., W. R. Turner, F. W. Larsen, T. M. Brooks, and C. Gascon. 2011. Global biodiversity conservation: The critical role of hotspots. In Biodiversity Hotspots: Distribution and Protection of Conservation Priority Areas, edited by F. E. Zachos and J. C. Habel, 3–22. Springer, Heidelberg.

Mogil, H. M., and K. L. Seaman. 2008. Florida's climate and weather. Weatherwise 61(6): 14–19.

Monk, C. D. 1965. Southern mixed hardwood forests of north central Florida. Ecological Monographs 35:335–354.

Moore, C. R., M. J. Brooks, D. J. Mallinson, P. R. Parham, A. H. Ivester, and J. K. Feathers. 2016. The Quaternary evolution of Herndon Bay, a Carolina bay on the Coastal Plain of North Carolina (USA): Implications for paleoclimate and oriented lake genesis. Southeastern Geology 51:145–171,

Moreira, B., and J. G. Pausas. 2012. Tanned or burned: The role of fire in shaping physical seed dormancy. PLoS ONE 7(12): e51523.

Morgan, G. S., and S. D. Emslie. 2010. Tropical and western influences in vertebrate faunas from the Pliocene and Pleistocene of Florida. Quaternary International 217:143–158.

Morgan, J. A., D. R. LeCain, E. Pendall, et al. 2011. C_4 grasses prosper as carbon dioxide eliminates dessication in warmed semi-arid grassland. Nature 476:202–205.

Morley, R. J., and K. Richards. 1993. Graminieae cuticle: A key indicator of Late Cenozoic climatic change in the Niger Delta. Review of Palaeobotany and Palynology 77:119–127.

Morrison, J. L., and S. R. Humphrey. 2001. Conservation value of private lands for Crested Caracaras in Florida. Conservation Biology 15:675–684.

Muhs, D. R., K. R. Simmons, R. R. Schumann, and R. B. Halley. 2011. Sea-level history of the past two interglacial periods: New evidence from U-series dating of reef corals from south Florida. Quaternary Science Reviews 30:570–590.

Murphy, M. J., and R. L. Holle. 2005. Where is the real cloud-to-ground lightning maximum in North America? Weather and Forecasting 20:125–133.

Mutch, R. W. 1970. Wildland fire and ecosystems—a hypothesis. Ecology 51:1046–1051.

Myers, J. A., and K. E. Harms. 2011. Seed arrival and ecological filters interact to assemble high-diversity plant communities. Ecology 92:676–686.

Myers, R. L. 1985. Fire and the dynamic relationship between Florida sandhill and sand pine scrub vegetation. Bulletin of the Torrey Botanical Club 112:241–252.

———. 1990. Scrub and high pine. In Ecosystems of Florida, edited by R. L. Myers and J. J. Ewel, 150–193. University of Central Florida Press, Orlando.

Myers, R. L., and S. E. Boettcher. 1987. Flowering response of cutthroat grass (*Panicum abscissum*) following fire. (Abstract). Bulletin of the Ecological Society of America 68:375.

Myers, R. L.,and J. J. Ewel, editors. 1990. Ecosystems of Florida. University of Central Florida Press, Orlando.

Negrón-Ortiz, V., and D. L. Gorchov. 2000. Effects of fire season and post-fire herbivory on the cycad *Zamia pumila* (Zamiaceae) in slash pine savanna, Everglades National Park, Florida. International Journal of Plant Sciences 161:659–669.

Neel, L., P. S. Sutter, and A. G. Way. 2010. The Art of Managing Longleaf: A Personal History of the Stoddard-Neel Approach. University of Georgia Press, Athens.

Newsom, L. A., and M. C. Mihlbachler. 2006. Mastodons (*Mammut americanum*) diet foraging patterns based on analysis of dung deposits. In First Floridians and Last Mastodons: The Page-Ladson Site in the Aucilla River edited by S. D. Webb, 263–331. Springer, Dordrecht, The Netherlands.

Nimmo, D. G., L. T. Kelly, L. M. Spence-Bailey, S. J. Watson, R. S. Taylor, M. F. Clarke, and A. F. Bennett. 2013. Fire mosaics and reptile conservation in a fire-prone region. Conservation Biology 27:345–353.

Nimmo, D. G., R. MacNally, S. C. Cunningham, A. Haslem, and A. F. Bennett. 2015. Vive la résistance: Reviving resistance for 21st century conservation. Trends in Ecology and Evolution 30:516–523.

North, M. P., S. L. Stephens, B. M. Collins, J. K. Agee, G. Aplet, J. F. Franklin, and P. Z. Fulé. 2015. Reform forest fire management: Agency incentives undermine policy effectiveness. Science 349:1280–1281.

Noss, C. F., and B. B. Rothermel. 2015. Juvenile recruitment of oak toads (*Anaxyrus quercicus*) varies with time-since-fire in seasonal ponds. Journal of Herpetology 49:364–370.

Noss, R. F. 1987a. From plant communities to landscapes in conservation inventories: A look at The Nature Conservancy (USA). Biological Conservation 41:11–37.

———. 1987b. Do we really want diversity? Whole Earth Review 55:126–128.

———. 1989. Longleaf pine and wiregrass: Keystone components of an endangered ecosystem. The Natural Areas Journal 9:211–213.

———. 1996. Ecosystems as conservation targets. Trends in Ecology and Evolution 11:351.

———. 2001. Beyond Kyoto: Forest management in a time of rapid climate change. Conservation Biology 15:578–590.

———. 2013. Forgotten Grasslands of the South: Natural History and Conservation. Island Press, Washington, DC.

Noss, R. F., and A. Cooperrider. 1994. Saving Nature's Legacy: Protecting and Restoring Biodiversity. Island Press, Washington, DC.

Noss, R. F., J. F. Franklin, W. L. Baker, T. Schoennagel, and P. B. Moyle. 2006. Managing fire-prone forests in the western United States. Frontiers in Ecology and the Environment 4:481–487.

Noss, R. F., M. Korosy, D. Breininger, J. Aldredge, and R. Bjork. 2008. An Investigation of Breeding and Non-breeding Season Ecology, Metapopulation Dynamics, and Recovery Options for the Florida Grasshopper Sparrow (*Ammodramus savannarum floridanus*). Final Report to the U.S. Fish and Wildlife Service. University of Central Florida, Orlando.

Noss, R. F., E. T. LaRoe, and J. M. Scott. 1995. Endangered Ecosystems of the United States: A Preliminary Assessment of Loss and Degradation. Biological Report 28. USDI National Biological Service, Washington, DC.

Noss, R. F., W. J. Platt, B. A. Sorrie, A. S. Weakley, D. B. Means, J. Costanza, and R. K. Peet. 2015. How global biodiversity hotspots may go unrecognized: Lessons from the North American Coastal Plain. Diversity and Distributions 21:236–244.

Nowacki, G. J., and M. D. Abrams. 2008. The demise of fire and "mesophication" of forests in the eastern United States. BioScience 58:123–138.

Nyman, J. A., and R. H. Chabreck. 1995. Fire in coastal marshes: History and recent concerns. Proceedings of the Tall Timbers Fire Ecology Conference 19:134–141.

O'Brien, J. J., J. K. Hiers, M. A. Callaham Jr., R. J. Mitchell, and S. B. Jack. 2008. Interactions among overstory structure, seedling life-history traits, and fire in frequently burned Neotropical pine forests. Ambio 37:542–547.

Odling-Smee, F. J., K. N. Laland, and M. W. Feldman. 2003. Niche Construction: The Neglected Process in Evolution. Princeton University Press, Princeton, NJ.

Odum, E. P. 1969. The strategy of ecosystem development. Science 164:262–270.

Olano, J. M., E. S. Menges, and E. Martinez. 2006. Carbohydrate storage in five resprouting Florida scrub plants across a fire chronosequence. New Phytologist 170:99–106.

Olivieri, I., J. Tonnabel, O. Ronce, and A. Mignot. 2015. Why evolution matters for species conservation: Perspectives from three case studies of plant metapopulations. Evolutionary Applications 9:196–211.

Orzell, S. L., and E. L. Bridges. 2006a. Species composition and environmental characteristics of Florida dry prairies from the Kissimmee River region of south-central Florida. In Land of Fire and Water: The Florida Dry Prairie Ecosystem, edited by R. F. Noss, 100–135. Proceedings of the Florida Dry Prairie Conference, 5–7 October 2004. E. O. Painter, DeLeon Springs, FL.

———. 2006b. Floristic composition of the south-central Florida dry prairie landscape. In Land of Fire and Water: The Florida Dry Prairie Ecosystem, edited by R. F. Noss, 64–99. Proceedings of the Florida Dry Prairie Conference, 5–7 October 2004. E. O. Painter, DeLeon Springs, FL.

Ostertag, R., and E. S. Menges. 1994. Patterns of reproductive effort with time since last fire in Florida scrub plants. Journal of Vegetation Science 5:303–310.

Ottmar, R. D., D. V. Sandberg, C. L. Riccardi, and S. J. Prichard. 2007. An overview of the Fuel Characteristic Classification System—quantifying, classifying, and creating fuelbeds for resource planning. Canadian Journal of Forest Research 37:2383–2393.

Outcalt, K. 2006. Prescribed burning for understory restoration. In Longleaf Pine Ecosystems: Ecology, Management, and Restoration, edited by S. Jose, E. J. Jokela, and D. L. Miller, 326–329. Springer, New York.

Overpeck, J. T., R. S. Webb, and T. Webb III. 1992. Mapping eastern North American vegetation change of the past 18 Ka: No-analogs and the future. Geology 20:1071–1074.

Palis, J. G. 1996. Flatwoods salamander (*Ambystoma cingulatum* Cope). Natural Areas Journal 16:49–54.

———. 1997. Breeding migration of *Ambystoma cingulatum* in Florida. Journal of Herpetology 31:71–78.

Palmquist, K. A., R. K. Peet, and A. S. Weakley. 2014. Changes in plant species richness following reduced fire frequency and drought in one of the most species-rich savannas in North America. Journal of Vegetation Science 25:1426–1437.

Panzer, R. 2002. Compatibility of prescribed burning with the conservation of insects in small, isolated prairie reserves. Conservation Biology 16:1296–1307.

Parr, C. L., and A. N. Andersen. 2006. Patch mosaic burning for biodiversity conservation: A critique of the pyrodiversity paradigm. Conservation Biology 20:1610–1619.

Patton, T. H. 1969. An Oligocene land vertebrate fauna from Florida. Journal of Paleontology 43:544–546.

Pausas, J. G. 2015a. Evolutionary fire ecology: Lessons learned from pines. Trends in Plant Science 20:318–324.

———. 2015b. Bark thickness and fire regime. Functional Ecology 29:315–327.

Pausas, J. G., and S. Fernández-Muñoz. 2012. Fire regime changes in the Western Mediterranean Basin: From fuel-limited to drought-driven fire regime. Climatic Change 110:215–226.

Pausas, J. G., and J. E. Keeley. 2009. A burning story: The role of fire in the history of life. BioScience 59:593–601.

———. 2014a. Abrupt climate-independent fire regime changes. Ecosystems 17:1109–1120.

———. 2014b. Evolutionary ecology of resprouting and seeding in fire-prone ecosystems. New Phytologist 204:55–65.

Pausas, J. G., and M. Verdú. 2008. Fire reduces morphospace occupation in plant communities. Ecology 89:2181–2186.

Pellegrini, A. F. A. 2016. Nutrient limitation in tropical savannas across multiple scales and mechanisms. Ecology 97:313–324.

Pellegrini, A. F. A., L. O. Hedin, A. C. Staver, and N. Govender. 2015. Fire alters ecosystem carbon and nutrients but not plant nutrient stoichiometry or composition in tropical savannas. Ecology 96:1275–1285.

Pennington, R. T., and C. E. Hughes. 2014. The remarkable congruence of New and Old World savanna origins. New Phytologist 2014:4–6.

Perkins, M. W., L. M. Conner, and M. B. Howze. 2008. The importance of hardwood trees in the longleaf pine forest ecosystem for Sherman's fox squirrels. Forest Ecology and Management 255:1618–1625.

Peterson, D. W., and P. B. Reich. 2001. Prescribed fire in oak savanna: Fire frequency effects on stand structure and dynamics. Ecological Applications 11:914–927.

Pickett, S. T. A., and J. N. Thompson. 1978. Patch dynamics and the design of nature reserves. Biological Conservation 13:27–37.

Pickett, S. T. A., and P. S. White, editors. 1985. The Ecology of Natural Disturbance and Patch Dynamics. Academic Press, Orlando, FL.

Pierce, J. L., G. A. Meyer, and A. J. T. Jull. 2004. Fire-induced erosion and millennial-scale climate change in northern ponderosa pine forests. Nature 432:87–90.

Pimm, S. L. 1984. The complexity and stability of ecosystems. Nature 307:321–326.

Pinter, N., S. Fiedel, and J. E. Keeley. 2011. Fire and vegetation shifts in the Americas at the vanguard of Paleoindian migration. Quaternary Science Reviews 30:269–272.

Platt, S. G., and C. G. Brantley. 1997. Canebrakes: An ecological and historical perspective. Castanea 62:8–21.

Platt, W. J. 1999. Southeastern pine savannas. In Savannas, Barrens, and Rock Outcrop Plant Communities of North America, edited by R. C. Anderson, J. R. Fralish, and J. M. Baskin, 23–51. Cambridge University Press, Cambridge.

Platt, W. J., and R. M. Gottschalk. 2001. Effects of exotic grasses on potential fine fuel loads in the groundcover of south Florida slash pine savannas. International Journal of Wildland Fire 10:155–159.

Platt, W. J., and M. W. Schwartz. 1990. Temperate hardwood forests. In Ecosystems of Florida, edited by R. L. Myers and J. J. Ewel, 194–229. University of Central Florida Press, Orlando.

Platt, W. J., D. P. Ellair, J. M. Huffman, S. E. Potts, and B. Beckage. 2016. Pyrogenic fuels produced by savanna trees can engineer humid savannas. Ecological Monographs 86:352–372.

Platt, W. J., G. W. Evans, and S. L. Rathbun. 1988a. The population dynamics of a long-lived conifer (*Pinus palustris*). American Naturalist 131:491–525.

Platt, W. J., G. W. Evans, and M. M. Davis. 1988b. Effects of fire season on flowering of forbs and shrubs in longleaf pine forests. Oecologia 76:353–363.

Platt, W. J., J. S. Glitzenstein, and D. R. Streng. 1991. Evaluating pyrogenicity and its effects on vegetation in longleaf pîne savannas. Proceedings of the Tall Timbers Fire Ecology Conference 17:143–162.

Platt, W. J., J. M. Huffman, M. G. Slocum, and B. Beckage. 2006. Fire regimes and trees in Florida dry prairie landscapes. In Land of Fire and Water: The Florida Dry Prairie Ecosystem, edited by R. F. Noss, 3–13. Proceedings of the Florida Dry Prairie Conference, 5–7 October 2004. E. O. Painter, DeLeon Springs, FL.

Platt, W. J., S. L. Orzell, and M. G. Slocum. 2015. Seasonality of fire weather strongly influences fire regimes in South Florida savanna-grassland landscapes. PLoS ONE 10(1): e0116952. doi:10.1371/journal.pone.0116952.

Ponisio, L. C., K. Wilkin, L. K. M'Gonigle, K. Kulhanek, L. Cook, R. Thorp, T. Griswold, and C. Kremen. 2016. Pyrodiversity begets plant-pollinator community diversity. Global Change Biology 22:1794–1808.

Poulson, T. L., and W. J. Platt. 1989. Gap light regimes influence canopy tree diversity. Ecology 70:553–555.

Power, M. J., F. E. Mayle, P. J. Bartlein, et al. 2012. Climatic control of the biomass-burning decline in the Americas after AD 1500. The Holocene 1–11. DOI: 10.1177/0959683612450196.

Pranty, B., and J. W. Tucker Jr. 2006. Ecology and management of the Florida Grasshopper Sparrow. In Land of Fire and Water: The Florida Dry Prairie Ecosystem, edited by R. F. Noss, 188–200. Proceedings of the Florida Dry Prairie Conference, 5–7 October 2004. E. O. Painter, DeLeon Springs, FL.

Prince, A., M. C. Chitwood, M. A. Lashley, C. S. DePerno, and C. E. Moorman. 2016. Resource selection by southeastern fox squirrels in a fire-maintained forest system. Journal of Mammalogy 97:631–638.

Provencher, L., B. J. Herring, D. R. Gordon, H. L. Rodgers, K. E. M. Galley, G. W. Tanner, J. L. Hardesty, and L. A. Brennan. 2001. Effects of hardwood reduction techniques on longleaf pine sandhill vegetation in northwest Florida. Restoration Ecology 9:13–27.

Putz, F. E. 2003. Are rednecks the unsung heroes of ecosystem management? Wild Earth 12(2/3): 10–14.

———. 2015. Finding Home in the Sandy Lands of the South: A Naturalist's Journey in Florida. Cypress Highlands Press of Florida, Gainesvile.

Pyne, S. J. 1982. Fire in America: A Cultural History of Wildland and Rural Fire. Princeton University Press, Princeton, NJ.

———. 2001. Fire: A Brief History. University of Washington Press, Seattle.

———. 2016. Florida: A Fire Survey. University of Arizona Press, Tucson.

Quarterman, E., and C. Keever. 1962. Southern mixed hardwood forest: Climax in the Southeastern Coastal Plain, U.S.A. Ecological Monographs 32:167–185.

Quintana-Ascencio, P. F., and E. S. Menges. 2000. Competitive abilities of three narrowly endemic plant species in experimental neighborhoods along a fire gradient. American Journal of Botany 87:690–699.

Quintana-Ascencio, P. F., R. W. Dolan, and E. S. Menges. 1998. *Hypericum cumulicola* demography in unoccupied and occupied Florida scrub patches with different time-since-fire. Journal of Ecology 86:640–651.

Quintana-Ascencio, P. F., E. S. Menges, and C. W. Weekley. 2003. A fire-explicit population viability analysis of *Hypericum cumulicola* in Florida rosemary scrub. Conservation Biology 17:433–449.

Raffa, K. F., B. H. Aukema, B. J. Bentz, A. L. Carroll, J. A. Hicke, M. G. Turner, and W. H. Romme. 2008. Cross-scale drivers of natural disturbances prone to anthropogenic amplification: The dynamics of bark beetle eruptions. BioScience 58:501–517.

Raman, S., A. Sims, R. Ellis, and R. Boyles. 2005. Numerical simulation of mesoscale circulations in a region of contrasting soil types. Pure and Applied Geophysics 162:1689–1714.

Ramirez, E. M., and H. K. Ober. 2014. Nest site selection and reproductive success of red-cockaded woodpeckers in Ocala National Forest. American Midland Naturalist 171:258–270.

Rao, L. A., E. D. Allen, and T. Meixner. 2010. Risk-based determination of critical nitrogen deposition loads for fire spread in southern California deserts. Ecological Applications 20:1320–1335.

Raymo, M. E., and J. X. Mitrovica. 2012. Collapse of polar ice sheets during the stage 11 interglacial. Nature 483:453–456.

Regan, H. M., J. B. Crookston, R. Swab, J. Franklin, and D. M. Larson. 2010. Habitat fragmentation and altered fire regime create trade-offs for an obligate seeding shrub. Ecology 91:1114–1123.

Rein, G., N. Cleaver, C. Ashton, P. Pironi, and J. L. Torero. 2008. The severity of smouldering peat fires and damage to the forest soil. Catena 74:304–309.

Reséndez, A. 2007. A Land So Strange: The Epic Journey of Cabeza de Vaca. Basic Books, New York.

Richardson, D. M., and P. W. Rundel. 1999. Biology and biogeography of *Pinus*: An introduction. In Ecology and Biogeography of *Pinus*, edited by D. M. Richardson, 3–46. Cambridge University Press, Cambridge.

Richmond, G. M., and D. S. Fullerton. 1986. Summation of Quaternary glaciations in the United States of America. Quaternary Science Reviews 5:183–196.

Rickey, M. A., C. W. Weekley, and E. S. Menges. 2013. Felling as a pre-treatment for prescribed fire promotes restoration of fire-suppressed Florida sandhill. Natural Areas Journal 33:199–213.

Ripley, B., V. Visser, P.-A. Christin, S. Archibald, T. Martin, and C. Osborne. 2015. Fire ecology of C_3 and C_4 grasses depends on evolutionary history and frequency of burning but not photosynthetic type. Ecology 96:2679–2691.

Robbins, L. E., and R. L. Myers. 1992. Seasonal effects of prescribed burning in Florida: A review. Misc. Publ. No. 8, Tall Timbers Research, Tallahassee, FL.

Robertson, K. M., and T. L. Hmielowski. 2014. Effects of fire frequency and season on resprouting of woody plants in southeastern US pine-grassland communities. Oecologia 174:765–776.

Robertson, K. M., and T. E. Ostertag. 2004. Problems with Schurbon and Fauth's test of effects of prescribed burning on amphibian diversity. Conservation Biology 18:1154–1155.

Robertson, W. B. 1962. Fire and vegetation in the Everglades. Proceedings of the Tall Timbers Fire Ecology Conference 1:67–80.

Robinson, G. S., L. P. Burney, and D. A. Burney. 2005. Landscape paleoecology and megafaunal extinction in southeastern New York state. Ecological Monographs 75:295–315.

Romans, B. C. 1775. A Concise Natural History of East and West Florida. Vol. 1. Reprinted by Pelican Publishing (1961), New Orleans.

Romps, D. M., J. T. Seeley, D. Vollaro, and J. Molinar. 2014. Projected increase in lightning strikes in the United States due to global warming. Science 346:851–854.

Rooney, B. 1993. Burnin' Bill and the Dixie Crusaders. American Forests 99(5/6): 35.

Ross, W. G., D. L. Kulhavy, and R. N. Conner. 1997. Stand conditions and tree characteristics affect quality of longleaf pine for red-cockaded woodpecker cavity trees. Forest Ecology and Management 91:145–154.

Rostlund, E. 1960. The geographic range of the historic bison in the Southeast. Annals of the Association of American Geographers 50:395–407.

Rowley, D. B., A. M. Forte, R. Moucha, J. X. Mitrovica, N. A. Simmons, and S. P. Grand. 2013. Dynamic topography change of the eastern United States since 3 million years ago. Science 340:1560–1563.

Roznik, E. A., and S. A. Johnson. 2009. Canopy closure and emigration by juvenile gopher frogs. Journal of Wildlife Management 73:260–268.

Rule, S., B. W. Brook, S. G. Haberle, C. S. M. Turney, A. P. Kershaw, and C. N. Johnson. 2012. The aftermath of megafaunal extinction: Ecosystem transformation in Pleistocene Australia. Science 335:1483–1486.

Rundel, P. 1981. Structural and chemical components of flammability. In Fire Regimes and Ecosystem Properties, edited by H. A. Mooney, T. M. Bonnicksen, N. L. Christensen, J. E. Lotan, and W. A. Reiners, 183–207. USDA Forest Service General Technical Report WO-86.

Runkle, J. R. 1985. Disturbance regimes in temperate forests. In The Ecology of Natural Disturbance and Patch Dynamics, edited by S. T. A. Pickett and P. S. White, 17–33. Academic Press, Orlando, FL.

Russell, D. A., F. J. Rich, V. Schneider, and J. Lynch-Stieglitz. 2009. A warm thermal enclave in the late Pleistocene of the south-eastern United States. Biological Reviews 84:173–202.

Ryan, K. C., E. E. Knapp, and J. M. Varner. 2013. Prescribed fire in North American forests and woodlands: History, current practice, and challenges. Frontiers in Ecology and the Environment 11:e15–e24. doi:10.1890/120329.

Sah, J. P. Sah, M. S. Ross, J. R. Snyder, S. Koptur, and H. C. Cooley. 2006. Fuel loads, fire regimes, and post-fire fuel dynamics in Florida Keys pine forests. International Journal of Wildland Fire 15:463–478.

Saha, S., A. Catenazzi, and E. S. Menges. 2010. Does time since fire explain plant biomass allocation in the Florida, USA, scrub ecosystem? Fire Ecology 6(2): 13–25.

Sandel, B., L. Arge, B. Dalsgaard, R. G. Davies, K. J. Gaston, W. J. Sutherland, and J.-C. Svenning. 2011. The influence of Late Quaternary climate-change velocity on species endemism. Science 334:660–664.

Sargent, C. S. 1884. Report on the Forests of North America (Exclusive of Mexico). Report to Department of Interior, Census Office. Government Printing Office, Washington, DC.

Schafale, M. P. 2012. Guide to the Natural Communities of North Carolina: Fourth Approximation. North Carolina Natural Heritage Program, Department of Environment and Natural Resources, Raleigh.

Schimper, A. F. W. 1903. Plant Geography upon a Physiological Basis. Translated by W. R. Fisher. Clarendon Press, Oxford.

Schmidtling, R. C., and V. Hipkins. 1998. Genetic diversity in longleaf pine (*Pinus palustris*): Influence of historical and prehistorical events. Canadian Journal of Forest Research 28:1135–1145.

Schrey, A. W., K. G. Ashton, S. Heath, E. D. McCoy, and H. R. Muschinsky. 2011a. Fire alters patterns of genetic diversity among 3 lizard species in Florida scrub habitat. Journal of Heredity 102:399–408.

Schrey, A. W., A. Fox, E. D. McCoy, and H. R. Mushinsky. 2011b. Fire increases variance in genetic characteristics of Florida Sand Skink (*Plestiodon reynoldsi*) local populations. Molecular Ecology 20:56–66.

Schrey, A. W., A. K. Ragsdale, E. D. McCoy, and H. R. Mushinsky. 2016. Repeated habitat disturbances by fire decrease local effective population size. Journal of Heredity 107:336–341.

Schurbon, J. M., and J. E. Fauth. 2003. Effects of prescribed burning on amphibian diversity in a southeastern U.S. national forest. Conservation Biology 17:1338–1349.

Schwilk, D. W. 2003. Flammability is a niche construction trait: Canopy architecture affects fire intensity. American Naturalist 162:725–733.

Schwilk, D. W., and D. D. Ackerly. 2001. Flammability and serotiny as strategies: Correlated evolution in pines. Oikos 94:326–336.

Schwilk, D. W., and B. Kerr. 2002. Genetic niche-hiking: An alternative explanation for the evolution of flammability. Oikos 99:431–442.

Scott, A. C. 2000. The pre-Quaternary history of fire. Palaeogeography, Palaeoclimatology, Palaeoecology 164:281–329.

Scott, A. C., D. M. J. S. Bowman, W. J. Bond, S. J. Pyne, and M. E. Alexander. 2014. Fire on Earth: An Introduction. Wiley Blackwell, West Sussex, UK.

Shea, J. P. 1940. Our pappies burned the woods. American Forests (April): 159–162, 174.

Shepherd, B. J., D. L. Miller, and M. Thetford. 2012. Fire season effects on flowering characteristics and germination of longleaf pine (*Pinus palustris*) savanna grasses. Restoration Ecology 20:268–276.

Shigo, A. L. 1984. Compartmentalization: A conceptual framework for understanding how trees grow and defend themselves. Annual Review of Phytopathology 22:189–214.

Shores, E. F. 2008. On Harper's Trail: Roland McMillan Harper, Pioneering Botanist of the Southern Coastal Plain. University of Georgia Press, Athens.

Shriver, W. G., P. D. Vickery, and S. A. Hedges. 1996. Effects of summer burns on Florida grasshopper sparrows. Florida Field Naturalist 24:68–73.

Shriver, W. G., P. D. Vickery, and D. W. Perkins. 1999. The effects of summer burns on breeding Florida grasshopper and Bachman's sparrows. Studies in Avian Biology 19:144–148.

Simon, M. F., and R. T. Pennington. 2012. Evidence for adaptation to fire regimes in the tropical savannas of the Brazilian Cerrado. International Journal of Plant Sciences 173:711–723.

Simon, M. F., R. Grether, L. P. de Queiroz, C. Skema, R. T. Pennington, and C. E. Hughes. 2009. Recent assembly of the Cerrado, a neotropical plant diversity hotspot, by in situ evolution of adaptations to fire. Proceedings of the National Academy of Sciences 106:20359–20364.

Slapcinsky, J. L., D. R. Gordon, and E. S. Menges. 2010. Responses of rare plant species to fire in Florida's pyrogenic communities. Natural Areas Journal 30:4–19.

Slocum, M. G., W. J. Platt, B. Beckage, S. L. Orzell, and W. Taylor. 2010. Accurate quantification of seasonal rainfall and associated climate-wildfire relationships. Journal of Applied Meteorology and Climatology 49:2559–2573.

Slocum, M. G., W. J. Platt, B. Beckage, B. Panko, and J. B. Lushine. 2007. Decoupling natural and anthropogenic fire regimes: A case study in Everglades National Park, Florida. Natural Areas Journal 27:41–55.

Slocum, M. G., W. J. Platt, and H. C. Cooley. 2003. Effect of differences in prescribed fire regimes on patchiness and intensity of fires in subtropical savannas of Everglades National Park, Florida. Restoration Ecology 11:91–102.

Small, J. K. 1929. From Eden to Sahara: Florida's Tragedy. Science Press, Lancaster, PA.

Smith, K. T., and E. K. Sutherland. 1999. Fire-scar formation and compartmentalization in oak. Canadian Journal of Forest Research 29:166–171.

———. 2001. Terminology and biology of fire scars in selected central hardwoods. Tree-Ring Research 57:141–147.

Smith, M. D. 2011. The ecological role of climate extremes: Current understanding and future prospects. Journal of Ecology 99:651–655.

Smouse, P. E., and L. C. Saylor. 1973. Studies of the *Pinus rigida-serotina* complex I: A study of geographic variation. Annals of the Missouri Botanical Garden 60:174–191.

Snyder, J. R. 1984. The role of fire: Mutch ado about nothing? Oikos 43:404–405.

———. 1991. Fire regimes in subtropical south Florida. Proceedings of the Tall Timbers Fire Ecology Conference 17:111–116.

Snyder, J. R., A. Herndon, and W. B. Robertson Jr. 1990. South Florida rockland. In Ecosystems of Florida, edited by R. L. Myers and J. J. Ewel, 230–277. University of Central Florida Press, Orlando.

Sorrie, B. A., and A. S. Weakley. 2001. Coastal Plain plant endemics: Phytogeographic patterns. Castanea 66:50–82.

Sousa, W. P. 1979. Disturbance in marine intertidal boulder fields: The nonequilibrium maintenance of species diversity. Ecology 60:1225–1239.

Souza, H. A. V. E., R. G. Collevatti, M. S. Lima-Ribeiro, J. P. de Lemos-Filho, and M. B.

Lovato. 2017. A large historical refugium explains spatial patterns of genetic diversity in a Neotropical savanna tree species. Annals of Botany 119:239–252.

Spier, L. P., and J. R. Snyder. 1998. Effects of wet- and dry-season fires on *Jacquemontia curtisii*, a south Florida pine forest endemic. Natural Areas Journal 18:350–357.

Stambaugh, M. C., R. P. Guyette, and J. M. Marschall. 2011. Longleaf pine (*Pinus palustris* Mill.) fire scars reveal new details of a frequent fire regime. Journal of Vegetation Science 22:1094–1104.

Stambaugh, M. C., J. M. Varner, R. F. Noss, D. C. Dey, N. L. Christensen, R. F. Baldwin, R. P. Guyette, B. B. Hanberry, C. A. Harper, S. G. Lindblom, and T. A. Waldrop. 2015. Clarifying the role of fire in the deciduous forests of eastern North America: Reply to Matlack. Conservation Biology 29:942–946.

Staver, A. C., and S. E. Koerner. 2015. Top-down and bottom-up interactions determine tree and herbaceous layer dynamics in savanna grasslands. In Trophic Ecology: Bottom-up and Top-down Interactions across Aquatic and Terrestrial Systems, edited by T. C. Hanley and K. J. La Pierre, 86–106. Cambridge University Press, Cambridge.

Staver, A. C., S. Archibald, and S. Levin. 2011. The global extent and determinants of savanna and forest as alternative biome states. Science 334:230–232.

Steen, D. A., L. M. Conner, L. L. Smith, L. Provencher, J. K. Hiers, S. Pokswinski, B. S. Helms, and C. Guyer. 2013a. Bird assemblage response to restoration of fire-suppressed longleaf pine sandhills. Ecological Applications 23:134–147.

Steen, D. A., L. L. Smith, L. M. Conner, A. R. Litt, L. Provencher, J. K. Hiers, S. Pokswinski, and C. Guyer. 2013b. Reptile assemblage response to restoration of fire-suppressed longleaf pine sandhills. Ecological Applications 23:148–158.

Stephens, S. L., and W. J. Libby. 2006. Anthropogenic fire and bark thickness in coastal and island pine populations from Alta and Baja California. Journal of Biogeography 33:648–652.

Stephens, S. L., J. D. McIver, R. E. J. Boerner, C. J. Fettig, J. B. Fontaine, B. R. Hartsough, P. L. Kennedy, and D. W. Schwilk. 2012. The effects of forest fuel-reduction treatments in the United States. BioScience 62:549–560.

Stevens, J. T., and B. Beckage. 2009. Fire feedbacks facilitate invasion of pine savannas by Brazilian pepper (*Schinus terebinthifolius*). New Phytologist 184:365–375.

Stevenson, H. M., and B. H. Anderson. 1994. The Birdlife of Florida. University Press of Florida, Gainesville.

Stevenson, J. A. 1993. Shifting to lightning season burns aids pineland restoration. Resource Management Notes: A DNR Newsletter on Natural Resource Management 5(4): 6–8.

Stoddard, H. L. 1931. The Bobwhite Quail, Its Habits, Preservation and Increase. USDA Bureau of Biological Survey. Charles Scribners' Sons, New York.

———. 1935. Use of controlled fire in southeastern upland game management. Journal of Forestry 33:346–351.

———. 1964. Bird habitat and fire. Proceedings of the Tall Timbers Fire Ecology Conference 3:163–175.

———. 1969. Memoirs of a Naturalist. University of Oklahoma Press, Norman.

Stork, W. 1769. An Account of East Florida. W. Nicoll, London.

Stowe, J. 2016. Whither the fire-oaks? Longleaf Leader 9(1): 40–41.

Streng, D. R., J. S. Glitzenstein, and W. J. Platt. 1993. Evaluating effects of season of burn in longleaf pine forests: A critical literature review and some results from an ongoing long-term study. Proceedings of the Tall Timbers Fire Ecology Conference 18:227–259.

Sugihara, N. G., J. W. Van Wagtendonk, and J. Fites-Kaufman. 2006. Fire as an ecological process. In Fire in California's Ecosystems, edited by N. G. Sugihara, J. W. Wagtendonk, K. E. Shaffer, J. Fites-Kaufman, and A. E. Thode, 58–74. University of California Press, Berkeley.

Swengel, A. B. 2001. A literature review of insect responses to fire, compared to other conservation managements of open habitat. Biodiversity and Conservation 10:1141–1169.

Swengel, A. B., and S. R. Swengel. 2007. Benefit of permanent non-fire refugia for Lepidoptera conservation in fire-managed sites. Journal of Insect Conservation 11:263–279.

Swihart, R. K., and N. A. Slade. 1984. Road crossing in *Sigmodon hispidus* and *Microtus ochrogaster*. Journal of Mammalogy 63:357–360.

Takahashi, M. K., L. M. Horner, T. Kubota, N. A. Keller, and W. G. Abrahamson. 2011. Extensive clonal spread and extreme longevity in saw palmetto, a foundation clonal plant. Molecular Ecology 20:3730–3742.

Talluto, M. V., and C. W. Benkman. 2013. Landscape-scale eco-evolutionary dynamics: Selection by seed predators and fire determine a major reproductive strategy. Ecology 94:1307–1316.

Talluto, M. V., and K. N. Suding. 2008. Historical change in coastal sage scrub in southern California, USA in relation to fire frequency and air pollution. Landscape Ecology 23:803–815.

Tansley, A. G. 1939. The British Islands and Their Vegetation. Cambridge University Press, Cambridge.

Tautenhahn, S., J. W. Lichstein, M. Jung, J. Kattge, S. A. Bohlman, H. Heilmeier, A. Prokushkin, A. Kahl, and C. Wirth. 2016. Dispersal limitation drives successional pathways in central Siberian forests under current and intensified fire regimes. Global Change Biology 22:2178–2197.

Taylor, R. S., S. J. Watson, D. G. Nimmo, L. T. Kelly, A. F. Bennett, and M. F. Clarke. 2012. Landscape-scale effects of fire on bird assemblages: Does pyrodiversity beget biodiversity? Diversity and Distributions 18:519–529.

Terando, A. J., J. Costanza, C. Belyea, R. R. Dunn, A. McKerrow, and J. A. Collazo. 2014. The southern megalopolis: Using the past to predict the future of urban sprawl in the southeast U.S. PLoS ONE 9(7): e102261.

Terborgh, J., and J. A. Estes, editors. 2010. Trophic Cascades: Predators, Prey, and the Changing Dynamics of Nature. Island Press, Washington, DC.

Thatcher, B. S., D. G. Krementz, and M. S. Woodrey. 2006. Henslow's sparrow winter-survival estimates and response to prescribed burning. Journal of Wildlife Management 70:198–206.

Thaxton, J. M., and W. J. Platt. 2006. Small-scale fuel variation alters fire intensity and shrub dominance in a pine savanna. Ecology 87:1331–1337.

Thomas, B. P., L. Law, and D. L. Stankey. 1979. Soil survey of Marion County, Florida.

USDA Soil Conservation Service in cooperation with the University of Florida Institute of Food and Agricultural Sciences, Agricultural Experiment Stations, Soil Science Department, Gainesville, FL.

Thompson, K., R. M. Ceriani, J. P. Bakker, and R. M. Bekker. 2003. Are seed dormancy and persistence in soil related? Seed Science Research 13:97–100.

Thurgate, N. Y., and J. K. Pechmann. 2007. Canopy closure, competition, and the endangered dusky gopher frog. Journal of Wildlife Management 71:1845–1852.

Tilman, D., P. B. Reich, and J. M. H. Knops. 2006. Biodiversity and ecosystem stability in a decade-long grassland experiment. Nature 441:629–632.

Troumbis, A. Y., and L. Trabaud. 1989. Some questions about flammability in fire ecology. Acta Oecologica–Oecologia Plantarum 10:167–175.

Tucker, J. W., Jr., W. D. Robinson, and J. B. Grand. 2004. Influence of fire on Bachman's sparrow, an endemic North American songbird. Journal of Wildlife Management 68:1114–1123.

———. 2006. Breeding productivity of Bachman's sparrows in fire-managed longleaf pine forests. Wilson Journal of Ornithology 118:131–137.

Turner, M. G. 2010. Disturbance and landscape dynamics in a changing world. Ecology 91:2833–2849.

Vale, T. R. 1998. The myth of the humanized landscape: An example from Yosemite National Park. Natural Areas Journal 18:231–236.

Van Lear, D. H., W. D. Carroll, P. R. Kapeluck, and R. Johnson. 2005. History and restoration of the longleaf pine-grassland ecosystem: Implications for species at risk. Forest Ecology and Management 211:150–165.

Van Wagner, C. E., and I. R. Methven. 1978. Discussion: Two recent articles on fire ecology. Canadian Journal of Forest Research 8:491–492.

Varner, J. M., D. R. Gordon, F. E. Putz, and J. K. Hiers. 2005. Restoring fire to long-unburned *Pinus palustris* ecosystems: Novel fire effects and consequences for long-unburned ecosystems. Restoration Ecology 13:536–544.

Varner, J. M., J. K. Hiers, R. Ottmar, D. R. Gordon, F. E. Putz, and D. Wade. 2007. Overstory tree mortality resulting from reintroducing fire to long-unburned longleaf pine forests: The importance of duff moisture. Canadian Journal of Forest Research 37:1349–1358.

Varner, J. M., J. M. Kane, E. M. Banwell, and J. K. Kreye. 2015a. Flammability of litter from southeastern trees: A preliminary assessment. In Proceedings of the 17th Biennial Southern Silvicultural Research Conference, edited by A. Gordon Holley, Kristina F. Connor, and James D. Haywood, 183–187. Shreveport, LA, 5–7 March 2013. Southern Research Station, Asheville, NC.

Varner, J. M., J. M. Kane, J. K. Hiers, J. K. Kreye, and J. W. Veldman. 2016. Suites of fire-adapted traits of oaks in the southeastern USA: Multiple strategies for persistence. Fire Ecology 12(2): 48–64.

Varner, J. M., J. M. Kane, J. R. Kreye, and E. Engber. 2015b. The flammability of forest and woodland litter: A synthesis. Current Forestry Reports. DOI 10.1007/s40725-015-0012-x.

Varner, J. M., F. E. Putz, J. J. O'Brien, J. K. Hiers, R. J. Mitchell, and D. R. Gordon. 2009.

Post-fire tree stress and growth following smoldering duff fires. Forest Ecology and Management 258:2467–2474.

Veldman, J. W., E. Buisson, G. Durigan, G. W. Fernandes, S. Le Stradic, G. Mahy, D. Negreiros, G. E. Overbeck, R. G. Veldman, N. P. Zaloumis, F. E. Putz, and W. J. Bond. 2015a. Toward an old-growth concept for grasslands, savannas, and woodlands. Frontiers in Ecology and the Environment 13:154–162.

Veldman, J. W., G. E. Overbeck, D. Negreiros, G. Mahy, S. Le Stradic, G. W. Fernandes, G. Durigan, E. Buisson, F. E. Putz, and W. J. Bond. 2015b. Where tree planting and forest expansion are bad for biodiversity and ecosystem services. BioScience 65:1011–1018.

Venable, D. L. 2007. Bet hedging in a guild of desert annuals. Ecology 88:1086–1090.

Wade, D., J. Ewel, and R. Hofstetter. 1980. Fire in south Florida ecosystems. USDA Forest Service General Technical Report SE-17. Asheville, NC.

Wahlenberg, W. G. 1946. Longleaf pine: Its use, ecology, regeneration, protection, growth and management. C. L. Pack Forestry Foundation and USDA Forest Service, Washington, DC.

Waldron, J. L., S. M. Welch, and S. H. Bennett. 2008. Vegetation structure and the habitat specificity of a declining North American reptile: A remnant of former landscapes. Biological Conservation 141:2477–2482.

Waldrop, T. A., and S. L. Goodrick. 2012. Introduction to Prescribed Fire in Southern Ecosystems. USDA Forest Service, Southern Research Station, Asheville, NC.

Waldrop, T. A., D. H. Van Lear, F. T. Lloyd, and W. R. Harms. 1987. Long-term studies of prescribed burning in loblolly pine forests of the southeastern Coastal Plain. USDA Forest Service, Southeastern Forest Experiment Station, General Technical Report SE-45.

Waldrop, T. A., D. L. White, and S. M. Jones. 1992. Fire regimes for pine-grassland communities in the southeastern United States. Forest Ecology and Management 47:195–210.

Walker, J., and R. K. Peet. 1984. Composition and species diversity of pine-wiregrass savannas of the Green Swamp, North Carolina. Vegetatio 55:163–179.

Walker, L. R. 2012. The Biology of Disturbed Habitats. Oxford University Press, Oxford.

Ware, S., C. Frost, and P. D. Doerr. 1993. Southern mixed hardwood forest: The former longleaf pine forest. In Biodiversity of the Southeastern United States: Lowland Terrestrial Communities, edited by W. H. Martin, S. G. Boyce, and A. C. Echternacht, 447–493. Wiley, New York.

Warner, R. R., and P. L. Chesson. 1985. Coexistence mediated by recruitment fluctuations: A field guide to the storage effect. American Naturalist 125:769–787.

Watts, A. C., and L. N. Kobziar. 2013. Smoldering combustion and ground fires: Ecological effects and multi-scale significance. Fire Ecology 9:124–132.

Watts, A. C., and G. Tanner. 2006. Restoration of dry prairie using fire and roller chopping. In Land of Fire and Water: The Florida Dry Prairie Ecosystem, edited by R. F. Noss, 225–230. Proceedings of the Florida Dry Prairie Conference, 5–7 October 2004. E. O. Painter, DeLeon Springs, FL.

Watts, A. C., L. N. Kobziar, and J. R. Snyder. 2012. Fire reinforces structure of pondcypress (*Taxodium distichum* var. *imbricarium*) domes in a wetland landscape. Wetlands 32:439–448.

Watts, W. A. 1971. Postglacial and interglacial vegetation history of southern Georgia and central Florida. Ecology 52:676–690.

———. 1975. A Late Quaternary record of vegetation from Lake Annie, south-central Florida. Geology 3:344–346.

———. 1980a. The Late Quaternary vegetation history of the southeastern United States. Annual Review of Ecology and Systematics 11:387–409.

———. 1980b. Late-Quaternary vegetation history at White Pond on the inner coastal plain of South Carolina. Quaternary Research 13:187–199.

Watts, W. A., and B. C. S. Hansen. 1994. Pre-Holocene and Holocene pollen records of vegetation history from the Florida peninsula and their climatic implications. Palaeogeography, Palaeoclimatology, Palaeoecology 109:163–176.

Watts, W. A., B. C. S. Hansen, and E. C. Grimm. 1992. Camel Lake: A 40,000-yr record of vegetational and forest history from northwest Florida. Ecology 73:1056–1066.

Way, A. G. 2006. Burned to be wild: Herbert Stoddard and the roots of ecological conservation in the southern longleaf pine forest. Environmental History 11:500–526.

Weakley, A. S. 2015. Flora of the Southern and Mid-Atlantic States. Working draft, 21 May 2015. University of North Carolina Herbarium, North Carolina Botanical Garden, Chapel Hill.

Weakley, A. S., and M. P. Schafale. 1991. Classification of pocosins of the Carolina Coastal Plain. Wetlands 11:355–375.

Webb, S. D. 1977. A history of savanna vertebrates in the New World. Part I: North America. Annual Review of Ecology and Systematics 8:355–380.

———. 1990. Historical biogeography. In Ecosystems of Florida, edited by R. L. Myers and J. J. Ewel, 70–100. University of Central Florida Press, Orlando.

———, ed. 2006. First Floridians and Last Mastodons: The Page-Ladson Site in the Aucilla River. Springer, Dordrecht, The Netherlands.

Webb, S. D., and K. T. Wilkins. 1984. Historical biogeography of Florida Pleistocene mammals. In Contributions in Quaternary Vertebrate Paleontology: A Volume in Memorial to John E. Guilday, edited by H. H. Genoways and M. R. Dawson, 370–383. Special Publication 8. Carnegie Museum of Natural History, Pittsburgh.

Webb, S. D., G. S. Morgan, R. C. Hulbert Jr., D. S. Jones, B. J. MacFadden, and P. A. Mueller. 1989. Geochronology of a rich early Pleistocene vertebrate fauna, Leisey Shell Pit, Tampa Bay, Florida. Quaternary Research 32:96–110.

Weekley, C. W., E. S. Menges, D. Berry-Greenlee, M. A. Rickey, G. L. Clarke, and S. A. Smith. 2011. Effects of mowing and burning on Florida scrub vegetation at two sites on the Lake Wales Ridge. Ecological Restoration 29:357–373.

Weigl, P. D., M. A. Steele, L. J. Sherman, J. C. Ha, and T. S. Sharpe. 1989. The ecology of the fox squirrel (*Sciurus niger*) in North Carolina: Implications for survival in the Southeast. Bulletin of the Tall Timbers Research Station 24:1–93.

Wells, B. W. 1928. Plant communities of the Coastal Plain of North Carolina and their successional relations. Ecology 9:230–242.

———. 1932. The Natural Gardens of North Carolina. University of North Carolina Press, Chapel Hill.

Wells, B. W., and I. V. Shunk. 1931. The vegetation and habitat factors of the coarser sands of the North Carolina Coastal Plain: An ecological study. Ecological Monographs 1:465–520.

Westerling, A. L., H. G. Hidalgo, D. R. Cayan, and T. W. Swetnam. 2006. Warming and earlier spring increase western U.S. forest wildfire activity. Science 313:940–943.

Whelan, A., R. Mitchell, C. Staudhammer, and G. Starr. 2013. Cyclic occurrence of fire and its role in carbon dynamics along an edaphic moisture gradient in longleaf pine ecosystems. PLoS ONE 8:e54045.

White, F. 1976. The underground forests of Africa: A preliminary review. Singapore Gardens' Bulletin 24:57–71.

White, P. S. 1979. Pattern, process, and natural disturbance in vegetation. Botanical Review 45:229–299.

White, P. S., and S. T. A. Pickett. 1985. Natural disturbance and patch dynamics: An introduction. In The Ecology of Natural Disturbance and Patch Dynamics, edited by S. T. A. Pickett and P. S. White, 3–13. Academic Press, Orlando, FL.

White, R., A. Murray, and M. Rohweder. 2000. Pilot Analysis of Global Ecosystems: Grassland Ecosystems. World Resources Institute, Washington, DC.

Whitehead, D. R. 1973. Late Wisconsin vegetational changes in unglaciated North America. Quaternary Research 3:621–631.

Whittaker, R. H. 1953. A consideration of climax theory: The climax as a population and pattern. Ecological Monographs 23:41–78.

———. 1975. Communities and Ecosystems. 2nd ed. Macmillan, New York.

Wiens, J. J., D. D. Ackerly, A. P. Allen, B. L. Anacker, L. B. Buckley, H. V. Cornell, E. I. Damschen, T. J. Davies, J.-A. Grytnes, S. P. Harrison, B. A. Hawkins, R. D. Holt, C. M. McCain, and P. R. Stephens. 2010. Niche conservatism as an emerging principle in ecology and conservation biology. Ecology Letters 13:1310–1324.

Willcox, E. V., and W. M. Giuliano. 2010. Seasonal effects of prescribed burning and roller chopping on saw palmetto in flatwoods. Forest Ecology and Management 259:1580–1585.

Williams, G. C. 1966. Adaptation and Natural Selection: A Critique of some Current Evolutionary Thought. Princeton University Press, Princeton, NJ.

Williams, J. W., B. N. Shuman, T. Webb III, P. J. Bartlein, and P. L. Leduc. 2004. Late-Quaternary vegetation dynamics in North America: Scaling from taxa to biomes. Ecological Monographs 74:309–334.

Williams, J. W., T. Webb III, P. H. Richard, and P. Newby. 2000. Late Quaternary biomes of Canada and the eastern United States. Journal of Biogeography 27:585–607.

Willig, M. R., and L. R. Walker. 1999. Disturbance in terrestrial ecosystems: Salient themes, synthesis and future directions. In Ecosystems of Disturbed Ground, Ecosystems of the World 16, edited by L. R. Walker, 747–767. Elsevier, Amsterdam.

Wilson, R. C. L., S. A. Drury, and J. L. Chapman. 2000. The Great Ice Age: Climate Change and Life. Routledge, London.

Winchester, B. H., J. S. Bays, J. C. Higman, and R. L. Knight. 1985. Physiography and vegetation zonation of shallow emergent marshes in southwestern Florida. Wetlands 5:99–118.

Wittkuhn, R. S., L. McCaw, A. J. Wills, R. Robinson, A. N. Andersen, P. van Heurck, G.

Liddelow, and R. Cranfield. 2011. Variation in fire interval sequences has minimal effects on species richness and composition in fire-prone landscapes of south-west Western Australia. Forest Ecology and Management 261:965–978.

Wonkka, C. L., W. E. Rogers, and U. P. Kreuter. 2015. Legal barriers to effective ecosystem management: Exploring linkages between liability, regulations, and prescribed fire. Ecological Applications 25:2382–2393.

Wright, B. R., and P. J. Clarke. 2008. Relationships between soil temperatures and properties of fire in feathertop spinifex (*Triodia schinzii* [Henrard] Lazarides) sandridge desert in central Australia. Rangeland Journal 30:317–325.

Wu, S.-Y. 2015. Changing characteristics of precipitation for the contiguous United States. Climatic Change 132:677–692.

Yoder, J., D. Engle, and S. Fuhlendorf. 2004. Liability, incentives, and prescribed fire for ecosystem management. Frontiers in Ecology and the Environment 2:361–366.

INDEX

Index includes frequently mentioned species (common names) and natural communities, as well as major historical figures in fire ecology of the region.

REED F. NOSS is lead scientist for the University of Florida's Center for Landscape Conservation Planning and Chief Science Advisor for the Southeastern Grasslands Institute. He is the author of several books, including *Forgotten Grasslands of the South: Natural History and Conservation.*

www.ingramcontent.com/pod-product-compliance
Lightning Source LLC
Chambersburg PA
CBHW021112270326
41929CB00009B/836

9780813080772